CHRIST, SCIENCE, AND REASON

Robert J. Spitzer, S.J., Ph.D.

Christ, Science, and Reason

What We Can Know about Jesus, Mary, and Miracles

IGNATIUS PRESS   SAN FRANCISCO

Cover image:
Christ Pantocrator by Viktor Mikhaylovich Vasnetsov, 1885–1896.
State Tretyakov Gallery, Moscow, © Bridgeman Images

Inset images:
Shroud of Turin, public domain
DNA image by julientromeur, Unsplash.com
Our Lady of Guadalupe, photographic reproduction, in the public domain
Solar eclipse, photo by Mathew Schwartz, Unsplash.com

Cover design by John Herreid

© 2024 by Ignatius Press, San Francisco
All rights reserved
ISBN 978-1-62164-743-0 (PB)
ISBN 978-1-64229-327-2 (eBook)
Library of Congress Catalogue number 2024937050
Printed in Canada ∞

In memory of my mother, Blanche Spitzer, whose tremendous faith inspired my devotion to Jesus Christ, the Holy Eucharist, and the Blessed Virgin Mary, and in memory of my father, Arthur Spitzer, whose love of electronics and radios opened the door to the world of science.

CONTENTS

ACKNOWLEDGMENTS

I am very grateful to Dr. August ("Gus") Accetta (surgeon and founder of the Shroud Center of Southern California) for introducing me to many of the scientists responsible for the Shroud of Turin Research Project.

I am most grateful to Joan Jacoby, who helped to bring this book into being by her invaluable work, dedication, editing suggestions, research, and preparation of the manuscript. I am also sincerely grateful to Jeena Rudy for her considerable help in research and preparation of the manuscript. I also want to thank Karlo Broussard and Kathleen Conway for their help in bringing the manuscript to its final form.

INTRODUCTION

In the previous volume, *Science at the Doorstep to God*,[1] we presented evidence and formulated conclusions about the *what* of God but were left with the question of *who* God really is. Therefore, we now investigate the evidence for Christ.

It may seem strange to address the topic of scientific evidence of Christ, because science is in the domain of reason while Christ, the domain of revelation. As we shall see, the two domains are not intrinsically distinct but rather touch upon, complement, and reinforce each other in many respects.

The intersection of Christ and science is so pronounced that one cannot help but come away with the belief that the Lord had providentially designed it over two thousand years ago—indeed at the inception of time—so that it might be revealed to a scientific age becoming increasingly more resistant and skeptical to the good news of salvation.

The scientific studies of the Shroud of Turin—particularly the new wide-angle X-ray scattering dating method, which places the Shroud between A.D. 55 and 74, and the growing consensus around the source of the image as particle radiation coming from the complete nuclear disintegration of the body—reinforce an already strong scientific case for the authenticity of the Shroud and the supernatural origin of its image. When these are combined with the physiological examination of the blood (and its indication of a crucifixion similar to that of Jesus') and other dating tests, as well as the statistical debunking of the 1988 carbon dating, they set out a strong scientific and reasonable foundation for the Passion and glorious Resurrection of Jesus Christ as depicted in the Gospel accounts.

The topic of Eucharistic miracles, long belittled by scientific audiences, has now gained scientific respectability because of the research

---

[1] Robert Spitzer, *Science at the Doorstep to God: Science and Reason in Support of God, the Soul, and Life after Death* (San Francisco: Ignatius Press, 2023).

of Dr. Ricardo Castañón Gomez and others on the Hosts of Tixtla, Mexico, in 2006, and Buenos Aires, Argentina, in 1996, as well as the research of Drs. Maria Elżbieta Sobaniec-Łotowska and Stanisław Sulkowski on the Host of Sokółka, Poland, in 2008. These Hosts, subjected to an electron microscope as well as histological, pathological, DNA, and other analyses, manifest an incredibly refined interweaving of the substance of the Host with the substance of living cardiac tissue, which cannot be duplicated by modern technology.

Though there has been considerable scholarly work surrounding the Guadalupe tilma image, the miracle of the dancing sun at Fatima, and the healings of Lourdes, a reexamination of these miracles in light of current scientific techniques shows them to be as scientifically inexplicable as ever—indeed, more so. When all this evidence is interrelated, it yields not only a tapestry of faith and science but also a remarkable probative corroboration of the strong historical and exegetical case for the Passion and glorious Resurrection of Jesus Christ as well as His Real Presence in the Eucharist. It also reveals the cosmic and salvific significance of the Blessed Virgin Mary and her answer to our prayers.

Some very credible scientists, physicians, and technicians have been involved with the intersection between Christian faith and science, mostly because they were confronted with scientifically inexplicable data connected with Jesus Christ. Some of those scientists were unbelievers when they started their investigations (e.g., Dr. Alexis Carrel at Lourdes, Dr. Ricardo Castañón Gomez investigating transformed Eucharistic hosts, and Mark Antonacci investigating the Shroud of Turin), and many scientists were confirmed in their faith. Since the evidence in support of Christian faith has been increasing, particularly over the last three decades, I decided to give a summary of some of the more important scientifically investigated findings so that the public might know the quality and quantity of scientific research that points to Christ's risen and supernatural presence in the past and present. The evidence for Christ's historicity, miraculous power, and Resurrection is quite probative (see Chapter 2) and does not need scientific corroboration to validate it. Nevertheless, there is, as we shall see, considerable scientific evidence pointing not only to the historical Christ but also to His Real Presence in the Holy Eucharist and through His mother in the present. This scientific evidence does in fact corroborate the historical and scriptural evidence given in Chapter 2, which strengthens the foundation of Christian faith in our scientific age.

Some readers may be thinking that most of the evidence presented in this volume points not only to Jesus Christ but also to controversial doctrines of the Catholic Church—Eucharistic and Marian miracles. Though it might be justly said that I am predisposed to "Catholic miracles" because I am a Catholic priest, that is not the real reason why I have spent three chapters on these Catholic themes. The real motivation arises out of the plain fact that the most remarkable and best scientifically researched manifestations of the miraculous in the twentieth and twenty-first centuries just happen to be related to the Holy Eucharist and the Blessed Virgin Mary. An examination of the depth and quality of this research moved me to the conviction that the evidence should not be hidden from the public just because it is Catholic; rather, it should be shared with the public so that people can make their own decisions about its significance for the Resurrection and divinity of Jesus as well as the veracity of the Church He founded. It should not be surprising that some of the most remarkable and scientifically well-researched manifestations of the miraculous concern Catholic doctrines, for if Jesus founded the Catholic Church, why would He not validate it for a scientific age? Besides, as we shall see in Chapter 7, there is no other church that even comes close to the Catholic Church in its contributions to the development of all branches of science as well as scientific education. It seems to me that this circumstance is just as providential as the seemingly miraculous phenomena being investigated by those disciplines. My purpose is simply to shine a light on this providential "conspiracy".

In writing this volume, I do not mean to imply that the scientific investigation of Jesus Christ, the Eucharist, and the Blessed Virgin Mary has been brought to a definitive conclusion. Scientific research must and will continue, new tests will be devised, new data will become available—as well it should. My suspicion is that this data will continue to corroborate not only the miraculous phenomena studied in this volume but others as well. Nevertheless, if readers want or need to withhold judgment until further scientific research is done, that is more than appropriate. That being said, the evidence presented in this volume is significant enough to form what Saint

John Henry Newman called "an informal inference"—an inference from a series of different sets of evidence, each of which is antecedently probable, that complement and corroborate one another toward a single conclusion.[2] The conclusion is not dependent on *every* implication being perfectly clear and precise, but rather in the totality of antecedently *probable* data pointing to the same conclusion. Hence, modifications, changes, and additions can be made without undermining the certitude of the general conclusion. My intention, therefore, is to provide sufficient evidence for a probative informal inference of the historicity, supernatural power, and Resurrection of Jesus as well as His Real Presence in the Eucharist and the presence of His mother in the economy of salvation.

Some readers will find the scientific terminology used in this book to be challenging, but it cannot be avoided in a book called *Christ, Science, and Reason*. Though I did try to make these terms and concepts accessible to general readers, I realize the virtual impossibility of doing this in a perfectly lucid way; nevertheless, I beg the reader's indulgence to make an effort at getting the general idea being conveyed, for it may not be perfectly clear. A general idea may be sufficient to enhance appreciation of the supernatural presence of Christ in history, the Holy Eucharist, and through His blessed mother. This may have the effect of deepening, enlivening, and strengthening the reader's faith.

This book has seven chapters:

1. *From Reason and Science to Revelation and Faith in Jesus Christ.* This chapter bridges the content from the previous volume (*Science at the Doorstep to God*) to this volume. The previous volume is concerned with the scientific and philosophical affirmation of a unique, uncaused, unrestricted, intelligent Creator as well as a transphysical soul capable of surviving bodily death (which can be accomplished through reason and science without making recourse to revelation). This chapter shows how and why to make the transition from reason to the revelation of Jesus Christ.

---

[2] John Henry Newman, *An Essay in Aid of a Grammar of Assent* (Notre Dame, Ind.: University of Notre Dame Press, 1992), pp. 259–342. See also Spitzer, *Doorstep to God*, Introduction, Section III.

2. *The Historical Evidence of Jesus.* This chapter shows the contemporary scriptural and historical evidence for Jesus' ministry, Passion, Resurrection, and gift of the Spirit. Since it is sufficient to ground belief in the Risen Jesus reasonably, it does not *need* scientific corroboration. Nevertheless, it provides a strong reasonable foundation that new scientific evidence can corroborate and strengthen.

3. *Science at the Doorstep to Jesus—The Shroud of Turin.* This chapter is primarily focused on the scientific investigation of the Shroud of Turin. There is considerable evidence for its authenticity from the vantage point of dating, bloodstains, anatomical integrity, pollen grains, and correlation with the customs of Roman crucifixion and weaponry. As we shall see, the only two scientific explanations of the image capable of explaining its many enigmas apparently entail some kind of transphysical/supernatural explanation. This evidence lends further credence not only to the historical Jesus but also to the eschatological (heavenly) Christ.

4. *Science and Eucharistic Miracles.* This chapter examines the transformation of three Eucharistic hosts manifesting living cardiac tissue growing out of the substance of the Host. The scientific examination of these hosts improved from that of Buenos Aires (1996) to Tixtla, Mexico (2006), and Sokółka, Poland (2008). The histopathological examination, electron microscope screening (Sokółka), and other analytical tests were performed by credible experts in adequate laboratory conditions implying that the living heart tissue with AB+ blood type and living white blood cells is real. The absence of DNA profile amid molecular indications of DNA is mysterious, but this does not invalidate other analytical tests. These scientifically inexplicable features are not fraudulent, indicating the body and blood of a victim of torture. The identity of the victim may be discerned through the eyes of faith—it is very likely the man who initiated the Eucharistic rite at the Last Supper: Jesus.

5. *Science at the Doorstep to Mary—Guadalupe and Fatima.* This chapter examines two apparently miraculous phenomena—the tilma of Guadalupe and the Miracle of the Sun at Fatima. The tilma has several scientifically inexplicable features—the longevity

of the material, the absence of natural pigment and brushstrokes, the longevity of the nonpaint substance through which
the image was produced, the triple reflection and proper curvature of the cornea, and the presence of multiple figures and the
detectable images of several individuals witnessing the revealing
of Juan Diego's tilma. The work of Dr. José Aste Tonsmann
(among other scientific experts) strongly suggests transphysical
(supernatural) causation of the image and the longevity of the
tilma. With respect to Fatima, the thorough historical and scientific examination of this phenomenon of October 13, 1917,
by physicist-priest Stanley Jaki gives a plausible natural explanation for the phenomenon predicted by Lucia three times
from July through September. This complex naturalistic explanation appears to be miraculous because the odds against the
occurrence of so many unusual natural phenomena at the exact
date, time, and place predicted by Lucia over three months is
thousands of trillions to one. Thus, it seems that some kind
of supernatural agency which Lucia identified as the Blessed
Virgin orchestrated this wholly improbable event to give credibility about the need for repentance, prayer (particularly the
Rosary), and fidelity to her Son.

6. *Science at the Doorstep to Mary—Lourdes*. This chapter continues the investigation of the appearance of Mary in our scientific age, focusing on the remarkable predictions and actions
of Saint Bernadette Soubirous that led to the founding of the
Lourdes Medical Bureau (now the Lourdes Office of Medical
Observations), organized to investigate scientifically inexplicable cures. Since that time, seven thousand inexplicable cures
involving hundreds of physicians have been documented by
that bureau.[3] Some of these cures, particularly those in which
large sections of atrophied bone and tissue were instantaneously
regenerated, seemed to defy scientific explanation, and three of
them occurred in conjunction with Eucharistic blessings. This
manifests the healing hand of both Christ and His mother in
our scientific age.

---

[3] "Miraculous Healings", Lourdes Sanctuaire (website), accessed February 21, 2024, https:
//www.lourdes-france.com/en/miraculous-healings/.

7. *The Catholic Church and Natural Science.* This chapter shows the close connection between science and the Catholic Church since the origin of the scientific method (Brother Roger Bacon, O.F.M.). The Catholic Church, far from being anti-science, was one of its champions, with 286 of her priests and clerics contributing to all branches of science. In light of this, we discuss why the Galileo affair happened, as well as the Church's position on the "Bible and science" and evolution. This close relationship indicates not only that science has been at the doorstep to the Church but that the Church has been at the doorstep to science.

In view of the above, some readers may have a new or renewed interest in the Catholic Church, and so an appendix is provided to help them understand the foundation of the Church founded by Christ, His continued presence in her, and the spiritual and emotional benefits coming from her.

Chapter One

From Reason and Science to Revelation and Faith in Jesus Christ

Introduction

My previous volume, *Science at the Doorstep to God*, was focused on the domain of reason and science. It first explored the significant scientific evidence for an intelligent Creator,[1] and then the philosophical/logical evidence for a unique, uncaused, unrestricted, intelligent reality, which is the Creator of everything else.[2] It then examined the peer-reviewed medical studies of near-death experiences and terminal lucidity that implied the existence of a transcendent soul capable of surviving bodily death,[3] which is corroborated by contemporary philosophical studies of human intelligence and self-consciousness.[4] Can reason and objective evidence go any further? We can discern some clues of what the next life might be like from near-death experiences (a heavenly domain with a loving white light), but we do not know much beyond what multiple witnesses indicate happened to them when they were clinically dead. Additionally, we might be able to obtain some clues about the goodness of the Creator from our conscience and the numinous experience,[5] but these are inferences from our subjective experience. Similarly, we may obtain insights into the intelligence, goodness, love, and beauty of the Creator from

---

[1] Robert Spitzer, *Science at the Doorstep to God: Science and Reason in Support of God, the Soul, and Life after Death* (San Francisco: Ignatius Press, 2023), Chapters 1 and 2.

[2] Ibid., Chapter 3.

[3] Ibid., Chapter 4.

[4] Ibid., Chapter 5.

[5] Ibid., Chapter 6.

the examination of the universe and its highly intelligible and mathematical laws, but again, these are subjective inferences and impressions from scientific data, which are themselves not scientific data. At
this point, the well of reason and science begins to run dry.

As might be inferred from the above, reason and objective evidence
alone leave many questions unanswered—questions relevant to our
purpose and priorities in life, as well as our principles and values, the
way we want to interact with others, and the intelligent Creator. For
example, is God perfectly good, partially good, or something else?
Does He love us unconditionally, or according to our just deserts, or
only when we have been good? What does this intelligent Creator
consider to be good—and evil? Does the idea of the good go beyond
the dictates of our conscience—does it have a specificity of which our
conscience is only dimly aware? If God is loving, why does He allow
suffering? Does this intelligent Creator see a purpose in suffering, and
is He willing to help us through this suffering? Is this intelligent Creator really interested in what happens to us in our day-to-day lives?
Would He deign to interact with us, respond to our prayers, inspire
us, guide us, and protect us, or is He the intelligent, but indifferent
God of Einstein, Aristotle, and the deists? How does God answer
our prayers? How does He guide us through suffering? If there is a
Heaven, as near-death experiences suggest, what is it really like? Is
there a Hell? If so, why would a loving God allow that?

There are literally hundreds of more nuanced questions impinging
on our lives both in this world and the next that reason and science
cannot answer. Unfortunately, the clues and impressions we might
obtain from reason's gaze on the universe and the soul do not give
us the specificity and definitiveness we need not only to live our
lives well but to enter into an eternal state of perfection implied by
our human nature and near-death experiences. Vague clues will not
answer the above questions sufficiently to fill us with purpose, guide
us with precision, and orient us toward perfection.

Can it be that the intelligent Creator has created us with intelligence, self-consciousness, moral and spiritual awareness, and free
will[6] sufficient to ask the above questions without at the same time
providing some path to obtain the answers? Did the intelligent

---

[6] For free will, see ibid., Chapter 6, Section V.

Creator create us to be universally and irresolvably frustrated, like a cosmic joke? If this seems to be inconsistent with the nature of the intelligent Creator who inspires our conscience and is numinously present to us—who incites us to desire perfect truth, love, goodness, beauty, and home[7]—then He must provide some other way of answering the above questions beyond reason and science. Inasmuch as the answers to the above questions come from the heart[8] as well as the mind, we may surmise that He has intended from the inception of time to reveal Himself to us—both His mind and His heart. Viewed from the opposite direction, if God did not reveal Himself to us—if He did not show us our true purpose, dignity, and destiny, revealing the good in its depth and specificity, and invite us into a relationship with Him—He would have left us for all intents and purposes, orphans—lost in frustration and given a nature that seeks Him but not given adequate information to pursue Him. We would, to paraphrase Saint Paul, be the most pitiable of all creatures (1 Cor 15:19).

The moment we bring up the idea of revelation, two questions arise. First, given that there are so many distinct religions around the world, all of which maintain they have some form of revelation from the Divinity, we must ask, which revelation most accurately represents the mind, heart, and nature of the Divinity—the intelligent Creator? Second, since revelation entails truths about morality, love, sacredness/holiness, beauty, and spiritual struggle, as well as our interior experience of the Divinity,[9] it is situated in the domain of *faith*, which gives rise to the question, what is meant by "faith"?

We will consider these questions in five sections:

1. Which Religion Most Accurately Represents the Self-Revelation of God? (Section I)

---

[7] See ibid., Chapter 6, Section III.

[8] By "heart" I refer to the consolidation of three capacities that go beyond (yet work with) conceptual intelligence and reason—the capacity for love (empathy leading to self-giving care), the capacity for moral rectitude, and the capacity for aesthetics (the recognition of beauty). Other capacities—such as the awareness of transcendence, the spiritual, and sacredness—belong to the perfectly loving, good, intelligent, beautiful, uncaused reality by His very nature.

[9] For an explanation of religious experience in most religions, see ibid., Chapter VI, Section I.

I. Which Religion Most Accurately Represents the Self-Revelation of God?

What religion most closely represents the true nature, goodness, and will of the intelligent Creator? We might begin with recognizing what all major religions have in common—though they specify and articulate these common characteristics quite differently. Friedrich Heiler, an important German historian of religion, has in concert with several scholars of different religions gleaned seven common characteristics among world religions:

1. The transcendent, the holy, the Divine, the Other is real.
2. The transcendent reality is immanent in human awareness.
3. This transcendent reality is the highest truth, highest good, and highest beauty.
4. This transcendent reality is loving and compassionate—and seeks to reveal its love to human beings.
5. The way to God requires prayer, ethical self-discipline, purgation of self-centeredness, asceticism, and redressing of offenses.
6. The way to God also includes service and responsibility to people.
7. The highest way to eternal bliss in the transcendent reality is through love.[10]

If this list of seven common characteristics is truly common to most major religions, we might infer that they really have come

---

[10] See Friedrich Heiler, "The History of Religions as a Preparation for the Cooperation of Religions", in The History of Religions, ed. Mircea Eliade and J. Kitagawa (Chicago: Chicago University Press, 1959), pp. 142–44.

from the intelligent Creator, revealed not only through chosen prophets, priests, and spiritual divines but also in the conscience and numinous awareness of every human being.[11] Furthermore, if the seventh characteristic represents the highest way to eternal bliss in the transcendent reality, then this presents a clue as to which religion most accurately represents the mind, heart, and will of the intelligent Creator—the transcendent reality. There can be little doubt that the most comprehensive, interior, exterior, and transcendent articulation of love in the history of religions was given by Jesus Christ (see below, Section III). People with a passing awareness of Christianity will probably recognize the central moral principle articulated by Jesus when He was asked about the greatest commandment in the Law. Jesus responded:

> You shall love the Lord your God with all your heart, and with all your soul, and with all your mind. This is the great and first commandment. And a second is like it, You shall love your neighbor as yourself. On these two commandments depend all the law and the prophets. (Mt 22:37–40)

When Jesus selected a "greatest commandment", He did something quite unique in the history of religions. No one prior to Jesus—neither Jewish nor any other religion—ever selected one principle to be the highest commandment. Further, no other religious figure claimed that all other commandments and religious writings were summed up in this one principle—love. Still further, no other religious figure in history spent so much of his preaching defining and articulating a single principle (such as "love")—for example, see the Beatitudes (Mt 5:3–11), the Parable of the Good Samaritan (Lk 10:25–37), about half the Sermon on the Mount (Mt 5–7), and a large portion of the Gospel of John. Finally, no other religious figure revealed that God's nature, mind, and heart is unconditional love (as He defined it)—for example, see the father of the prodigal son

---

[11] There is significant evidence for this in the work of Rudolf Otto, Mircia Eliade, John Henry Newman, C. S. Lewis, and others. For an extensive review of this evidence, see Robert Spitzer, *The Soul's Upward Yearning* (San Francisco: Ignatius Press, 2015), Chapters 1 and 2, as well as Robert Spitzer, *The Moral Wisdom of the Catholic Church* (San Francisco: Ignatius Press, 2022), Chapter 5, III.A.

(Lk 15:11–32), Jesus' address to God as "Abba"—affectionate, compassionate, forgiving, trustworthy, father/"daddy" (Mk 14:36)—and Jesus' commandment about laying down one's life for his friends (Jn 15:13).[12] If love truly is the highest way to eternal bliss in the transcendent reality (as recognized in all major religions), and Jesus has given us by far the deepest and most comprehensive articulation of this principle and capacity, then Jesus would seem to be accurately reflecting the heart of the Creator—going beyond any other religion. A bold statement, but logically sound. Did Jesus really intend this? In addition to the above, His teaching that love is the greatest commandment implies this, because the rabbis held that the heart of God is reflected in the Torah (the Law), and if the commandment to love sums up the Torah and is its highest principle, then Jesus is saying that the heart of God is love, which is consistent with what all major religions believe is the highest way to eternal bliss in the transcendent reality.

As will be seen in Chapters 2 and 3, there is other evidence of Jesus giving us the most accurate and comprehensive self-revelation of God—His Resurrection in glory, His gift of the Holy Spirit with its charismatic power of healing and prophecy, His miracles by His own authority, and above all, His self-sacrificial death, which He intended as an unconditional act of love in imitation of His Father's unconditional love. Jesus not only defined and identified love as the highest commandment and the nature of His Father but also identified with it concretely and historically in His Passion and Crucifixion. If Jesus did not give us the *fullest* articulation of God's self-revelation, then it would imply that love is not the highest way to eternal bliss in the transcendent reality, because He could not have given a deeper or more comprehensive revelation of love by His teaching, His love of sinners, His healing of the sick, His love of the poor, His Holy Eucharist, His complete self-sacrificial Passion and death, and His Resurrection and the gift of His Spirit. In sum, by the criterion given by all major religions, Jesus should be the *fullness* of God's self-revelation, because His words and actions are love and He has demonstrated that, like His Father, He is love itself.

---

[12] See Robert Spitzer, *God So Loved the World* (San Francisco: Ignatius Press, 2016), Chapters 1–3.

II. From Belief in God to Revelation in Christ: Six-Question Method

We might summarize the argument in the previous section as follows:

- Science and philosophy strongly imply the existence of an intelligent Creator and a transphysical soul capable of surviving bodily death. This in turn implies that we have a transcendent dignity and destiny that is inseparable from the transcendent intelligent Creator.
- Though science and philosophy can give evidence about the fact and powers of this Creator and our souls (the *what* of God and the soul), they cannot enlighten us on the heart of that intelligent Creator (the *who* of God and His full purpose in creating our souls); therefore, He must reveal Himself to us.
- The consensus among all major world religions is that love truly is the highest way to eternal bliss in the transcendent reality. Jesus Christ focused the whole of His moral and theological teaching on this particular virtue and attribute, saying that love is the highest commandment in which the whole Law and the Prophets are summarized and that God by nature is unconditional love (*caritas/agapē*—see below, Section III). He also demonstrated through His self-sacrificial Passion and death that He shares in His father's unconditionally loving nature. Therefore, Jesus Himself is, according to the criteria of all major religions, the highest way to eternal bliss in the transcendent reality.
- In view of this, if we seek to know and follow Jesus and His revelation of love, we will understand the path to the perfection of our nature and our relationships with others. This in turn will reveal the transcendent eternal dignity and destiny intended for us (free intelligent creatures) by the loving intelligent Creator from all eternity.

Many of us may ask the question, if we believe in an intelligent Creator and a transcendent soul capable of surviving bodily death, why do we need Jesus—why is He relevant? By now the answer might be apparent—we need not only revelation, but specifically a revelation about God's nature; and if His nature is love, then we need a revelation about the nature of love as it exists in God and how it should exist in us. Yet as we have seen, science and philosophy will not get us to that revelation by themselves. So how do we help others (and even ourselves) to move logically from the domain of science and reason to that of religion and faith? I discovered a method several years ago that uses six questions to lead inquiring minds through the transition from reason to revelation. The following story illustrates its effectiveness.

When I was teaching at Georgetown University, I was privileged to direct a science and philosophy student (Steve) on an Ignatian retreat. He was exceptionally bright and good-willed, and had the capacity to express what was on his mind in a very straightforward way. At the beginning of our first conference he said, "Could I ask you something very elementary which has been bothering me for several years? I don't have any real difficulty believing in God, because I think the evidence of physics points to the finitude of past time—implying a beginning and a creation. My real problem is Jesus—I don't get it. If I believe in God, why do I need anything more—like Jesus? Can't we just stick with a 'Creator outside of space-time asymmetry'?"

I thought about it for a couple of minutes and said to him, "Jesus is about the *unconditional love* of God. He is about God's desire to be with us in a perfect act of empathy; about God wanting to save us from evil and sin—the contrary of love—and to bring us to His own life of unconditional love. A Creator alone, indeed, even a Creator with infinite power, could be tantamount to Aristotle's God. Once He has fulfilled His purpose of ultimate, efficient, and final causation, He is detached from the affairs of rather base and boring human beings. The God of Jesus Christ is about the desire to be intimately involved in the affairs of human beings made in His image, affected by the unloving influence of evil, and destined for His eternity—and that makes all the difference."

He said in reply, "This all seems a bit too good to be true. I would like the Creator to be the God of Jesus Christ, but do you have any evidence that this is not just wishful thinking—evidence showing that this is really the way God is? Is there any reason why we would think that God is loving instead of indifferent?" I responded by noting that it would be better for *him* to answer six questions rather than have *me* give an extended discourse, because the six questions could reveal not only what was in his mind, but more importantly, what was in his heart—what he thought about love, life's purpose, others, and his highest imaginable state of existence. If he answered these six questions (from his heart) in a manner commensurate with "the logic of love", then the unconditional love and divinity of Jesus (Jesus being Emmanuel, "God with us") would become evident.

1. What is the most positive and creative power or capacity within you? Steve thought about it for a while, and he said, "Though I value my intelligence, energy, and athleticism, I have a feeling that they are not my most positive powers, because I can use them for evil and manipulative ends." I said, "That's pretty good, so if they are not your most positive powers, what do you think might fit the bill?" I really don't know whether he was raised in a Christian household, but after thinking about it, he said, "Well—is it love—or something like that?", and I enthusiastically replied, "That's what I would have said." I briefly summarize here what I discussed with him after he identified love as a potential candidate for his highest power.

At first glance, one might want to respond, as Steve did, that this power is intellect, creativity, wisdom, or artistic or literary genius, but further reflection shows that the capacity to apprehend truth or knowledge, or to create beauty, *in and of itself*, is not necessarily positive. Knowledge and beauty can be misused, and therefore be negative, destructive, manipulative, inauthentic, and thus undermine both the individual and the common good. There is but one human power that contains its own end of "positivity", one power that is directed toward the positive by its very nature, and therefore one power that directs intellect and artistic creativity to their proper, positive end. As may by now be evident, that power is love *(agapē)*. Love's capacity for empathy, its ability to enter into a unity with others leading to a natural "giving of self", forms the fabric of the common good and the human community, and so seeks as its end the good of both individuals and the community.

The highest form of love (called *caritas/agapē*) seeks the good of *the other* and derives its power from looking for the intrinsic goodness, lovability, and transcendent mystery of *the other*. For this reason, it needs no rewards like the mutuality of friendship or the romantic dimensions of *eros*. The good of the other *is* its own reward. Thus, it is not deterred by the appearance of the other, whether the other is a stranger, or even whether the other has been offensive or harmful. This enables *agapē* to be the dynamic of forgiveness, compassion, and self-sacrifice—for anyone and everyone.

Agapē by its very nature unifies, seeks the positive, orders things to their proper end, finds a harmony amid diversity, and gives of itself in order to initiate and actualize this unifying purpose. This implies that love (*agapē*) is naturally oriented toward perfect positivity and perfect fulfillment.

Furthermore, love (*agapē*) would seem to be the one *virtue* that can be an end in itself. Other virtues do not *necessarily* result in positivity or culminate in a good for others. So, for example, courage, left to itself, might be mere bravado or might lead to the persecution of the weak. Self-discipline, left to itself, might lead to a disdain for the weak or a sense of self-sufficiency that is antithetical to empathy. Even humility can be overbearing and disdainful if it is not done out of love. Even though these virtues are necessary means for the actualization of love (i.e., authentic love cannot exist without courage, self-discipline, and humility), they cannot be ends in themselves, for they can be the instruments of "unlove" when they are not guided by the intrinsic goodness of love. Love seems to be the only virtue that can be an end in itself and therefore can stand by itself.

If we affirm this power within ourselves and further affirm that it is the guiding light of both intellect and creativity, that its successful operation is the only way in which all our other powers can be guided to a positive end, that it is therefore the only way of guaranteeing positivity for both ourselves and others, and that it therefore holds out the promise of authentic fulfillment, purpose in life, and happiness, then we will have acknowledged *love* to be the highest of all powers and the central meaning of life. We will then want to proceed to the next question.

2. *If love is the one power that seeks the positive in itself, and we are made to find our purpose in life through love, could God (the intelligent Creator of the universe and our transphysical souls), who created us with this loving nature, be devoid of love?* When I asked this question, Steve readily agreed that if the Creator of the universe was the transcendent Creator of our transphysical souls, then God would have to be self-contradictory to

create us with a soul that yearns for love as its highest most positive fulfillment while Himself being fundamentally unloving. He thought this was inconsistent not only with the nature of an *intelligent Creator* but also with his and others' desire for love and to love. The following summarizes our subsequent discussion.

If the Creator were devoid of love, why would that Creator create human beings not only with the capacity for love but for them to be fulfilled only when they are loving? If the Creator is devoid of love, why make love the fulfillment of all human powers and desires, and therefore of human nature? If the Creator is not loving, then the creation of "beings meant for love" seems absurd. However, if the Creator *is* love, then creating a loving creature (i.e., sharing His loving nature) would seem to be both intrinsically and extrinsically consistent with what (or perhaps better, "who") He is. Could the Creator be any less loving than the "loving nature" He has created? Furthermore, if the Creator is perfectly intelligent—a unique unrestricted act of thinking[13]—wouldn't that perfection extend to the highest perfection (love)?

If we affirm the love of the Creator from the above, then we can proceed to the third question.

3. Is your desire to love and to be loved merely conditional or unconditional? At this point, Steve bristled, saying, "Look—I am a realist. I really don't believe that a desire for unconditional love can be fulfilled, and therefore, I don't believe that I *truly* have a desire for unconditional love—I know better." I asked him, "Well—before your realist viewpoint stifled your desire for unconditional love, do you think that you desired unconditional love? Did you ever have a girlfriend who you sort of wanted to love you unconditionally—to be perfectly understanding, perfectly responsive, perfectly authentic, and perfectly desirous to be a perfect soul mate? Over the course of your relationship with her, did you get progressively more frustrated when she was not *perfectly* understanding, authentic, and a soul mate? Finally, did her imperfections cause you to give up in frustration and declare, 'She's not the one!', which resulted in your breaking up with her?"

---

[13] See Spitzer, *Doorstep to God*, Chapter 3, Section II (the philosophical proof of God), and Section III (proof that the uncaused reality is an unrestricted act of thinking).

Steve candidly responded, "I have done precisely that!" To which I retorted, "Well, what were you looking for?" He responded, "Uhhh—unconditional love?" I said, "I think so." The following summarizes our discussion.

In *Science at the Doorstep to God*,[14] we showed that we not only have the power to love (i.e., the power to be naturally connected to another human being in profound empathy, care, self-gift, concern, and acceptance); we have a "sense" of what this profound interpersonal connection would be like if it were perfect. This sense of perfect love has the positive effect of inciting us to pursue ever more perfect forms of love. However, it has the drawback of inciting us to *expect* ever more perfect love from others. This generally leads to frustrated expectations of others and consequently to a decline of relationships that can never grow fast enough to match this expectation of perfect and unconditional love.

The evidence for our awareness of and desire for perfect love can be seen in our capacity to recognize *every* imperfection of love in others and in ourselves.[15] How could we have this seemingly unlimited capacity to recognize imperfection in love without having some sense of what perfect love would be like? Without at least a tacit awareness of perfect love, we would be quite content with any manifestation of affection that just happens to come along.

If we have a capacity to recognize imperfection of love in others and ourselves—seemingly unlimitedly—and sense in this ability our tacit awareness of what perfect love would be like, then we must ask ourselves whether we would be satisfied with anything less than unconditional love. If not, proceed to the next question.

4. If our desire for love can be ultimately satisfied only by unconditional love, then could the Creator of this desire be anything less than unconditional love? At this point, Steve simply said, "I can see where you are going, and I am not sure that I want to go there, but it seems likely that if I have a desire for unconditional love, which is beyond anything I could manufacture or experience, then perhaps the Creator of my desire for unconditional love is itself unconditional love." I indicated that his

<hr/>

[14] See ibid., Chapter 6, Section III, for a discussion about the five transcendental desires.
[15] Ibid.

tentative agreement was enough to proceed with the argument, but I tried to supplement his thinking with the following thoughts.

If we assume that the Creator does not intend to frustrate our desire for unconditional love, it would seem that His creation of the desire would imply an intention to fulfill it, which would, in turn, imply the very presence of this quality within Him. This would mean that the Creator of the desire for unconditional love is Himself unconditional love.

Recall from above that we have the capacity to recognize every imperfection of love in others and ourselves, revealing at least a tacit *awareness* of perfect love that brings these imperfections to light. This tacit awareness of unconditional love seems to be beyond any specifically known or concretely experienced love, because every manifestation of love we encounter is *imperfect*. How can we have an awareness of unconditional love that we have not experienced? How can we even extrapolate to it if we do not know what we are looking for? So it seems that there must be some source of our awareness of unconditional love that is capable of unconditional love. This source would seem to be the Creator of our transcendental nature—God.

If we are in agreement with "God being unconditional love", then proceed to the next question.

5. If the Creator is unconditional love, would He want to be with us and enter into a personal empathetic relationship with us—face-to-face? Would he be Emmanuel ("God with us")? Would it be typical of an unconditionally loving God to want to become incarnate? When Steve heard the question, he responded right away, "I think I get the 'Jesus thing'. An unconditionally loving God not only would want to reveal Himself to us but to enter into a relationship with us on our own terms— through felt empathy manifest in voice, eyes, touch, and feeling; in this way, He could show us the way to unconditional love through word and action. I get it!" The following summarizes our subsequent discussion.

If one did not attribute unconditional love to God, then the idea of God wanting to be with us would be implausible. If God were not loving, He would not bother to relate to creatures, let alone actually be among them and enter into empathetic relationship with them.

However, in the logic of love, or rather, in the logic of unconditional love, all this changes.

If we attribute the various parts of the definition of *agapē* to an unconditionally loving Creator, we might obtain the following result: God would be focused on what is uniquely good, lovable, and mysterious in each one of us, and in seeing this perfectly would enter into a perfect empathetic relationship with us—whereby doing the good for us would be just as easy if not easier than doing the good for Himself. Thus, God would empathize with us and do the good for us unconditionally—without expecting a "reward". He would love us unconditionally even if we did not love Him—even if we resented and rejected Him. He would love us unconditionally even if we had sinned terribly—so terribly that we had no hope of being excused, but only forgiven. His unconditional love would seek as deep a relationship with us as *we*, in our freedom, would allow. He would not only want to be with us in deepest intimacy; He would even sacrifice Himself for us—sacrifice Himself *unconditionally* for us, even if we did not deserve it, *particularly* if we did not deserve it. If God were unconditional love, then God's love would naturally extend itself to us in an unmitigated act of compassion and affection, irrespective of our transgressions. If we open ourselves and respond to His love, He will deepen it until He brings us into the fullness of relationship with Him, which is perfect joy. If God truly is unconditional *agapē*, then it would be perfectly consistent with His nature (and heart) to want to be perfectly present to us—as Emmanuel.[16]

If God is truly unconditional love (*agapē*), then He is also unconditional empathy; and if He is unconditional empathy, He would want to enter into a perfectly empathetic relationship with us—"face-to-face" and "peer-to-peer"—where the Lover and beloved would have an equal access to the uniquely good and lovable personhood and mystery of the other through empathy. A truly unconditionally

---

[16] Even though an unconditionally loving God would never stop loving us, He would give us the freedom to reject His love, because He would not want to force it upon us. Therefore, He would have to make some accommodation for those who wanted to live without Him and without love—even eternally. So, an unconditionally loving Creator would not create Hell to torment sinners, but rather to give people who want to reject Him (and to live a life without this kind of love) a *separate* place for their eternal dwelling—away from the love, God, and others they do not want to be near.

loving Being would want to give *complete* empathetic access to His heart and interior life in a way that was proportionate to the receiving apparatus of the weaker (creaturely) party. Thus, it seems that an unconditionally loving Creator would want to be Emmanuel in order to give us complete empathetic access to that unconditional love through voice, face, touch, action, concrete relationship, and in every other way that love, care, affection, home, and felt response can be concretely manifest and appropriated by us. If God really is unconditional love, and *agapē* is the perfection of love, then we might expect that this God would want to be perfectly present to us as Emmanuel. If this resonates with your thoughts and feelings, proceed to the next question.

6. Inasmuch as the unconditionally loving God would want to be perfectly present to us, is Jesus the One? Is He Emmanuel? Steve responded, "I really don't know the answer to that question, and I will have to do a lot of study to figure it out. However, if God really is unconditional love, then Jesus seems like a very good candidate to be the full expression of that love." The following is for the reader. Steve spent a good part of his retreat studying some of it.

To answer this question, we will want to examine three dimensions of Jesus' teaching and life:

- Jesus' proclamation of the unconditional love of God (see Section III below)
- His life of unconditional love (see Section III below)
- His manifestation of divine authority and power—in His Resurrection, gifts of the Spirit, and miracles (see Chapters 2 and 3)

An examination of the remainder of this chapter as well as Chapters 2 and 3 can show us that Jesus really is unconditional love—sharing this nature with His divine Father—and that He holds the key not only to the definitive revelation of our ultimate dignity and destiny but also to the eternal love promised by His Father.

As we reflect on these three dimensions of Jesus' revelation in both our minds and hearts, it will impart a gradual freedom to believe. Instead of thinking, "God with us is too good to be true", we begin to think that an unconditionally loving God would *really* want to be with us—and if He were to come, that Jesus would be His perfect presence to us.

We might use Pascal's words to describe the above questioning process—the heart's process. If we affirm the validity of this reasoning, it now remains we must enter with our hearts into the mystery of Jesus' teaching and life of love. As we do so, we are likely to feel a deeper affinity for Him and recognize His love for us personally. This will galvanize the truth about *agapē*, Jesus' identity, and the Father's love within our hearts.

III. How Jesus' Words and Acts of Love Changed the History of Religions

We ended the six questions with, "Is Jesus the one—the Emmanuel—the unconditional love of God with us?" To answer this question we will first explore Jesus' distinctive meaning of "love" and how it changed the world (this section) and then explore the divine power He manifested in His Resurrection in glory, miracles, and gift of the Holy Spirit (Chapters 2 and 3). The synthesis of these two studies will show that divine power is unconditional love and that unconditional love is divine power—the two come together not only in the teachings and actions of Jesus but more importantly in Jesus Himself.

This section will be divided into three subsections:

1. Jesus' Distinctive Teaching on Love (Section III.A)
2. Jesus' Distinctive Teaching on the Unconditional Love of God: The Father of the Prodigal Son (Section III.B)
3. The Effects of Jesus' Teaching on World History (Section III.C)

A. Jesus' Distinctive Teaching on Love

As noted above, all major world religions (according to Heiler) believe that the supreme transcendent power is both good and loving. This general belief is interpreted quite differently in each of the major religions. Some subordinate love to justice and the moral law, some hold that the deity's love is oriented toward a group rather than toward individuals, and some define love in a very restricted way.

Jesus' teaching on the love of God is quite distinct. First, He proclaims the *unconditional* love of God and places it at the center of His teaching, making all other teachings and doctrines subordinate to it. He also defines love in a special way that requires Christians to find a distinct word to describe it (*agapē*). In these two respects, Jesus appears to be quite distinctive in the history of religions. Jesus proclaims the unconditional love of God through several distinct teachings. First, He teaches His disciples to address God as He does—as "*Abba*", which means "affectionate, understanding, trustworthy father" with connotations of childlike delight, for example, "daddy" (see below, Section III.B). Second, He identifies God the Father with the father in the Parable of the Prodigal Son (Lk 15:11–32)—who is unconditionally forgiving, compassionate, and humble (see below, Section III.B). Third, He says that the whole Law and the Prophets are summed up in the commandments to love God and neighbor. Inasmuch as the Torah (the Jewish law) reflects the heart of God, love must be the essence of God's heart.

Jesus places His radical doctrine of love (His definition of love and His proclamation of God's unconditional love) at the very center of His teaching, making all other teachings subordinate to it. The combined effect of these proclamations is a distinctive recasting of love into the primary end or goal of every individual—and even of history and culture.

With respect to Jesus' definition of love, He is primarily concerned with the *interior heart* of love. This is most manifest in the Beatitudes (interior attitudes of love), which are placed at the *beginning* of Jesus' Sermon on the Mount (Mt 5:3–11)—showing the priority of the interior disposition of the heart, in care and compassion. We may summarize them as follows:

- "Poor in spirit", which means humble-hearted (5:3)
- "Meek" (5:5), which means gentle-hearted
- "Hunger and thirst for righteousness" (5:6), which means the ardent desire to follow the way of God
- "Merciful" (5:7), which means unqualified forgiveness as well as compassion for the weak, marginalized, and poor

- "Pure in heart" (5:8), which means true to oneself and others
- "Peacemaker" (5:9), which means bringing reconciliation and peace to everyone we touch

Jesus gives several examples in His parables and actions to illustrate this love. The most notable parable, the Parable of the Good Samaritan, portrays a foreigner whose *heart* is moved with *care* and *compassion* toward a Jewish man (an enemy of the Samaritans) who has been beaten severely by robbers (Lk 10:25–37). The most notable action in Jesus' life is His love for sinners and His self-sacrificial death on the Cross.

In the concluding passages of the first part of the Sermon on the Mount, Jesus then shows how these interior attitudes should manifest themselves in exterior actions—turning the other cheek (Mt 5:39) as well as loving our enemies and praying for those who hate us (5:44). The combination of these teachings in a single doctrine of love is distinctive in the history of religions. Jesus' transformation of the reality of love goes even further—to the heart of God Himself.

B. Jesus' Distinctive Teaching on the Unconditional Love of God: The Father of the Prodigal Son

Jesus illustrates His revelation of the Father's love in the well-known Parable of the Prodigal Son (Lk 15:11–32). This parable may be considered one of Jesus' primary revelations of God the Father's *unconditional* love.

Three preliminary considerations should be made before retelling the parable as a first-century audience would have understood it. First, Jesus intends that the father in the story be a revelation of the heart of God the Father. The parable would be more aptly named the Parable of the *Father* of the Prodigal Son. Second, notice that the younger son has committed just about every sin imaginable according to the mindset of Second Temple Judaism (the religious context in which Jesus was living), and so he has absolutely no basis or merit for asking the father to receive him back into the household—even as one of the servants. Third, the older son in this story represents the Pharisees and those who are trying to remain righteous according to their understanding of the Jewish law. So we can see that Jesus has not abandoned them, but He desires to give them everything He has—so long as they come back into the house.

Now we may proceed to a retelling of the parable. A father had two sons, the youngest of whom asked for his share of the inheritance. This would have been viewed as an insult to the father, which would have shamed both father and family (because the son is asking not only for the right of possession but for the right of disposal of the property, which legally does not occur until the death of the father[17]). Nevertheless, the father hears the son's request and acquiesces to it. He divides his property and lets his son go. Remember, the father in the story is Jesus' revelation of God the Father.

The son chooses to go to a foreign land—probably a Gentile land, indicated by his living on a Gentile farm with pigs. Whether he started there or simply ended there is of little consequence. His actions indicate a disregard for (if not a rejection of) his election and his people, and a further shaming of the family from which he came.

Then the son adds further insult to injury by spending his father's hard-earned fortune on dissolute living (violations of the Torah) in the Gentile land. This shows the son's callous disregard for (if not rejection of) God's law, God's revelation, and perhaps God Himself. Furthermore, he manifests his callous disregard for his people, the Law, and God before the entire Gentile community—bringing shame upon them all.

Just when it seems that the son could not possibly sin any more egregiously, the foreign land finds itself in a famine. The son has little money left and is constrained to live with the pigs, which were considered to be highly unclean animals. The son incurs defilement not only from working with the pigs but from actually living with them! He even longs to eat the food of the pigs, which would defile him both inside and outside. This reveals the son's wretched spiritual state, which would have engendered both disgust and revulsion from most members of Jesus' first-century audience.

The son experiences a "quasi-change" of heart, not so much because of what he's done to his family, country, people, election, law, religion, and God, but because of the harshness of his condition

[17]Joachim Jeremias, *The Parables of Jesus*, 2nd ed. (London: SCM Press, 1972), pp. 128–29.

("How many of my father's hired servants have bread enough and
to spare, but I perish here with hunger!"—15:17). He decides to
take advantage of what he perceives to be his father's merciful nature
by proffering an agreement to accept demotion from son to servant
(even though it was the father's right to reject and even disown him
altogether). The son then makes his way back home.

The father (who represents God—the Abba figure in Jesus' parable) sees him coming while he is still on his way (possibly indicating
that the father is looking for him) and is so completely overjoyed that
he runs out to meet him (despite the fact that the son has so deeply
injured and shamed both him and his family). When he meets his
son, he throws his arms around him and kisses him. The kiss is not
only an act of affection but also a sign of forgiveness.[18] The son's list
of insults, injuries, and sins is incapable of turning the father's heart
away from him. The father is almost compelled to show unrestrained
affection toward him. The son begins to utter his prepared speech
of quasi-repentance/quasi-negotiation: "Father, I have sinned against
heaven and before you; I am no longer worthy to be called your
son" (15:18–19, 21). But before he can say, "Treat me as one of
your servants" (15:19), the father tells the servants to get him a robe,
which not only takes care of his temporal needs but is also a mark
of high distinction.[19] He then asks that a ring be put on his hand.
New Testament scholar Joachim Jeremias indicates that this ring is
very likely a signet ring,[20] having the seal of the family. This would
indicate not only belonging to the family but also the authority of the
family (showing the son's readmission to the family in an unqualified
way). He then gives him shoes, which again takes care of his obvious
temporal need, and inasmuch as they are luxuries, signifies a free man
who no longer has to go about barefoot like a servant or slave.[21] He
then kills the fatted calf (reserved only for very special occasions) and
holds a feast. This is a further indication of the son's readmission to
the family by being received at the festal family table.[22]

Jesus' audience probably felt conflicted (if not angered) by the
father's "ridiculously merciful" treatment of his son, because it

[18] Ibid., p. 130.
[19] Ibid.
[20] Ibid.
[21] Ibid.
[22] Ibid.

ignored (and even undermined) the "proper" structures of justice. The father's love and mercy seem to disregard the justice of the Torah. This does not deter Jesus, because He is convinced that God the Father treats sinners—even the most egregious sinners—in exactly the same fashion, that is, with a heart of unconditional love.

Jesus continues the story by turning His attention to the older son, who reflects a figure of righteousness according to the Old Covenant. He has stayed loyal to his father, family, election, country, religion, law, and God. Furthermore, he has been an incredibly hard worker and seems to accept patiently the father's frugality toward him, saying, "You never gave me a kid [goat]" (15:29). Most of Jesus' audience probably sympathized with this older son's plight when the father demonstrated his extraordinary generosity to his younger son. By all rights, the father should have either rejected or disowned the younger son, and if not that, he certainly should have accepted the younger son's offer to become a servant—but an unqualified readmittance to the family appeared to be an injustice (if not a slap in the face) to his loyal son.

The father understands the son's difficulty with his actions and goes outside to "entreat" (15:28) his son, virtually begging him to come back into the house (an almost unthinkable humiliation for a father at that time). He begins by giving his older son all his property, addressing his older son's need for justice, telling him, "You are always with me, and all that is mine is yours" (15:31). Then he gives him an explanation that did not fall within the mainstream interpretation of the Law: mercy must take precedence over justice and love take precedence over the Law, for that is the only way that the negativity of sin and evil can be redressed and overcome—"Your brother was dead, and is alive; he was lost, and is found" (15:32).

If we view the father in this parable as Jesus' definitive revelation of the heart of God His Father, we can also see by contrast some false notions of God—the "payback God", the "disgusted God", the "indifferent God", the "stoic God", and "the intransigently angry God". We can then use the portrait of the father and the prodigal son to counter these false notions of God as they pop up in our lives. The mystery of Jesus' transformation of love goes even further because practicing it transforms the people and culture around us. As we shall see, it had an exceedingly positive effect upon the Middle East, Europe, and even the world.

C. The Effects of Jesus' Teaching on World History

Most religions emphasize the Silver Rule.[23] It can be roughly translated as, "Do not do a harm to others that you would not want done to you." This is generally termed "ethical minimalism" because it places the emphasis on avoiding harm rather than doing good.

When Jesus removed the "nots" from the Silver Rule, He converted it from "ethical minimalism" to "ethical maximalism". The emphasis is no longer on merely avoiding harm but also on doing good (beyond the avoidance of harm)—doing every good to another that you would want done to you. There is really no limit to these goods, and so the Golden Rule might be viewed as "open-ended altruism".

Jesus not only emphasized the Golden Rule but also made it *central* to His ethical doctrine. He superseded the Silver Rule and rejected the *lex talionis*:

> You have heard that it was said, "An eye for an eye and a tooth for a tooth." But I tell you, Do not resist one who is evil. But if any one strikes you on the right cheek, turn to him the other as well. (Mt 5:38–39; Jesus was quoting Ex 21:24)

Jesus' positive ethical maximalism has its origins in His doctrine on love. He asks us to imitate the Father's love of enemies (Mt 5:44–48), to forgive everyone from the heart (Mt 6:12), not to judge others negatively (Mt 7:1–5), to consider *everyone* our neighbor—worthy of compassionate love (like the Good Samaritan—Lk 10:25–37). When we look at these teachings collectively, we can see Jesus' underlying viewpoint that love, mercy, and compassion are higher than justice (which is derived from the Silver Rule). Love and mercy (from which the Golden Rule is derived) go beyond justice (the Silver Rule) and encourage a positive, altruistic, and compassionate social order. This emphasis had a profound effect on the development of sociopolitical ethics throughout the world, particularly with respect to the development of universal public healthcare and

---

[23] See Spitzer, *Moral Wisdom*, Chapter 6, Section I.A.1. Note: The Silver Rule is mentioned explicitly in the Old Testament two times (Tob 4:15; Sir 31:15).

welfare and universal public education, as well as on the development of inalienable rights and social responsibility.

Inasmuch as Jesus is the fullness of God's self-revelation (as the divine Son manifest in His glorious Resurrection, miraculous power, self-revelation, gift of the Spirit, and self-sacrificial love on the Cross), we should give serious consideration to following Jesus' words and actions, which we may reasonably infer come from God Himself.

This revelation has changed the course of history and the world. The Christian Church created such large missions to help the needy, cure the sick, and educate all classes of people, she ultimately undermined the barbarity, social stratification, and slavery of Rome.[24] Throughout her history, the Catholic Church has been the largest public educational system, healthcare system, and public welfare system in the world. She remains so today:

- With respect to nonstate education, recent data shows that the Church provides services in approximately fifty thousand secondary schools and one hundred thousand primary schools.[25]
- With respect to healthcare, the Church oversees approximately a quarter of all worldwide healthcare facilities and hospitals.[26]
- With respect to public welfare, the Church provides services in approximately fifteen thousand homes for the elderly, chronically

[24] For a summary of how the Catholic Church progressively weakened the institution of slavery and social elitism and built the largest international education, healthcare, and welfare systems, see ibid., Chapter 6. See also Helmut Koester, "The Great Appeal: What Did Christianity Offer Its Believers That Made It Worth Social Estrangement, Hostility from Neighbors, and Possible Persecution?", Frontline, WGBH Educational Foundation, April 1998, https://www.pbs.org/wgbh/pages/frontline/shows/religion/why/appeal.html; and Christopher Dawson, The Formation of Christendom (New York: Sheed & Ward, 1965), pp. 111–37.

[25] Quentin Wodon, Global Catholic Education Report 2023: Transforming Education and Making Education Transformative (Washington, D.C.: Global Catholic Education, 2022), p. 98, https://www.globalcatholiceducation.org/_files/ugd/b9597a_b54239f33dec48ddb4f2d735d10cba7c.pdf.

[26] Matt Moran, "The Role of the Catholic Church in Healthcare Provision Globally", Independent Catholic News, October 11, 2023, https://www.indcatholicnews.com/news/48212#:~:text=The%20contribution%20of%20the%20Catholic,services%20throughout%20Africa%20is%20inestimable. This same percentage was reported in 2010; see Catholic News Agency, "Catholic Hospitals Comprise One Quarter of World's Healthcare, Council Reports", February 10, 2010, https://www.catholicnewsagency.com/news/catholic_hospitals_represent_26_percent_of_worlds_health_facilities_reports_pontifical_council.

ill, and disabled, as well as approximately ten thousand orphanages, marriage counseling centers, and nurseries—not including
any healthcare facilities.[27]

Inasmuch as these institutions arising out of the teaching of Jesus
have literally transformed history and the world, we should seriously
consider the truth and efficacy of Jesus' revelation about the unconditional love of God, the definition of love (as highest commandment),
and His foundation of the Catholic Church on these principles.

IV. Faith and Our Interior Awareness of God

Up to now, we have addressed the transition from reason and science to revelation and faith through a consensus of world religions
(Section I), a series of *logical* deductions on love (Section II), and the
distinctiveness and effectiveness of Jesus' teaching on love and His
divine power (Section III). Readers may be thinking that this analysis places priority on the mind (rather than the heart) and the outer
world (rather than our inner world). Does our interior life—intuition,
choices, and feelings—have anything to do with the movement to
revelation and faith? By now, it may be evident that the discussion of
love in Sections II and III is filled with strong interior intuitions and
feelings because love cannot be divorced from them. In this section,
we will move beyond the intuition and feelings of human love to
our sense of divine love in the numinous experience (Rudolf Otto)
and the "fatherliness" of the morally authoritative voice within our
conscience (Saint John Henry Newman).

The above analysis of world religions, deductions on love, and
Jesus' distinctive view of love implies that Jesus Christ holds out
the opportunity to reach our true dignity, destiny, fulfillment, and
perfection—eternally. When this rationale is filled with the interior
heartfelt awareness of faith, it speaks loudly that knowledge—even
the highest, most important knowledge—is not enough; we need

[27] Agenzia Fides (Vatican news agency), "Vatican—Catholic Church Statistics 2023",
October 20, 2023, https://fides.org/en/news/74319-VATICAN_CATHOLIC_CHURCH
_STATISTICS_2023.

a real loving relationship, a real loving community, and real loving service to bring that knowledge to its fulfillment. This requires *faith*, which is our next subject.

Faith begins not with us but with God's presence and invitation to us. When we talk about our sense of the moral, the holy, and love, we are not speaking only of feelings but about something relational—a connection between two minds or hearts.[28] Thus, as Newman so brilliantly articulates, our sense of morality comes not from feelings of the superego but rather from God's presence to our conscience (the presence of another authoritative yet fatherly being). As he states it:

> This is Conscience, and, from the nature of the case, its very existence carries on our minds to a Being exterior to ourselves; or else, whence did it come? And to a being superior to ourselves; else whence its strange, troublesome peremptoriness? ... Its very existence throws us out of ourselves and beyond ourselves, to go and seek for Him in the height and depth, whose voice it is.[29]

Furthermore, our sense of the holy comes not only from feelings but from God's numinous presence within our consciousness—what Rudolf Otto calls the "numinous experience"—a preconceptual awareness of the mysterious, fascinating, overpowering, yet inviting, wholly Other.[30]

Finally, our sense of an obligation to love originates not simply with our isolated feelings of affection, but with an experience of empathy—experiencing unity in feeling and being with another. As Newman and Otto show, we have empathy not only with other human beings but also primordially with a divine Other. The obligations of conscience are imbued with empathy and love of the fatherly authority, and the numinous experience is also imbued not only with awareness of the overpowering and mysterious but also with fascinating and inviting love.[31]

[28] See Spitzer, *Doorstep to God*, Chapter 6, Sections I–III.

[29] John Henry Newman, "Proof of Theism", in *The Argument from Conscience to the Existence of God according to J. H. Newman*, ed. Adrian Boekraad and Henry Tristram (Louvain: Editions Nauwelaerts, 1961), pp. 114–15. (The essay was not previously published.)

[30] Rudolf Otto, *The Idea of the Holy: An Inquiry into the Non-Rational Factor in the Idea of the Divine and Its Relation to the Rational* (New York: Oxford University Press, 1958).

[31] Spitzer, *Soul's Upward Yearning*, pp. 50–62.

Inasmuch as Newman and Otto are correct about the interior presence of God within our conscience, our sense of the numinous, and our sense of primordial empathy, then we might say that faith begins with this threefold presence of God within us. However, this is not yet faith, for faith requires our response—our openness to this numinous, sacred, wholly Other, and His authoritative precepts within our conscience. If we do not want to be responsible to a moral authority beyond ourselves, if we do not want to be a creature before the wholly Other, if we want to be completely autonomous—an authority unto ourselves (responsible only to ourselves)—if we want to be free from any authority and responsibility beyond ourselves, then we will never move into the domain of faith. No evidence will ever convince us of a creative and transcendent authority, and no power in the world will be able to compel us to "cross the line".

However, if we are open to the interior presence of the authoritative and numinous wholly Other, then it is quite natural for us not only to acknowledge Him but also to be filled with wonder, desire, and awe. To the extent that we are open to (accepting of) the divine reality within us, we will sense fear, respect, and creatureliness, as well as being accepted, invited deeper, and loved. This felt relationship with the wholly Other is filled with anticipation—anticipation of fulfillment of our desires, fulfillment of our natures, absolute meaning, absolute groundedness—not only for this world but also for the life to come. Most of the time, this anticipation of absolute meaning, fulfillment, and completeness is quite subtle, but sometimes the source of the fulfillment breaks out for a fleeting moment with His sacredness, joy, love, and unity with everything—what the mystics would call "bliss".[32]

Let us now return to the topic of faith. We might describe the first stage of faith as a divine invitation to open ourselves to the sacredness and moral precepts manifest in our numinous awareness and conscience. If we respond by opening ourselves to this sacred wholly Other, He initiates the second stage of faith by giving us a sense of anticipation of absolute fulfillment, meaning, and completeness, which fills us with wonder and desire. Subtle though this anticipation often is, we feel drawn to praise and thank Him and to follow His sacred way. If we follow through on this desire to give thanks, praise,

[32] Otto, *Idea of the Holy*, pp. 16–18, 38–44.

and obedience, we will naturally want to know more about Him and find the best and truest way to worship and follow Him. This provokes us to seek an authoritative declaration of the Divinity's self-revelation, as well as an authoritative public expression of worship. This, in turn, leads almost invariably to a commitment to a religious authority and community—a church.

About 84 percent of the world's population associates with a public religious community,[33] not only because of cultural traditions but also because of the profound awareness of divine holiness and moral authority within us. The idea that someone can be religious on their own—without an authoritative source of God's revelation and worship—is unlikely. People who follow through the second stage of faith want *authoritative* revelation and worship, and they know that they are *not* this authoritative source. Hence, they look outside of themselves for that authority and are naturally drawn to religious communities led by seemingly authentic representatives of the Divine. Those who do not belong to a church are unlikely to be in the second stage of faith, because the desire to thank, praise, and obey the divine authority almost invariably leads to the desire for authentic and authoritative divine revelation and worship, which in turn leads to an authoritative religious community.

When we sincerely begin to follow the precepts of a religious community in our desire to thank, praise, and obey the divine authority (the third stage of faith), the Divinity responds by providing inspiration, guidance, and protection, as well as absolute meaning and a path to eternal fulfillment, completeness, and bliss. This heightened relationship with the Divinity and the graces it brings elevate us to a whole new level of being. We are no longer merely profane beings, but share in the sacredness, moral goodness, and transcendence of the Divinity. Our souls are no longer dormant, but fully charged with the presence of God. Feelings of spiritual emptiness, loneliness, alienation,

[33] Pew Research Center, "Religious Composition by Country, 2010–2050", December 21, 2022, https://www.pewresearch.org/religion/interactives/religious-composition-by-country-2010-2050/. This same percentage was reported by the Pew Research Center in 2012, when it stated that "a comprehensive demographic study of more than 230 countries and territories conducted by the Pew Research Center's Forum on Religion & Public Life estimates that there are 5.8 billion religiously affiliated adults and children around the globe, representing 84% of the 2010 world population of 6.9 billion." Pew Research Center, "The Global Religious Landscape", December 18, 2012, https://www.pewresearch.org/religion/2012/12/18/global-religious-landscape-exec/.

dread, and guilt are overcome and replaced by a sense of completeness and incorporation into perfect truth, love, goodness, beauty, and
home. Several psychiatric studies show that nonreligiously affiliated
people experience significantly higher rates (doubling or more) of
depression, anxiety, malaise, substance abuse, familial tensions, suicidal
ideation, and suicides.[34] Evidently, the feelings of emptiness, loneliness, and alienation coming from nonreligious affiliation (nonaffilia-
tion with the Divinity) undermine us emotionally and relationally as
well as spiritually. All of this is reversed when we commit ourselves
to following the Divinity through religious affiliation and we become
transformed in being—categorically different from the merely profane
person, for we now have access to the only fulfillment and meaning
that will satisfy and complete us: perfection in truth, love, goodness,
beauty, and being, which is found only in the Divinity Himself.[35] As
Saint Augustine notes at the beginning of his *Confessions*, "You have
made us for yourself, and our heart is restless until it rests in you."[36]

When the relationship with God (with the invitation and free
response of faith) is infused with Jesus' distinctive idea of love (*caritas/*

[34] Kanita Dervic et al., "Religious Affiliation and Suicide Attempt", *American Journal of Psychiatry* 161, no. 12 (December 2004): 2303–8, https://ajp.psychiatryonline.org/doi/full /10.1176/appi.ajp.161.12.2303. See also Harold Koenig, "Research on Religion, Spirituality and Mental Health: A Review", *Canadian Journal of Psychiatry* 54, no. 5 (2009): 283–91, https:// pubmed.ncbi.nlm.nih.gov/19497160/; Raphael Bonelli et al., "Religious and Spiritual Factors in Depression: Review and Integration of the Research", *Depression and Research Treatment* (2012): 1–8, https://www.hindawi.com/journals/drt/2012/962860/; Stefano Lassi and Dan-iele Mugnaini, "Role of Religion and Spirituality on Mental Health and Resilience: There Is Enough Evidence", *International Journal of Emergency Mental Health and Human Resilience* 17, no. 3 (2015), 661–63, https://www.omicsonline.org/open-access/role-of-religion-and -spirituality-on-mental-health-and-resilience-there-is-enough-evidence-1522-4821-1000273. pdf; Harold Koenig, "Religion, Spirituality, and Health: A Review and Update", *Advances in Mind-Body Medicine* 29, no. 3 (2015): 19–26, https://pubmed.ncbi.nlm.nih.gov/26026153/; and Corina Ronenberg et al., "The Protective Effects of Religiosity on Depression: A 2-Year Prospective Study", *Gerontologist* 56, no. 3 (2016): 421–31, https://academic.oup.com /gerontologist/article/56/3/421/2605601.

[35] The philosophical proof of the one uncaused, unrestricted reality being perfect truth, perfect love, perfect goodness, and perfect beauty is found in Robert Spitzer, *New Proofs for the Existence of God: Contributions of Contemporary Physics and Philosophy* (Grand Rapids, Mich.: Eerdmans, 2010), Chapter 7.

Note also that the identification of perfect truth, goodness, and beauty with the Divinity is one of the common characteristics of most major religions—Heiler's third characteristic (see Section I above).

[36] Saint Augustine, *Confessions* 1, 1, ed. and trans. Henry Chadwick (New York: Oxford University Press, 1991), p. 3.

agapē) and directed toward Jesus' divine Father (portrayed in the Parable of the Prodigal Son), it naturally moves toward a *Christian* community through which that relationship can deepen and grow in worship, prayer, and learning. At this point, Christ's love moves beyond a mere idea to a living spiritual reality, opening upon a life of goodness and service, culminating in eternal salvation for us and those we touch. Thus, faith transforms intellectual belief into the reality of divine love.

V. Why Christian Faith?

In the Dogmatic Constitution on the Church, the Second Vatican Council Fathers state:

> Those also *can attain* to salvation who through no fault of their own do not know the Gospel of Christ or His Church, yet sincerely seek God and moved by grace strive by their deeds to do His will as it is known to them through the dictates of conscience. Nor does Divine Providence deny the helps necessary for salvation to those who, without blame on their part, have not yet arrived at an explicit knowledge of God and with His grace strive to live a good life.[37]

If all authentically religious people are eligible for eternal salvation, we might ask, why pursue specifically Christian faith? Four reasons stand out among others:

1. *The completeness of Jesus' revelation of God.* As noted above (Section I), Jesus' revelation in its development and primacy of love is apparently the most accurate and comprehensive manifestation of God's self-revelation, and therefore, the most comprehensive, deepest, and most nuanced way to know, love, and serve God—and through this to be brought into eternal salvation. Furthermore, as will be shown in Chapters 2 and 3, there is significant evidence for Jesus' Resurrection in glory, gift of the Spirit, miracles by His own authority, and self-sacrificial

---

[37] Vatican Council II, Dogmatic Constitution on the Church *Lumen Gentium* (November 21, 1964), no. 16, https://www.vatican.va/archive/hist_councils/ii_vatican_council/documents/vat-ii_const_19641121_lumen-gentium_en.html (emphasis added).

Passion and death, which He intended as an unconditional act of love for mankind. These events substantiate the truth of His claim to be the only begotten Son of His divine Father—and therefore to be divine. This makes His revelation absolutely significant and makes His self-sacrificial death absolutely efficacious for the salvation of the world.

2. *The goodness of Christianity*. As we saw in Chapter 1, Jesus' definition of love is the best candidate for the full actualization of human beings. This view of love also brings to fulfillment the dictates of our conscience and our highest religious aspirations. Furthermore, Jesus' specification of what this love entails in moral life, marriage, and relationships with others is the most comprehensive, consistent, and deep articulation of morality and goodness ever proposed in human history.[38]

3. *The path to salvation*. Inasmuch as Jesus has the most complete and comprehensive expression of the truth and goodness of God's self-revelation, the path He gives us to salvation will not have inaccuracies, incompleteness, pitfalls, and other problems that could distract from or negatively affect our path to His eternal life. Other paths may be ultimately successful, but could be less direct, more lengthy, and require changes of belief and habits. C. S. Lewis' *Pilgrim's Regress* (his allegorical autobiography)[39] details the long and arduous road taken by Lewis through every imaginable non-Christian philosophy and spirituality. Near the end of the book, he reaches what he believes to be his goal, but it turns out to be an immense untraversable chasm beyond which lies his goal. He suddenly realizes that he has been going in the wrong direction, and the only way of getting to the goal is to return from whence he came and approach the goal from the opposite direction—his Christian origins that he had rejected. His non-Christian journey took him on a much lengthier route and was filled with distractions, falsities, and dangers to his soul.

4. *Increasing joy, hope, fulfillment, and moral rectitude in this life*. Inasmuch as Jesus' teaching is the most complete and accurate expression of God's self-revelation, it will provide a path to true

[38] See Spitzer, *Moral Wisdom*, Chapters 5 and 6.
[39] See C. S. Lewis, *Pilgrim's Regress* (London: J. M. Dent and Sons, 1933).

happiness in this life as well as the most direct path to eternal life. This can be assessed in three subpoints:

a. *Suffering.* Though this life will have its crosses, the Christian outlook on suffering will fill that suffering with hope, meaning, and purification of our love as we move toward eternal life. The Christian interpretation of suffering is by far the most comprehensive and meaningful proposed in human history. It shows not only why the Son of God entered into our suffering but how suffering can help us purify our love, faith, and authenticity. It also helps us use our faith to suffer well and serve others.[40]

b. *Christianity and restoration of emotional health.* Christianity combines well with other psychotherapeutic remedies in holistic therapy.[41] This is confirmed in several statistical studies, such as that of Dr. Harold Koenig and others: only those in Christian therapy, combined with normal psychotherapy, experienced significantly lower posttreatment depression, compared to those who did not have a religious component as part of their psychotherapy.[42] In view of this, Christianity can play an important role along with other therapeutic techniques to redress the huge increase in depression, anxiety, suicides, and homicides among young people.

c. *Christianity, morality, and emotional health.* We are most fulfilled and content when we are living a morally upright life. There is significant evidence from secular surveys that following Jesus' moral teachings leads to considerably lower rates of anxiety, depression, substance abuse, familial tensions, and suicides.[43] In addition to emotional health, Christian moral life significantly enhances relational health and marital health as well as spiritual health (according to secular surveys).[44]

---

[40] See Robert Spitzer, *The Light Shines on in the Darkness: Transforming Suffering through Faith* (San Francisco: Ignatius Press, 2017), Chapters 4–10.

[41] See Hsin-Nan Lin, "Dealing with Depression: A Christian Perspective", *Taiwanese Journal of Psychiatry* (Taipei) 25, no. 4 (2011): 224–32, https://www.sop.org.tw/sop_journal/Upload_files/25_4/02.pdf.

[42] Bonelli et al., "Religious and Spiritual Factors".

[43] I have collected multiple secular surveys from studies showing these correlations very probatively. See Spitzer, *Moral Wisdom*, Chapters 1–4.

[44] Ibid.

If you the reader believe that Jesus' view of love holds out the possibility of not only the highest meaning of life but also the highest expression of God's self-revelation, and therefore the truest, the most direct, and the most comprehensive path to His eternal life, then the above four reasons for moving from "faith in general" to Christian faith will not be arbitrary or off-putting because they logically follow from this view of love. If Jesus' view of love really is the highest commandment, the nature of God, and our highest meaning and dignity, and is the highest truth, the highest good, the most direct path to salvation, and the highest joy, fulfillment, and interpretation of suffering and morality in this life, then it warrants serious consideration of faith in Him.

Christian faith is not an infinite leap of the will to overcome the nonrational or irrational. Rather, it is built on a strong foundation of reason. In *Science at the Doorstep to God*, we show how science and reason support an intelligent Creator, a transcendent soul capable of surviving bodily death, and the Creator as the cause of this soul. In this volume, we will also apply the methods of science and reason to Jesus Christ in five areas:

1. Historical and scriptural evidence for the historicity of Jesus and the Gospel accounts (Chapter 2)
2. Scientific evidence of Jesus' Passion, Crucifixion, and glorious Resurrection on the Shroud of Turin (Chapter 3)
3. Scientific evidence that the Holy Eucharist is the real Body and Blood of Christ in the transubstantiated Eucharistic Host (Chapter 4)
4. Scientific evidence of the cosmic and salvific significance of the Blessed Virgin Mary in the tilma of Guadalupe and the Miracle of the Sun at Fatima (Chapter 5)
5. Scientific evidence of the healing power of the Blessed Virgin Mary and the Holy Eucharist at Lourdes (Chapter 6)

The Shroud of Turin gives us *direct* scientific access to Jesus' Crucifixion, glorious Resurrection, and supernatural cause of the image; the Guadalupe tilma gives us *direct* scientific access to the Blessed Mother's appearance to Juan Diego and the supernatural character of her image; the transubstantiated Eucharistic hosts (integrated with live cardiac tissue) give us *direct* scientific access to the real Body and Blood of Christ in a consecrated Host; the Miracle of the Sun at Fatima gives us access to a historical event manifesting the cosmic significance and power of the Blessed Mother; and the scientifically inexplicable cures of Lourdes give us access to the Blessed Mother's ongoing presence with her Son in the work of salvation.

As we shall see, God has not left us bereft of rational evidence for His Son, the Eucharist, and the Blessed Mother. On the contrary, He provides substantially probative evidence for these Christian revelations which can be affirmed through the lens of science and the eyes of faith.

This book will conclude with a chapter on science and the Catholic Church (Chapter 7), examining the Church's important role in the development and ongoing pursuit of science, as well as her views on the Bible and science, and evolution of intelligent life.

The focus placed on scientific evidence in this volume is not meant to deemphasize other rational, intuitive, and faith-based approaches to Christian revelation, but only to complement and corroborate them. We do not need scientific evidence given in this volume or elsewhere to validate Christian faith. The apostolic testimony, the historical, exegetical, and philosophical evidence of Jesus' life and teaching, and the collective witness of the Catholic Church are more than enough to establish the truth and reasonableness of Christianity. Nevertheless, the method and light of science can deepen our understanding and rationale for this most important revelation, and so we present it here in the hopes that it will open the eyes of believers and skeptics alike in this scientific age.

Chapter Two

The Historical Evidence of Jesus

Introduction

As we have seen, there is considerable evidence for the truth and goodness of Jesus' view of love and the effects it has had in world history as a sign of Jesus' claim to be the exclusive Son of His divine Father and therefore the pinnacle of God's self-revelation. Indeed, many Christian converts came to believe in Jesus solely for this "reason of the heart". Though this reason of the heart is substantial and foundational, Jesus provided us with more than this—reasons of the mind. In this chapter, we will examine the *nonscientific* reasonable evidence for the historicity and Resurrection of Jesus, and in Chapters 3–6, we will examine the *scientific* evidence for His Resurrection and Passion, as well as His Real Presence in the Eucharist, and the presence of His mother in the work of salvation. This chapter will be divided into six sections:

1. The Evidence of the Heart (Section I)
2. The Evidence of Jesus from Three Hostile Extratestamental Sources (Section II)
3. The Historical Ground of the Gospels (Section III)
4. The Evidence of Jesus' Resurrection (Section IV)
5. The Miracles of Jesus (Section V)
6. Jesus' Divinity and the Cost of Proclaiming It (Section VI)

I. The Evidence of the Heart

Those who have read the New Testament with care will probably have discovered the worthiness of its authors. There are several indications of their faithfulness to sources, such as their apologetical self-restraint, their honesty about themselves, their care to report the words of Jesus faithfully, and their concern for the salvation of readers. A brief overview of these indicators, which stand in contrast to later Gnostic Gospels, provides an intuitive sense—evidence of the heart—that the Gospel writers intended to convey the truth about the life, activities, and words of Jesus.

As we read the Gospel stories, we will probably notice that their authors could have embellished the accounts of miracles and the Resurrection beyond their rather prosaic form. Indeed, they seem to underplay these "deeds of power" so much that the miraculous event or Resurrection appearance appears somewhat anticlimactic. When we compare the exorcism stories (which are dramatic) with the miracle stories (which are quite subdued), we get the impression that an editor went through the miracle stories to take out the exciting parts. Moreover, when we compare the four canonical Gospels to the later second-century Gnostic Gospels,[1] the former are quite reserved in their presentation of miracles and the Resurrection, while the latter are hyperbolic and frequently ridiculous. Additionally, the canonical Gospels portray Jesus as the beloved Son of the Father, who came into the world to teach us, heal us, sacrifice Himself completely for us, and give us His Spirit and the Church for the sole purpose of saving us. In contrast to this, the Gnostic Gospels frequently portray Jesus as self-interested, impatient, and unduly concerned with power.

The authors of the canonical Gospels are respectful, reserved, and humble—even to the point of being self-critical (to follow their teacher, Jesus). The inclusion of insults leveled at Jesus by the religious authorities—for example, "He casts out demons by the prince

---

[1] The Gnostic Gospels are a set of apocryphal works attributed falsely to Jesus' disciples and friends. They were written several decades after the four canonical Gospels (Matthew, Mark, Luke, and John) during the second half of the second century to the fourth century. Their authors are not accepted authorities within the apostolic Church (as the four canonical Gospels), but rather spiritual writers who were heavily influenced by Gnostic philosophy (which attempts to achieve spiritual freedom through special knowledge or enlightenment). The so-called Christian Gnostics who wrote these texts departed from apostolic Christianity by advocating salvation not only through Jesus Christ but through enlightenment proposed by its spiritual leaders. As can be seen from their miracle stories, their view of salvation and miracles was considerably different from that of Jesus', and in some cases, ridiculous and fantastic.

of demons" (Mt 9:34; cf. 12:24; Mk 3:22; Lk 11:15)—and the failings and weaknesses of the apostles—such as Peter's denials, Thomas' doubts, Matthew's former profession as tax collector (a traitor to Israel), Phillip's naiveté, and the accusation that the apostles stole Jesus' body—show the Gospel writers' interest in the truth before the reputation of Christianity's foundational leaders. If those leaders had not had the humility to tell the whole truth, wouldn't they have asked the evangelists to use their editorial pens a little more assiduously? Humility speaks convincingly about the reliability of witnesses and authors.

Furthermore, the *tone* of the Gospel texts is "just right". The Gospels manifest an almost exclusive interest in our *salvation*, telling us the whole truth even if it is difficult to bear, such as the inevitability of the Cross, the power of the devil, the difficulties we are likely to encounter on our journeys, and warnings about the possibility of Hell. They are the opposite of the "prosperity gospel", which tries to mitigate difficulties and challenges in the Gospel message to make it more palatable and easier to accept. They were written not to gain readers' approval but rather in a challenging, almost "off-putting" way, to help us toward salvation—to call us out of self-delusion and darkness into the light of Christ's love. "Tough love" can dissuade more converts than it persuades. If the evangelists had been more interested in "winning converts" instead of "helping souls", the Gospels would have been written quite differently—avoiding "tough love".

Now ask yourself some questions. If you were a Gospel writer who was interested solely in winning converts rather than faithfully presenting the truth about Jesus, would you have presented the miracles and Resurrection accounts soberly or with exaggeration and flashiness? Would you have presented insults leveled against Jesus and the moral weakness and mistakes of the apostles—or would you have tried to omit them? Would you have presented the really tough dimensions of Jesus' preaching—about the necessity of the Cross and the difficulties of Christian life—or left them out? Would you have accentuated Jesus' humility and persecution on the Cross—or emphasized His powerful deeds and rhetoric? In light of this, do you think that the Gospel writers had an interest in presenting the truth about Jesus? We now proceed to what might be called "objective evidence"—publicly accessible evidence about Jesus that can be experienced and corroborated by the scholarly community and others.

II. The Evidence of Jesus from Three Hostile Extratestamental Sources

Is there any testimony to the historical existence of Jesus outside of the New Testament? There are three very credible first-century sources whose authors were hostile to both Jesus and Christianity. Despite this, they report not only the existence of Jesus but also His Crucifixion, authoritative words, and miracles. Given their hostility to Christianity, we might infer that their report of Jesus' authoritative words and miracles is accurate. Why would they have done this if they did not believe some accounts of His miracles and powerful rhetoric? Let's examine the sources.

A. Cornelius Tacitus

Cornelius Tacitus was a famous Roman historian who wrote the *Annals* for the Roman emperor and politicians around A.D. 80 to 90—close to the time of Christ. He makes explicit reference to the Crucifixion of Jesus in the *Annals* when speaking about Nero's blaming the Christians for the burning of Rome:

> Consequently, to get rid of the report [that Nero ordered the burning of Rome], Nero fastened the guilt and inflicted the most exquisite tortures on a class hated for their abominations, called Christians by the populace. Christus, from whom the name had its origin, suffered the extreme penalty [Crucifixion] during the reign of Tiberius at the hands of one of our procurators, Pontius Pilatus, and a most mischievous superstition, thus checked for the moment, again broke out not only in Judaea, the first source of the evil, but even in Rome, where all things hideous and shameful from every part of the world find their center and become popular.[2]

[2] Cornelius Tacitus, *Annals* 15, 44, in *The Complete Works of Tacitus*, trans. Alfred John Church and William Jackson Brodribb (New York: Random House, 1942), edited for Perseus Digital Library, http://www.perseus.tufts.edu/hopper/text?doc=Perseus%3Atext%3A1999.02.0078%3Abook%3D15%3Achapter%3D44.

B. Flavius Josephus

Flavius Josephus, a Jewish historian writing a history of the Jewish people for a Roman audience in approximately A.D. 93, provides the most impressive and detailed evidence for the historical Jesus outside Christian Scripture. He reported the following in his *Jewish Antiquities*:

> Now there was about this time, Jesus, a wise man, for he was a doer of wonderful works [i.e., miracles], and a teacher.... He drew over to him both many of the Jews and many of the Gentiles. And Pilate, at the suggestion of the principal men amongst us, condemned him to the cross. And the tribe of Christians, so named from him, are not extinct at this day.[3]

C. The Babylonian Talmud

The Babylonian Talmud refers to Jesus in several references that can be dated between A.D. 70 and 200. In view of the fact that the passages indicate rabbinical hostility toward Jesus and cast His Crucifixion in a negative light, they probably refer to Jesus as a real historical figure of considerable significance. One of the passages states that Jesus was accused of "witchcraft", indicating that Jesus was known to have some kind of extraordinary and other-worldly power.[4]

Given these three testimonies from hostile authors reporting near the time of Jesus, it is reasonable to believe that He existed, performed miracles, spoke authoritatively, was crucified, and that the religious movement He started (the Christians) grew rapidly beyond Judea to

---

[3] Flavius Josephus, *Jewish Antiquities* 18, 3, 3, trans. Louis H. Feldman, Loeb Classical Library (Cambridge, Mass.: Harvard University Press, 1965). This rendition of the text is the one that most scholars agree on as the precise words of Josephus. Luke Timothy Johnson, Raymond Brown, and John P. Meier have provided significant justification for the version. See Luke Timothy Johnson, *The Gospel according to Luke*, vol. 3 of Sacra Pagina Series, ed. Daniel J. Harrington, S.J. (Collegeville, Minn.: Liturgical Press, 1991), pp.113–14; Raymond Brown, *An Introduction to New Testament Christology* (New York: Paulist Press, 1994), pp. 373–76; and John P. Meier, *A Marginal Jew: Rethinking the Historical Jesus*, vol. 2, *Mentor, Message, and Miracles* (New York: Doubleday, 1994), pp. 592–93.

[4] See Babylonian Talmud: Tractate Sanhedrin 43a.

Rome itself. There is another external source currently under scientific investigation that provides probative evidence of Jesus' Passion, death, and Resurrection—the Shroud of Turin. We will discuss this in the next chapter.

III. The Historical Ground of the Gospels

How were the Gospels formed and why would scholars believe that the accounts of Jesus' words and actions are accurate? The Gospels are one of the most exhaustively studied subject areas by contemporary scholars, who generally agree that the Gospels underwent three major stages of composition:

1. From eyewitness accounts to documented oral traditions (A.D. 33 to 40)
2. From documented oral traditions to written accounts and collections of written accounts (A.D. 40 to 55)
3. From written collections to the Gospels (A.D. 55 to 85)

1. *From eyewitness accounts to documented oral traditions (A.D. 33 to 40).* Richard Bauckham, senior scholar at Cambridge University, explains how eyewitness accounts generated oral traditions linked to those eyewitnesses:

> Gospel traditions did not, for the most part, circulate anonymously but in the name of the eyewitnesses to whom they were due. Throughout the lifetime of the eyewitnesses, Christians remained interested in and aware of the ways the eyewitnesses themselves told their stories. So, in imagining how the traditions reached the Gospel writers, not oral tradition but eyewitness testimony should be our principal model.[5]

As Bauckham shows, the Gospel traditions were not anonymous. The oral traditions and written documents were linked to well-known named persons, and it was the early Jewish practice to memorize

---

[5] Richard Bauckham, *Jesus and the Eyewitnesses: The Gospels as Eyewitness Testimony*, 2nd ed. (Grand Rapids, Mich.: Eerdmans, 2017), p. 8.

sacred traditions so they could be passed on faithfully. As evidence from early Church writers shows, Christians were intensely interested in the eyewitnesses linked to particular accounts of Jesus' life, and often interviewed these witnesses or the persons who had contact with them.[6]

2. *From documented oral traditions to written accounts and collections of written accounts (A.D. 40 to 55).* Within a decade of Jesus' Resurrection, documented oral traditions were committed to writing still linked to the eyewitnesses from which the oral traditions were derived. These initial accounts of Jesus' words and deeds were placed in early collections at which point the names of the eyewitnesses were preserved *in* the accounts where these witnesses are specifically mentioned.[7] Many of these written accounts were witnessed by apostles who are named in those accounts, such as Peter, Matthew, James, John, Nathaniel, Phillip, Thomas, and Andrew. Many accounts were witnessed by nonapostles, such as Mary, the mother of Jesus; Mary Magdalene; Martha; Mary, the sister of Martha; Lazarus; Bartimaeus; Simon the Pharisee; Jairus; Joses; Mary, the mother of Clopas; Nicodemus; Joseph of Arimathea; and Zacchaeus.[8]

3. *From written collections to the Gospels (A.D. 55 to 80).* Written collections eventually gave rise to three major compilations, two of which are Gospels—the Gospel of Mark and the Gospel of John—and Q, referring to *Quelle*, meaning "source" in German; Q is an early collection of Jesus' sayings originally in Aramaic and translated into Greek.[9] Matthew and Luke base their Gospels on Mark and Q, adding traditions from their own special sources. Thus, the four Gospels have five major sources, all of which were initially linked to eyewitnesses:

a. The Gospel of Mark, whose primary eyewitness is Peter with the addition of other apostolic and nonapostolic eyewitnesses[10]

[6] Ibid., pp. 210–65.
[7] Ibid., pp. 37–39.
[8] Ibid., p. 56.
[9] Most scholars believe that Q is a single written source, though some hold that it is a plurality of sources. We do not know who the editor(s) was.
[10] Bauckham, *Jesus and Eyewitnesses*, pp. 219–46.

b. The Gospel of John, whose primary eyewitness was the Apostle John, son of Zebedee,[11] with the addition of other apostolic witnesses, particularly Peter and other named nonapostolic witnesses

c. Q, which had multiple nonnamed eyewitnesses, many of whom were apostles[12]

d. The special sources of Matthew, mostly unnamed eyewitnesses[13]

e. The special sources of Luke, mostly named eyewitnesses[14]

How can scholars be sure that these eyewitness accounts of Jesus' deeds and words were accurate? There are three major ways of determining this:

1. Formulation of eyewitness accounts into oral traditions, widely circulated within living memory of Jesus

2. Application of the criteria of historicity

3. Confirmation of geographical and architectural features of Gospel narratives by archeological specialists

First, as we saw above, the Gospel narratives can be traced back to oral traditions linked to named eyewitnesses. These eyewitness oral traditions were circulating widely in the regions where they originated within ten years of Jesus' Resurrection. Most of the witnesses to Jesus' *public* deeds and words (exorcisms, healings, raisings of the dead, debates with Jewish authorities, and compassion toward sinners), as well as to events of His Passion, Resurrection, and gift of the Spirit, *were still alive* at the time these eyewitness oral traditions were circulating, and these living witnesses could have challenged those oral traditions (and the eyewitnesses whose names were attached to them) if they were inaccurate. Furthermore, many traditions mentioned specific villages (and places in those villages), and a spectacular event such as raising the dead or curing blindness

[11] Andreas Köstenberger and Stephen Stout, " 'The Disciple Jesus Loved': Witness, Author, Apostle—A Response to Richard Bauckham's Jesus and the Eyewitnesses", *Bulletin for Biblical Research* 18, no. 2 (2008): 209–31.

[12] Bauckham, *Jesus and Eyewitnesses*, pp. 41–42.

[13] Ibid.

[14] Ibid.

or a paralytic would certainly be known in those particular villages. If one of these unforgettable events were reported to have occurred in a particular town, but people in that town had not heard of it, they would have challenged the story, and the eyewitnesses connected with that story would have been embarrassed and disgraced. The very fact that these eyewitness oral traditions seemingly survived unchallenged from A.D. 35 to 55 strongly suggests that they were unchallenged by those who could have identified fabrications. In view of this, it is likely that these eyewitness oral traditions (which later became the written Gospel traditions) were reasonably accurate.

We now proceed to the second way of determining the accuracy of the Gospels—criteria of historicity. There are six criteria used by historians to ascertain the accuracy of historical accounts, which apply to the Gospel narratives:

1. Multiple attestation
2. Embarrassment
3. Coherence with first-century Palestinian culture
4. Coherence with Jesus' unique style
5. Semitisms
6. Specific identifiable names and places

Multiple attestation. This refers to the principle that the more often a story or saying appears in *independent* traditions, the more probable its historicity. Note that the converse statement cannot be deduced from the former ("The less often a story or saying appears in independent traditions, the less probable its historicity"). This is the logical fallacy of negating the antecedent.[15] Appearance in a multiplicity of independent traditions strongly suggests that those traditions go back to a common source, which would presumably be either the early Palestinian community or Jesus Himself. However, an absence of multiple attestation does not *necessitate* nonhistoricity, because

[15] Since the time of Aristotle, it has been widely known that negating the antecedent is fallacious. It takes the following form: "If A, then B. Not A; therefore, not B." This applies to the following syllogism: "If there are multiple attestations of a story, then it is historically probable. There are no multiple attestations; therefore, it is not historically probable." This conclusion is fallacious because it negates the antecedent.

sometimes the author(s) of particular traditions may not have heard about a particular story or may have chosen to ignore it for theological or apologetical reasons.

What are these independent traditions? Recall from above that Mark was very likely the first Gospel, and that Matthew and Luke relied very heavily upon it. Recall also that Matthew and Luke shared a common source (which Mark did not use or know)—namely, Q. Luke and Matthew had their own special sources that are not found in either Mark or Q. We know that these special sources are not mere inventions of the evangelists because many of them have the characteristics of an eyewitness oral tradition developed prior to *any* literary tradition, and many of them do not follow the literary proclivities of the evangelists (e.g., some of Luke's sources write in a far less sophisticated and stylized way than Luke himself—and the fact that Luke does not correct them indicates that he is being respectful of his sources). Finally, recall that the Johannine source has long been recognized to be independent of the Synoptics (Matthew, Mark, and Luke). Thus, contemporary biblical criticism has been able to identify five *independent* traditions for the four Gospels—namely, Mark, Q, M (Matthew special), L (Luke special), and J (the independent Johannine tradition).

So what does multiple attestation mean? Given the above five original independent sources, historians define multiple attestation as the more often a story appears in the five *independent* Gospel traditions, the more probable its historicity. Thus, if a story appears in all five traditions (Mark, Q, M, L, and J), it is probable that it originated with an early common Palestinian oral tradition or Jesus' ministry itself. If it appears in three or four independent traditions, it is still quite probable. If a story appears in only one or two traditions, it does *not* indicate nonhistoricity, but only that we cannot confirm historicity through multiple attestation.[16]

Embarrassment. This criterion refers to actions or sayings that the early Church would have found embarrassing, apologetically unappealing, disrespectful to Jesus, or disrespectful to the apostles. Evidently, no evangelist would want to include such statements in the

---

[16] See Harvey K. McArthur, "Basic Issues: A Survey of Recent Gospel Research", in *In Search of the Historical Jesus* (London: Charles Scribner's Sons, 1969), pp. 139–40.

Gospels (which are written to instruct and edify the community and potential converts), because they would undermine the Gospels' purpose. Therefore, we assume they are included in the Gospel only because they were historically true. For example, with respect to the empty tomb, Matthew reports the accusation of the religious authorities that the disciples of Jesus stole His body (28:13). Why would Matthew have reported such an accusation—with all of its negative implications—unless it were true? Again, the Gospels report that the Pharisees accused Jesus of casting out demons through the power of Beelzebul (Mt 9:34; 12:24; Mk 3:22; Lk 11:15). Why would they do this unless the Pharisees had really made the accusation? This leads to the further question of why the Pharisees would have made such a self-contradictory (and unintelligent) accusation unless Jesus' exorcisms were successful, prolific, and well known by the public.

Coherence with first-century Palestinian culture. Béda Rigaux in 1958[17] recognized that the evangelists' accounts conform almost perfectly with the Palestinian and Jewish milieu of the period of Jesus, as confirmed by history, archeology, and literature. René Latourelle summarizes several of Rigaux's examples as follows:

> The evangelical description of the human environment (work, habitation, professions), of the linguistic and cultural environment (patterns of thought, Aramaic substratum), of the social, economic, political and juridical environment, of the religious environment especially (with its rivalries between Pharisees and Sadducees, its religious preoccupations concerning the clean and the unclean, the law and the Sabbath, demons and angels, the poor and the rich, the Kingdom of God and the end of time), the evangelical description of all this is remarkably *faithful* to the complex picture of Palestine at the time of Jesus.[18]

The environment of the early Church, with its post-Resurrection faith and extensive ministry to the Gentiles, became progressively

---

[17] Béda Rigaux, "L'historicité de Jésus devant l'exégèse récente", *Revue Biblique* 68 (1958): 481–522.

[18] René Latourelle, *Finding Jesus through the Gospels: History and Hermeneutics* (New York: Alba House, 1979), p. 227 (emphasis added).

detached from the ethos of Palestine at the time of Jesus, and by the writing of the Gospels, much of this ethos was obscure to many Christians. Remarkably, the Gospel narratives preserve not only the customs and actions of Palestinian Judaism but also expressions (such as "Son of David"[19] or "Rabbi"[20] or "He is a prophet"[21]) that would have been superseded by other more suitable titles or expressions in the post-Resurrection Church. This speaks to the historicity not only of the Gospels in general but of the specific narratives within the Gospels where these anachronisms occur.

Coherence with Jesus' unique style. This occurs when some expressions, attitudes, and actions of Jesus depart significantly from those of the milieu in which he lived and constitute a style that is distinctive or unique to Him. For example, the way in which Jesus worked miracles is completely different from that of Jewish or Hellenistic miracle workers. This unique style of miracle working is present in all five independent sources, which leads to the question, if the evangelists did not derive this *unique* style from the teachings, expressions, and actions of an original common tradition about *Jesus*, how could it occur so consistently in every independent tradition? This leads to the inference of a common source for this common tradition—the most probable of which is Jesus Himself.

Semitisms. The Gospels were written in Greek. However, the oral and written traditions underlying their many narratives were formulated in Aramaic. If these traditions can be identified from the Greek text, it shows a probable origin within a Palestinian community near the time of Jesus. Aramaic does not translate perfectly into Greek, so when linguists identify strange or awkward Greek expressions, they look for possible underlying Aramaic expressions. Much of the time, a *strange Greek* expression reveals a very *common Aramaic* expression of Palestinian origin. Additionally, there are Palestinian expressions that are virtually unknown to Gentile audiences, and so their occurrence in, say, a Gospel written by a Gentile for Gentile audiences (e.g., Luke) shows an earlier Palestinian origin.

[19] Mt 1:1; 1:20; 9:27; 12:3; 12:23; 15:22; 20:30–31; 21:9; 21:15; 22:42; Mk 10:47; 12:35; Lk 3:31; 18:38–39; 20:41.

[20] Mt 25:25; 26:49; Mk 9:5; 10:51; 11:21; 14:45; Jn 1:38; 1:49; 3:2; 3:26; 4:31; 6:25; 9:2; 11:8.

[21] Mt 14:5; 16:14; 21:11; 21:46; Mk 6:4; 6:15; 8:28; Lk 13:33; 24:19; Jn 4:19; 4:44; 6:14; 7:40; 7:52; 9:17.

Specific identifiable names and places. As noted above, when specific names, places, and topographical features are present in a Gospel narrative, they are probably retained from an early Palestinian tradition. Furthermore, many of the proper names are likely those of the eyewitnesses responsible for the formulation and transmission of that tradition. Details are frequently lost during a tradition's transmission, and so their inclusion indicates a retention of them (from an early source). Recall also that many of these details can be checked by individuals within living memory of Jesus, because the people mentioned are known within the community. Furthermore, a spectacular event such as raising the dead or curing blindness or a paralytic would certainly be known and remembered by people in a particular small town or village.

Many of these six criteria are applicable to all the traditions concerned with Jesus raising the dead as well as many of His exorcisms and healings, His empty tomb, His disputes with religious authorities, most of the Passion accounts, His outreach to sinners, His parables, His love of friends and disciples, and many of His sayings that solidify and often validate these traditions. There are other important historical criteria used to validate Jesus' Resurrection (discussed below in Sections IV and V). In view of this, we can make a reasonable and responsible judgment of not only the truth of the specific traditions validated by these criteria but also the general validity of the Gospels themselves.

We now proceed to the third way of ascertaining the accuracy of the Gospels—archaeology. Archaeology cannot be used to validate a specific tradition or Gospel account, but it can be used to validate geographical and topographical features given in those accounts. James Charlesworth (George L. Collord Professor of New Testament Language and Literature Emeritus and the director of the Dead Sea Scrolls Project at Princeton Theological Seminary) gives a summary of the archaeological evidence substantiating geographical and topographical features in the Gospel accounts.[22] With respect to the Synoptic

[22] James H. Charlesworth, ed., *Jesus and Archaeology* (Grand Rapids, Mich.: William B. Eerdmans Publishing, 2006); see esp. "Jesus Research and Archaeology: A New Perspective", pp. 11–63.

Gospels, he lists the following validatable geographical, topographical, and cultural-political features (among others): the tomb of Jesus, which is likely accurately located at the Church of the Holy Sepulchre; the house of Peter in Capernaum; the gate to the town of Nain (whose location was formerly unknown); the presence of Pontius Pilate as procurator; the presence of Herod Antipas as provincial ruler; descriptions of the Temple area, as well as descriptions of cultural practices (such as table fellowship, the shunning of sinners, disputes between rival Jewish camps, and disputes with the Romans).[23]

John's Gospel is particularly important because it gives many geographical and topographical features that can be validated by contemporary archaeological evidence. Charlesworth (following Urban C. von Wahlde[24]) cites studies on the tomb of Jesus; the Pool of Siloam (discovered in 2006), where Jesus sent the blind man; the Pool of Bethesda (discovered in the late nineteenth century), where Jesus cured the paralytic who could not move himself into the stirred waters; Jacob's Well, near Sychar (Samaria); Golgotha, where Jesus was crucified; the judgment seat of Pontius Pilate; and detailed descriptions of the Temple area. Additionally, John's Gospel gives several geographical details in various regions that can be validated today, specifically in Galilee (Bethsaida, Nazareth, Cana, Capernaum, Sea of Galilee, and Tiberius), Samaria, the Jordan River, the Judean countryside, the vicinity of Jerusalem (Bethlehem, Mount of Olives, Bethany, garden across the Kidron Valley, Golgotha, garden where the tomb of Jesus was located), as well as the city and Temple of Jerusalem, the Upper Room, the High Priest's house, the praetorium, and the house where Jesus' disciples met.[25] Charlesworth concludes that where the Gospels mention geographical and topographical features, it is very accurate.

In conclusion, the New Testament is the most studied historical document in the history of mankind. The majority of scholars agree

[23] For additional information, see "Archaeology and the Synoptic Gospels: Which Way Do the Rocks Roll", Bible.org, 2024, https://bible.org/seriespage/archaeology-and-synoptic -gospels-which-way-do-rocks-roll.

[24] See Urban C. von Wahlde, "Archeology and John's Gospel", in Charlesworth, *Jesus and Archaeology*, pp. 523–86.

[25] Ibid.

that the traditions underlying the Gospel accounts are quite credible.[26] As we shall see below, Gary Habermas' survey of New Testament scholarship shows that a majority of scholars—both liberal and conservative—think that Jesus' followers did in fact have genuine experiences of Him risen from the dead, and that these experiences are the foundation of the early Church's conviction about Jesus' divine Sonship.[27] As we have seen above, the traditions underlying the Gospels were connected with the names of eyewitnesses, many of which are preserved in the Gospel narratives. These traditions could have been challenged by those still alive when those traditions were being initially circulated, but there is a striking absence of such challenges even from small towns and villages where spectacular events (such as raising the dead) had taken place. Furthermore, many parts of the Gospel narratives have been validated by the use of historical criteria throughout the last 120 years. This validation along with considerable archeological evidence shows that the Gospel narratives are more accurate than most other ancient historical texts.[28] When we combine this evidence with the external testimony of Roman and Jewish historians, we can see that the case for the historical Jesus and the accuracy of the traditions underlying the Gospel accounts is quite strong. Further historical validation can be ascertained by specific studies of the Resurrection and miracle narratives (see Sections IV and V below).

IV. The Evidence of Jesus' Resurrection

While belief in Jesus as "God with us" requires a movement of the heart to open ourselves to acknowledge our need for moral guidance

[26] See McArthur, In Search of the Historical Jesus; Joachim Jeremias, New Testament Theology, vol. 1, The Proclamation of Jesus (New York: Charles Scribner's Sons, 1971); Meier, Mentor, Message, and Miracles; John P. Meier, "The Present State of the 'Third Quest' for the Historical Jesus: Loss and Gain", Biblica 80 (1999): 459–87.

[27] See Gary R. Habermas, "Mapping the Recent Trend toward the Bodily Resurrection Appearances of Jesus in Light of Other Prominent Critical Positions", in The Resurrection of Jesus: John Dominic Crossan and N. T. Wright in Dialogue, ed. Robert B. Stewart (Minneapolis: Fortress Press, 2006), pp. 78–92.

[28] See McArthur, In Search of the Historical Jesus; Jeremias, Proclamation of Jesus; Meier, Mentor, Message, and Miracles; and Meier, "'Third Quest' for the Historical Jesus".

and to affirm Jesus' primary commandment to love, it can be supported by evidence for the mind that Jesus is actually divine. Much of this evidence was manifest in the early Church, which is why early Christians placed it at the foundation of their teachings. Saint Paul even says that if Jesus didn't rise from the dead, then the Christians' faith and the apostles' preaching would both be in vain (1 Cor 15:13–14), and all those who died in Christ would have died in their sins (15:17). Is there any historical way of verifying whether the Resurrection happened? We'll look at four main areas of evidence in this section:

1. The common elements in the Gospel narratives about Jesus' risen appearance to the apostles (Section IV.A)
2. The historical evidence of the Resurrection in the writings of Saint Paul (Section IV.B)
3. N. T. Wright's historical analysis of the Resurrection (Section IV.C)
4. The historical status of the empty tomb (Section IV.D)

Before examining the evidence, it will be helpful to review briefly what scholars currently think about the Resurrection. A recent survey of exegetes (scholars who analyze the Scriptures) by Gary Habermas[29] indicates that most agree the early disciples had experiences that they perceived as the Risen Christ:

> The latest research on Jesus' resurrection appearances reveals several extraordinary developments. As firmly as ever, *most* contemporary scholars agree that, after Jesus' death, his early followers had *experiences* that they at least believed were appearances of their risen Lord. Further, this conviction was the chief motivation behind the early proclamation of the Christian gospel. These basics are rarely questioned, even by more radical scholars. They are among the most widely established details from the entire New Testament.[30]

What do scholars believe was the cause of the apostles' experiences of the Risen Jesus? A few exegetes propose *natural causes* for these experiences (e.g., subjective visions caused by "religious intoxication"

[29] Habermas, "Mapping Recent Trend", pp. 74–89.
[30] Ibid., p. 79 (emphasis added).

and "enthusiasm"). According to Habermas, "In the twentieth century, critical scholarship has largely rejected wholesale the naturalistic approaches to the resurrection."[31] In contrast to this, the *vast majority of scholars* conclude that there was a *supernatural cause* of Jesus' Resurrection and that something transformative and glorious happened to Him after His body was placed in the tomb. The consensus is that Jesus' body was supernaturally transformed. He didn't just appear to the apostles as a vision, a ghost, or a resuscitated corpse, but rather as a gloriously transformed manifestation of His former self—what scholars call "transformed corporeality".[32] What evidence has led them to this consensus?

A. Common Elements in the Gospel Accounts of Jesus' Risen Appearances

The Gospel accounts of Jesus's risen appearances to the apostles share many common elements, and one of the most telling of these is that the apostles weren't expecting what they saw, which shocked and amazed them. They were so overwhelmed by His transformation that they thought they were witnessing a divine appearance—not the same corporeal Jesus they had worked with during His ministry. In the midst of His glorious appearance, He reveals His former embodied appearance with the wounds of His Crucifixion. Far from something the apostles were expecting, let alone planning, this risen Christ that appeared to them was so transformed that Saint Paul would have to coin a new word to describe it: a "spiritual body [Greek: *soma pneumatikon*]" (1 Cor 15:44).

What are the clues from the Gospel texts that Jesus was radically transformed in spirit, power, and glory? Consider the following:

- Jesus possessed transphysical capacities, such as the ability to pass through closed doors (Jn 20:19–20).
- He possessed spirit-like qualities, which caused the disciples to think He was a spirit (Lk 24:37).

[31] Ibid., p. 86.
[32] See ibid., pp. 81–88.

- When the apostles saw Him, they bowed down and worshiped, thinking they were seeing God Himself (Mt 28:17).
- Jesus reveals Himself to have divine power, telling the apostles, "All authority [power] in heaven and on earth has been given to me" (Mt 28:18).
- When the apostles saw the resurrected Jesus, they identified Him as "the Lord [*ho Kurios*, the Greek translation of the divine name Yahweh]" (Jn 20:20, 28; 21:7–8, 12, 15–17, 20–21).

The Gospel texts are unanimous in proclaiming that Jesus is transformed in divine power and glory, which causes shock and amazement on the part of the apostles. Jesus is so transformed in appearance that He has to manifest His bodily features and wounds of the Crucifixion to assure the apostles that it is really He.

B. Paul's Testimony to the Resurrection of Jesus

One of the earliest New Testament traditions from the First Letter of Paul to the Corinthians contains a list of witnesses to the Resurrection:

> For I delivered to you as of first importance what I also received, that Christ died for our sins in accordance with the Scriptures, that he was buried, that he was raised on the third day in accordance with the Scriptures, and that he appeared to Cephas, then to the Twelve. Then he appeared to more than five hundred brethren at one time, most of whom are still alive, though some have fallen asleep. Then he appeared to James, then to all the apostles. Last of all, as to one untimely born, he appeared also to me. (15:3–8)

Paul offers this list with an eye to its value as legal evidence. He mentions the five hundred, "most of whom are still alive", implying that His audience could still consult the witnesses to corroborate the story. The reason the list does not include the women at the tomb—the earliest witnesses mentioned in the Gospel narratives—is because Jewish law of the time did not acknowledge them to be valid witnesses in a court of law. Thomas Oden notes with respect to the Pauline list that since the women's testimony was not permitted in the official court of law, the list was deliberately shortened to provide the best evidence for the Christian faith.[33]

Having offered the evidence of these witnesses, Paul probes the value of their evidence by laying out a dilemma in 1 Corinthians 15. Either the witnesses believe in God or they do not believe in God. In either case, Paul argues, they had everything to lose and nothing to gain by falsely claiming to have witnessed the Resurrection.

For a *believer* to lie publicly that he had witnessed the resurrected Christ would jeopardize his own salvation by bearing false witness that undermined and caused apostasy to the Jewish faith. If Paul really believed in the God of his forefathers, he would have viewed lying about the Resurrection (that undermine the Jewish faith) to be the worst possible crime that would seriously jeopardize his salvation. As he himself said, "We are even found to be misrepresenting God" (1 Cor 15:15), a most grave offense in Judaism and Christianity. He had everything to lose and nothing to gain by lying about the Resurrection if he were a believer.

Those who did *not believe* in God would also have to make great sacrifices to preach the Resurrection, which would make lying about it virtually unintelligible, because such a lie would bring needless suffering onto oneself. The Resurrection doctrine set the early Christians at odds with the Jewish tradition, causing them to get expelled from their synagogues and ostracized by their communities. Soon, it led to harassment by the Roman authorities as well. Ultimately, it meant active persecution, torture, and death. Indeed, Paul himself suffered repeatedly from these harassments and would soon die for the faith. If the Resurrection was a lie, all that suffering would be for nothing. As he himself indicates, "If Christ has not been raised, . . . we are of all men most to be pitied. . . . Why am I in peril? . . . 'Let us eat and, drink, for tomorrow we die'" (1 Cor 15:17, 19, 30, 32).

Paul uses this dilemma to show (in a legal fashion) that he and the other witnesses have everything to lose and nothing to gain by bearing false witness to the Resurrection of Christ. If they believe in

---

[33] Thomas Oden, *The Word of Life*, vol. 2, *Systematic Theology* (New York: Harper-One, 1992), pp. 497–98. See also Brian Chilton, "Resurrection Defense Series: The Testimony of Women", Crossexamined.org (blog), March 29, 2021, https://crossexamined.org /resurrection-defense-series-the-testimony-of-women/; Josephus, *Jewish Antiquities* 4, 8, 15: "Do admit the testimony of women."

God, they condemn themselves by falsely instigating apostasy, and if they don't believe in God, they suffer persecution for nothing. Their testimony is more reliable since it goes against their own self-interest.

C. N. T. Wright's Two Arguments for the Historicity of Jesus' Resurrection

N. T. Wright argues from the historical record that two aspects of the early Church are difficult to account for without the Resurrection—the remarkable success of Christian messianism and the Christian mutations of Second Temple Judaism.

1. The Remarkable Success of Christian Messianism

There were many messianic movements in the time of Christ. Wright lists several of these: "Judas the Galilean, Simon, Athronges, Eleazar ben Deinaus and Alexander, Menahem, Simon bar Giora, and bar-Kochba".[34] In every case, a charismatic leader would attract an enthusiastic following, the leader would die (usually at the hands of the authorities), and the followers would scatter and the messianic movement would shortly die. One example of this pattern is even found in the Gospels with John the Baptist—whose disciples either faded away or joined the disciples of Jesus.

Christianity is the one dramatic exception to this pattern. After the public humiliation and execution of their leader, the disciples don't fade away, but instead begin preaching throughout the surrounding countries that their crucified leader is not only the Messiah but has somehow succeeded in fulfilling the ancient prophecies, is risen from the dead, and in fact is divine. Even more shockingly, the messianic movement grew exponentially, and in a few generations, Christianity would be the dominant religion of the Roman Empire.[35]

[34] N. T. Wright, *Christian Origins and the Question of God*, vol. 2, *Jesus and the Victory of God* (Minneapolis: Fortress Press, 1996), p. 110. An extensive consideration of all these figures is given in N. T. Wright, *Christian Origins and the Question of God*, vol. 1, *The New Testament and the People of God* (Minneapolis: Fortress Press, 1992), pp. 170–81.

[35] Wright, *Victory of God*, p. 110.

Where did this momentum come from? What inspired them with such conviction?

Wright believes that the apostles would have had no credibility with this message among the Jewish or Gentile people were it not for two extraordinary occurrences: (1) the apostles' ability to perform healings and miracles on a regular basis through the Holy Spirit, and (2) the fact that they worked these miracles through the name of *Jesus*. As the apostles imply, if Jesus is not risen from the dead as they had preached, then how could they work miracles through His Spirit *in His name*? In view of the fact that no other messianic movement worked regular miracles in the name of their messiah (including the movement of John the Baptist), it explains how Christianity's preaching of the resurrection of the body was so credible and therefore how the early Church grew so rapidly. This gives evidence not only of the power of the Holy Spirit in the ministry of the apostles but also of the Resurrection of Jesus. We might all ask, why would God have worked miracles through the hands of the apostles, in the name of Jesus, if the apostles were lying about Jesus' Resurrection and divinity? This seems to have occurred to many early converts in the Church, catalyzing Christianity's exponential growth.

2. The Christian Mutations of Second Temple Judaism

The other historical anomaly is in the teaching of the early Church. Whenever possible, the early Christians tried to maintain continuity in teaching with the broader Jewish community whose teachings were developed in a period called Second Temple Judaism (516 B.C. to A.D. 70—the time from the construction to the destruction of the Second Temple). They did so because Jesus cited these teachings, and they believed they were generally the will of God. Given this great reluctance to depart from the doctrines of Second Temple Judaism, we must ask why the early Christians made such an explicit exception to their doctrinal loyalty in one area: the resurrection of the body. It should be noted that this doctrinal departure was in part responsible for the Christians' expulsion from the synagogues (which they definitely did not want). The following are the specific changes that early Christians made to Second Temple Judaism's doctrine of the resurrection of the body:

Second Temple Judaism	Christianity
Resurrection is a return to the same kind of physical body (e.g., like a resuscitated corpse).	Resurrection is transformation into a spiritual and glorified body (the *soma pneumatikon*—1 Cor 15:44).
No one will rise before the end times—Parousia (the Second Coming).	Jesus and others are risen before the Parousia.
Everyone will rise together.	People will rise individually.[36]
The Messiah is not connected with the resurrection.	The hope for a Messiah and the hope of a resurrection are both fulfilled in Jesus.
Resurrection is a minor doctrine.	Resurrection is the central doctrine that justifies and connects all other doctrines.

Historians have tried to theorize where these new ideas came from—perhaps from paganism, or the Christians' own desire to come to terms with the death of their leader. The problem with these theories is that the ideas are unprecedented; no one had proposed them even outside of Judaism, so it's hard to find a plausible source other than the one given in the Gospels—namely, that Jesus really did rise from the dead and appeared to the disciples in a spiritual, powerful, glorified body. (Remember that Paul even had to make up a new term to describe the idea of a spiritual and glorified body, the *soma pneumatikon*; they literally didn't have a word for it prior to Him). Combine the novelty of the doctrine with how uniquely unprecedented it was—the only area where Christians opposed the prevailing Jewish doctrine—and again we must ask, where did this conviction come from, if not from a powerful, glorious risen appearance of Jesus that the early *Christians witnessed*?

---

[36] This doctrine seems to have evolved in Paul's thought. In his earlier letters, he leans toward the side of imminent Parousia and group resurrection but later moves toward a later Parousia and individual resurrection. See Joseph A. Fitzmyer, S.J., "Pauline Theology", in The New Jerome Biblical Commentary (Englewood Cliffs, N.J.: Prentice Hall, 1990), p. 1392.

In short, Wright's two arguments give us evidence from history *outside the New Testament* for the Resurrection of Jesus:

- The exponential growth of Christianity after the humiliation and execution of its Messiah is unintelligible outside of the early Church's capacity to perform miracles in the name of Jesus. This provokes the question of why God would work miracles through the hands of the Christian leaders in the name of Jesus if those Christian leaders were lying about Jesus' Resurrection and divinity.
- Given the early Church's desire to stay within the synagogue and remain loyal to the doctrines of Judaism, we must ask the question why the Church departed radically from those doctrines in the one area of the resurrection of the body, proclaiming that the resurrection was spiritual and glorious (rather than a resuscitation of the physical body as in Judaism), that the resurrection was already occurring (rather than deferred to the end of time as in Judaism), and that it was central to all other doctrines (rather than minor and peripheral as in Judaism).

If we cannot find a reasonable explanation of the above two historical anomalies other than the one given in the Gospels and by Saint Paul—that Jesus rose in glory, power, and spirit—then we have probative evidence from outside the New Testament to show that Jesus did rise from the dead in glory. God vindicated the apostles' and disciples' preaching of the resurrection of the body by allowing them to work miracles through the Holy Spirit in the name of Jesus.

D. The Empty Tomb

The empty tomb does not give direct evidence of the Resurrection but does provide indirect corroboration of it. When the apostles began gaining converts by preaching the Resurrection, it was in the interests of the Jewish authorities to undermine this claim by producing the body of Jesus, which they apparently could not do. This points to the likelihood that the body was not in its burial place. We know this from the charge the Jewish authorities made (reported in the Gospel of Matthew) that the apostles stole the body (28:13)—an accusation that would not be necessary unless there was an identifiable burial site that was now empty.

We are quite sure about the historical reliability of the Jewish authorities' claim that the apostles had stolen the body because it is unthinkable that the Christians would have reported such a damaging claim to their own credibility in Matthew's Gospel unless it were really *true*. The empty tomb also indicates continuity between the original body and the glorified body of the Resurrection—that is, it was the same body that was buried that was later raised and left the tomb transformed in glory.

E. Conclusion

In conclusion, the early Church proclaimed Jesus to be risen in glory, taking on a spiritual, powerful, divine-like appearance—an appearance revealing His divine identity. This experience was so powerful and undeniable that the apostles and disciples proclaimed it at great cost to themselves. Fully aware that they would be responsible for falsely undermining the religion of their forefathers if their proclamation of Jesus' glorious Resurrection were untrue, they boldly asserted it, enduring every form of hardship, ostracization, and persecution. As Saint Paul queried, what could they have possibly gained by making such a proclamation if it were false? (see 1 Cor 15:12–19). If they believed in God, they jeopardized their own salvation, and if they did not, they endured hardship and persecution for nothing.

The Lord did not leave them alone in their bold and risky proclamation. He bestowed on them the Holy Spirit with immense power to work miracles in the name of Jesus. This power, combined with the apostles' remarkable and courageous oratorical skill (absent before the Resurrection), led to one of the most confounding historical anomalies—the exponential growth of the Christian Church after the humiliation and execution of her Messiah.

When we combine these historical facts with the early Church's departure from Second Temple Judaism (almost solely with respect to the resurrection of the body), the empty tomb, and the unanimity of the Gospels and Saint Paul on Jesus' glorious Resurrection,

we can see why the vast majority of historical and scriptural scholars believe that the apostles and early Christians really did have an
experience of the Risen Christ transformed in power and glory—
a transformation entailing a supernatural cause. In Chapter 3, we
will examine the considerable scientific evidence on the Shroud
of Turin, validating the Gospel accounts of Jesus' Crucifixion and,
most impressively, His glorious Resurrection. Before considering
this evidence, we will complete our nonscientific investigation for
the historicity of Jesus.

V. The Miracles of Jesus

Can Jesus' miracles be corroborated historically? We have already
examined the evidence for the Gospel eyewitnesses and the criteria
of historicity for the Gospel traditions (above in Section III). Can
this general historical certitude be focused on the miracle narratives
themselves? They can, and the well-known historical exegete John P.
Meier has done one of the best studies of them.[37]

Jesus' miracles are quite central to His mission. As acts of healing and deliverance, they were part of His mission to initiate God's
Kingdom in the world. And by performing these miracles by His
own power (unlike the earlier prophets, who invoked the power of
God), Jesus demonstrates his divine authority and validates his claim
as Emmanuel, "God with us". Given the importance of the miracles
in Jesus' mission and for showing Jesus' divinity before the Resurrection, we will want to explore the evidence for their historicity.

A. The Purpose and Distinctiveness of Jesus' Miracles

At the outset, recall that two non-biblical sources acknowledge Jesus'
miracles—the Jewish historian Josephus (who puts them in a positive
light) and the Babylonian Talmud (which attributes them to witchcraft). Since these Jewish sources had nothing to gain and everything
to lose by attributing miracles to Jesus, it is very likely that Jesus had

[37] Meier, *Mentor, Message, and Miracles*, Chapters 17–23.

a widespread reputation for a prolific ministry of miracles (see above, Section II).

Furthermore, the Gospels report that the Pharisees attributed Jesus' exorcisms to demonic power, "the power of Beelzebul" (Lk 11:15; cf. Mt 9:34; 12:24; Mk 3:22). The idea of associating Jesus with Satan would have been utterly repugnant to any Christian, and it is scarcely imaginable that Christian writers would have included this accusation in the text of the Gospels unless it were true. This provokes the question of why the Jewish authorities accused Jesus of casting out demons by the power of Beelzebul. If Jesus had not had an undeniably successful and prolific ministry of exorcism, there would have been no need for the Pharisees to make a self-refuting accusation (casting out demons by demons)—they simply could have denied that He had such a successful ministry. Therefore, it is likely that Jesus' reputation for successfully exorcising demoniacs was so widespread that a denial would have seemed patently false and ineffective, requiring that they make this self-contradictory accusation.

We may now examine the miracle stories in the New Testament. Miracles are documented extensively in the New Testament, with specific verifiable references to public places and times, and are mentioned in the earliest *kerygmas* of the Church. These documented miracles are distinctive in several ways:

1. Jesus does miracles by His own authority. Every other Old Testament prophet appealed to God for the power to work a miracle, but Jesus claimed and showed that the divine power to heal comes from within Himself alone—for example, "*I* say to you, arise" (Mk 5:41; emphasis added). This not only differentiates Jesus from the Old Testament prophets but implies His divinity—divine power originates from within Him without asking God for His divine power.

2. Jesus' miracles have the purpose not so much of showing His glory as actualizing the coming of the Kingdom and the vanquishing of evil. Jesus' miracles always free someone from affliction (which Judaism viewed as originating from evil), and in the case of exorcisms, they free people explicitly from the bondage of Satan.

3. Jesus is not a wonder-worker or magician in either the pagan or Jewish sense. Raymond Brown has made an extensive study of

the rare miracles of pagan magicians and Jewish wonder-workers that are oriented toward astonishment and self-aggrandizement. Jesus' miracles are distinct from this, arising out of compassion with a view to dispelling Satan and establishing the Kingdom of God. Moreover, Jesus frequently asks that the miraculous feat be kept secret when possible, for the sake of awaiting the true revelation of who He is in the Passion and Resurrection.[38]

4. Jesus combines teaching with His miracles. Lessons on faith, forgiveness of sins, and giving thanks to God are integrated into His ministry of miracles.

5. The faith and freedom of the recipient is integral to the miraculous deed. Before working a miracle, Jesus frequently asks the recipient if he believes He can heal him—He involves the recipient's free participation in seeking healing, just as his free participation is needed to receive God's deeper healing and salvation.

These five unique aspects of Jesus' miracle working are reflected in how the Gospels report on them. As noted above in Section I, the Gospels show marked restraint in reporting miracles—there is no hyperbolic aggrandizement, no frivolous or punitive miracles as are frequently found in the Gnostic Gospels. In some of those late second-century "gospels", we see a young Jesus bringing clay sparrows to life and striking down a child that bumped into Him—among many other bizarre stories. Jesus' miracles have a clear purpose— to deliver people from suffering and evil and to introduce God's Kingdom.

B. The Historicity of Jesus' Exorcisms and Healings

Exorcisms and healings are complementary aspects of Jesus' mission. Exorcisms emphasize the vanquishing of evil, while healings emphasize the presence of God's redeeming love—both of which actualize God's Kingdom in the world.

Exorcisms are frequently narrated throughout the Gospels (seven individual cases are narrated, in addition to several "summary" mentions of exorcisms), indicating that they formed a significant part

[38] Brown, *Introduction to New Testament Christology*, pp. 61–64.

of Jesus' ministry. The accounts are consistent and restrained, featuring Jesus' unique use of His own power, as He Himself commands the demons to come out of the people they are afflicting. The seven narratives are as follows:

1. The possessed boy (Mk 9:14–29)
2. A passing reference to the exorcism of Mary Magdalene (Luke special—Lk 8:2)
3. The Gerasene demoniac (Mk 5:1–20)
4. The demoniac in the Capernaum synagogue (Mk 1:23–28)
5. The mute and blind demoniac in the Q tradition (Mt 12:24; Lk 11:14–15)
6. The mute demoniac (Matthew special—Mt 9:32–33)
7. The Syrophoenician woman's daughter (Mt 15:21–28; Mk 7:24–30)

References to healing are even more frequent in the Gospels. Fifteen unique cases are described, in addition to dozens of references in other contexts, including the accusations of the scribes that Jesus performs miracles by the power of demons and the account of Jesus conferring the power of healing on His disciples. There are fifteen narratives:

- Mark relates eight miracle accounts: cure of a paralytic (2:1–12); cure of a man with a withered hand (3:1–6); two cures of blindness (8:22–26; 10:46–52); cure of leprosy (1:40–45); and three concerned with various diseases mentioned only once (fever of Peter's mother-in-law in 1:29–31, the woman with a hemorrhage in 5:24–34, and the deaf-mute in 7:31–37).[39]
- Q relates only one account of a healing miracle, which is the cure of a centurion's servant—at a distance (Mt 8:5–13; Lk 7:1–10). Matthew calls this a cure of a paralytic, but Luke calls it a cure of someone with a grave illness. John (independently) also recounts this tradition (Jn 4:46–54). The presence of this miracle in both Q and John indicates multiple attestation of sources for a single healing account. Q also has a list of miracles (Mt 11:2–6;

_____

[39] This reflects Meier's list given in Meier, *Mentor, Message, and Miracles*, p. 678.

Lk 7:18–23), which includes healing of the blind, the lame, lepers, and the deaf.

- L (Luke special) relates four healings: one paralytic (13:10–17), one concerned with leprosy (17:11–19), and two cures of various ailments mentioned only once (the man with dropsy in 14:1–6 and the ear of the slave of the High Priest in 22:49–51).
- John relates two healings: one concerned with the cure of a paralytic (5:1–9), and one concerned with the man born blind (9:1–41).

Evidently, healing narratives have multiple attestation, and specifically, healings of paralytics, the blind, and lepers also have independent multiple attestation. Healings are mentioned in a variety of other contexts outside of narratives:

- Allusions to miracles that are not narrated in full (e.g., Mk 6:56: "Wherever he came, in villages, cities, or country, they laid the sick in the marketplaces, and begged him that they might touch even the fringe of his garment; and as many as touched it were made well")
- In sayings implying His fulfillment of prophetic expectation (e.g., Lk 4:17–18, 21: "He opened the book [of Isaiah] and found the place where it was written, 'The Spirit of the Lord is upon me, because he has anointed me ... to proclaim ... recovering of sight to the blind' ")
- The disciples performing miracles (Lk 9:6; 10:17–20; Mk 3:15) or failing to perform them (Mk 9:18; 28, 38)
- Giving the power to heal to the disciples (Mt 10:1, 8)

Again, the accounts of these miracles illustrate Jesus exercising His own power. Miracles occur typically in response to someone's petition to help him, so the miracle is worked through the faith of the petitioner. It is often further linked to a relevant spiritual teaching.

Note that many exorcism and healing miracles have additional corroboration through living memory and historical criteria (see above, Section III), because these stories sometimes have place names, personal names, and unusual details that would be easy for a contemporary reader to verify or refute. It is difficult to imagine that anyone in the Gerasene territory would forget Jesus' exorcism of Legion, which caused an entire herd of swine to rush down the hill and drown in the sea (Mt 8:28–32; Mk 5:1–13; Lk 8:26–33). Additionally, exorcisms of demoniacs in the synagogues of small towns would also have been unforgettable. Likewise, the healing of Bartimaeus, the blind man who begged for years on the outskirts of Jericho who was healed amid a large crowd (Mk 10:46–52), and the paralytic who was let through the roof of a house in Capernaum amid protestations from the Pharisees (Mk 2:1–12) would have also been well remembered and "checkable". As we shall see, these place names and specific witnesses are significant in all three raisings of the dead.

C. The Historicity of Jesus Raising the Dead

Raising the dead is more rare in the Gospel accounts than other kinds of miracles; there are three specific stories that are reported in three different narrative traditions: the raising of Jairus' daughter (Mt 9:18–19, 23–26; Mk 5:21–24, 35–43; Lk 8:40–42, 49–56), the raising of the son of the widow of Nain (Luke special—Lk 7:11–17), and the raising of Lazarus (Jn 11:38–44). These particular miracles have considerable historical substantiation:

1. Two out of three narratives have either Semitisms or pre-Christian Palestinian expressions in the Gospel texts—for example, "talitha cumi" (Mk 5:41) and "a great prophet has arisen among us!" and "God has visited His people" (Lk 7:16). These expressions indicate Palestinian origins (showing composition near the time of Jesus).[40]

2. All three narratives contain either proper names of people, names of places, or both. With respect to the names of people, Mark (probably reporting Peter's testimony) and Luke indicate that the daughter of Jairus (a synagogue leader) was raised (Mk 5:22; Lk 8:41). A Galilean synagogue leader named Jairus would have been well known in the region, and a raising from the dead would have been widely remembered. Additionally, John

[40] See ibid., pp. 780–97.

reports three names in connection with the raising of Lazarus—Lazarus and his sisters, Martha and Mary. He also mentions that the raising occurred in Bethany—a small town in which this miracle would have been remembered by almost everyone. Luke special reports the raising of the son of the widow of Nain. Though no name is mentioned in connection with the miracle, the town of Nain is so small that again almost everyone would have known and remembered this remarkable miracle. Interestingly, the town of Nain was unidentified until the actual gate of the town (where Jesus performed the miracle) was discovered in a recent archeological excavation.[41]

3. The raising of Jairus' daughter has several features confirmable by the criterion of embarrassment—Jesus being "laughed at" (Mt 9:24; Mk 5:40; Lk 8:53), and Jesus healing a synagogue leader's daughter (which would not have been popular at the time of the Gospels' writing because of persecution).

These stories of raising the dead should be distinguished from the story of Jesus' Resurrection, because they are not a transformation in glory and divine power like the Resurrection of Jesus but only a restoration of the person's former embodied state. Furthermore, raisings of the dead are only temporary, while spiritual resurrection is eternal. Despite these differences, the incidents of raising the dead are significant in revealing Jesus' divinity, because the power over life and death was thought to belong to God alone[42] and Jesus demonstrates that this power comes from within Himself and not by appealing to God—for example, Luke 7:14: "Young man, *I say* to you, arise" (emphasis added), and Mark 5:41: "Talitha cumi"—"Little girl, *I say* to you, arise" (emphasis added). Jesus' Resurrection and glory and His gift of the Holy Spirit (called "the power of God") are definitive manifestations of His divinity, but since they are not manifest until after His

---

[41] Ibid., p. 795.

[42] From Jesus' vantage point, sharing in power is similar to sharing in life (implying divine Sonship). John L. McKenzie notes that the Old Testament does not distinguish between life as a principle or power of vitality and life as living (the concrete experience of vitality), and so he states, "Its language is concrete rather than abstract, and life is viewed as the fullness of power [which is God]." John L. McKenzie, *Dictionary of the Bible* (New York: Macmillan, 1965), p. 507 (emphasis added).

Resurrection, it might be thought that Jesus received divine power at that time. However, the raisings of the dead (occurring in His own name and by His own authority) show that He had divine power even before the Resurrection. His self-proclamation of divinity (see below, Section VI) points convincingly to His divinity before the Incarnation—indeed, from all eternity.

D. The Holy Spirit Is Still Active Today

As noted above, Jesus gave the Holy Spirit to His apostles, who in turn worked similar miracles in His name throughout their ministry. The Acts of the Apostles mentions many of these miracles—exorcisms, healings, and raising the dead—throughout the Jewish and Gentile worlds, and Saint Paul explains these charisms and miracles in several of his letters.

The Holy Spirit is just as active today as in apostolic times. One does not have to look far to see the millions of testimonies to the charismatic manifestation of the Spirit (with literally millions of internet search results devoted to the Holy Spirit, healings, miracles, prophesy, and tongues) that resemble those recounted by Luke and Paul almost two thousand years ago.[43] Additionally, several scholars have chronicled hundreds of modern, medically documented miracles occurring through the power of the Holy Spirit in Jesus' name.[44] With so many accounts of *visible* manifestations of the Holy Spirit (i.e., modern miracles) in the *United States*, how much greater would be the accounts of the *interior* gifts of the Holy Spirit, and how much greater still when both the charismatic and interior gifts of the Spirit are seen throughout the *entire world*? It seems evident that the Holy Spirit is truly alive and well in any individual or culture that wants the Spirit's help, guidance, inspiration, peace, and above all, love.

There are also interior charisms of the Holy Spirit—faith, hope, love, zeal, peace, joy, guidance, inspiration, and protection—and for practical service, teaching, exhortation, works of mercy, almsgiving,

[43] A simple Google search on the internet for "Holy Spirit healing" currently yields 58,900,000 results; for "Holy Spirit miracles", 21,400,000 results; for "Holy Spirit prophecy", 31,400,000 results; and for "Holy Spirit tongues", 12,100,000 results.

[44] See, for example, the two-volume work of Craig Keener, *Miracles: The Credibility of the New Testament Accounts* (Grand Rapids, Mich.: Baker Academic Publishing, 2011).

and governance (see especially Rom 12:4–11; 1 Cor 12:7–13). These interior gifts are also as present today as when Saint Paul so eloquently wrote about them in his letters.[45]

E. Conclusion

As we consider the overwhelming evidence for Jesus' miracles, it is a virtual certainty that He not only performed successful exorcisms, healings, and raisings of the dead but also had a prolific ministry of doing so, which was central to His mission. Let us review for a moment that evidence:

- The testimony about His miracles in non-Christian sources written near the time of Jesus (e.g., Flavius Josephus and the Babylonian Talmud)
- The accusations against Jesus (e.g., "It is only by Beelzebul" that He "casts out demons"—Mt 12:24; cf. 9:34; Mk 3:22; Lk 11:15), implying that His adversaries acknowledged His miraculous power
- Multiple attestations of exorcisms, healings, and raisings of the dead in all five original Gospel sources (Mark, Q, John, Luke special, and Matthew special), which show a multiplicity of eyewitness traditions from different towns and villages that would have been remembered and checkable by witnesses living within memory of Jesus' ministry. Note that many of these traditions contain Semitisms, indicating an early Palestinian origin close to the time of Jesus. Note also that many of the traditions mention the towns and names of witnesses where miracles took place, particularly those concerned with raising the dead. This would have allowed for accurate verification of the events near the time when they occurred.
- The virtual certainty that the apostles performed miracles in the name of Jesus (exactly as recounted in the Acts of the Apostles and the Letters of Saint Paul), because without them, the

---

[45] See Robert Spitzer, *Five Pillars of the Spiritual Life: A Practical Guide to Prayer for Active People* (San Francisco: Ignatius Press, 2008), Chapters 4 and 5. See also Robert Spitzer, *Finding True Happiness: Satisfying Our Restless Hearts* (San Francisco: Ignatius Press, 2015), Chapter 8.

exponential growth of the early Church (after the humiliation and execution of her Messiah) is virtually inexplicable (see N. T. Wright's study of early Christian messianism cited above in Section V.C, note 82). Why would the apostles be able to perform miracles in Jesus' name that Jesus Himself did not perform?

- The charismatic power of the Holy Spirit still visibly present today in thousands of miracles manifest throughout the international Christian Church. The Spirit is also evidently present in the interior lives of those who seek the Spirit's guidance, inspiration, and protection.

In view of the above, it is reasonable and responsible to believe that Jesus had a prolific ministry of miracles, including raising the dead. Furthermore, Jesus performed these miracles by His own authority, differentiating Himself from every previous prophet who needed to invoke the power of Yahweh through prayer.[46] When the apostles and evangelists reflected on this, they realized that Jesus possessed the power of God within Himself to exorcise, heal, and raise the dead. These powers were considered divine, and so the early Church was certain that Jesus did not become divine after His Resurrection, but was divine throughout His ministry. More than this, Jesus' testimony to be the exclusive Son of the Father (see below, Section VI) and Mary's testimony to have conceived by the power of the Holy Spirit led to the inevitable conclusion that Jesus was divine from all eternity with the Father. Let us now examine that claim.

VI. Jesus' Divinity and the Cost of Proclaiming It

We now approach the question of why the early Church proclaimed Jesus to be divine. At first glance, one might think, "Well, that's obvious—the Church wanted to aggrandize Jesus." However, this makes no sense; because proclaiming a man to be divine was completely repugnant to the Jewish audience that the early Church

---

[46] Recall that this unique characteristic of Jesus' miraculous power is attested by all original Gospel sources and applies to His exorcisms and healings, as well as the multiple people He raised from the dead. Therefore, it is very likely historical.

was trying to attract. Quite frankly, the early Church could not have picked a more apologetically *unappealing* proclamation for this audience than "Jesus is the Lord." This proclamation led to the Christians' loss of religious status, social status, and financial status, within the Jewish community inside and outside Jerusalem. Ultimately, it led to the Christians being expelled from the synagogue and being actively persecuted by both Jewish and Roman authorities. Why would the apostolic Church have selected a doctrine that was viewed so unfavorably by the very audience she wanted to attract, leading to her active persecution?

As Joachim Jeremias remarks, this was wholly unnecessary, because the apostolic Church did not have to proclaim or even imply that Jesus was divine in order to bestow great favor upon Him within the religious culture of the day. The early Church could have proclaimed Him to be a "martyr prophet", which would have allowed converts to worship at His tomb and to pray through His intercession.[47] This more modest claim would have made Him acceptable to Jewish audiences who would then have ranked Him high among the "holy ones". Why then did the leaders of the apostolic Church go so unapologetically and dangerously far to proclaim that "Jesus is the Lord"? Why did they suffer social and financial loss, religious alienation, and even persecution and death, when it all could have been avoided by simply giving up the implication of His divinity? The most likely answer is that they really believed Him to be divine. So why did the apostolic Church believe Him to be divine (and even to share a unity with the Father throughout all eternity)? How could they be so sure of this radical proclamation which had so many negative consequences, when they could have taken the "easier and safer route" in proclaiming Him to be a martyr-prophet? Was it simply because of Jesus' claim to be the exclusive Son of the Father—or something more? The Church proclaimed the "something more" in her earliest creedal statements called *kerygmas*.[48] These *kerygmas* were written shortly after Jesus' Resurrection and gift of the Spirit (A.D. 34–35) and are identifiable

---

[47] Joachim Jeremias, *Heiligengräber in Jesu Umwelt* (Göttingen: Vandenhoeck & Ruprecht, 1958).

[48] *Kerygmas* were initially elucidated for the English world by C.H. Dodd in his seminal work *The Apostolic Preaching and Its Development* (New York: Harper and Brothers, 1962).

in the letters of Saint Paul and the Acts of the Apostles through form-critical methods. These *kerygmas* identify the foundation upon which the Church proclaimed that Jesus is the Lord:

1. His glorious Resurrection, manifesting a divine quality
2. His gift of the Holy Spirit (identified as "the power of God")
3. His miracles by His own power and authority
4. His unconditional self-sacrificial love that saves us from our sins

When these events are combined with Jesus' preaching about Himself as the exclusive Son of the Father, and early Church leaders' experience of the power to perform miracles in the name of Jesus, they had little doubt that Jesus truly shared in His Father's divinity as He Himself proclaimed to His apostles. They must have thought to themselves, "Why would God work miracles in the name of *Jesus* if Jesus was not really His Son as He proclaimed it?"

Did Jesus really claim to be divine? Most scholars agree that the following Q source passage is of ancient Palestinian origin, most probably from Jesus Himself, because it is filled with His vocabulary and style (explained below):

> In that same hour [Jesus] rejoiced in the Holy Spirit and said, "I thank you, Father [*pater*], Lord of heaven and earth, that you have hidden these things from the wise and understanding and revealed them to infants; yes, Father [*ho pater*], for such was your gracious will. All things have been delivered to me by my Father [*tou patros mou*]; and no one knows who the Son is except the Father [*ho pater*], or who the Father [*ho pater*] is except the Son and any one to whom the Son chooses to reveal him." (Lk 10:21–22; cf. Mt 11:25–27)

Why do so many scholars believe that this passage came from Jesus Himself? First, this passage is not a redaction of the Gospel authors (Luke and Matthew), but comes from Q (an early source of Jesus' sayings used by Matthew and Luke—see above, Section III). Second, the presence of Semitisms indicates a Palestinian origin—either the Palestinian community or Jesus Himself. Third, there are five mentions of *pater*—*pater* in a prayer and *ho pater*, *tou pater*, and *tou patros mou* within the context of a prayer. Jeremias notes in this regard, "We have every reason to suppose that an *Abba* underlies every instance of *pater (mou)* or *ho patēr in his words of prayer.*"[49] This means that there are five uses of "Abba" in this short passage. Jeremias further indicates that "Abba" as an address to God is virtually unique to Jesus in all Palestinian literature.[50] Since this passage is virtually filled with "Abba" addresses and references, and since this address is virtually unique to Jesus, it is very likely that it was uttered by Jesus Himself.

What does this passage mean? The expression "all things" (*panta*) refers not only to the whole of creation but also to the Divine itself. In the next sentence, "no one knows who the Son is except the Father", the word "knows" has a Semitic meaning that goes beyond abstract knowing. It refers to knowledge of the person interiorly and completely. Matthew's Semitic audience would understand this, and so he stays with the simpler expression, "no one knows the Son except the Father; and no one knows the Father except the Son" (11:27). Luke has to translate this expression for his Gentile audience, and so he adds "who" and "is", rendering the proclamation as follows: "No one knows *who* the Son *is* except the Father, or *who* the Father *is* except the Son" (10:22; emphasis added). This slight change in the Q text suggests personal knowledge ("who") as well as complete knowledge ("is"—being). Thus, Luke's Greek translation of Jesus' Hebrew prayer and teaching might be, "No one knows the Father through and through (interiorly and completely) except the Son."

Thus, we might translate the passage as follows:

> The whole of creation and divinity has been given to me by my Father, and no one knows the Son through and through (interiorly and completely) except the Father, and no one knows the Father through and through (interiorly and completely) except the Son.

This passage strongly implies that Jesus shares in His Father's divinity and has the same complete intimate knowledge of the Father as the Father has of Him. Jesus' apostles and disciples would have understood the implications of divine status and equality of knowledge between the Father and the Son, and so would have naturally concluded to His divinity. N. T. Wright asserts that Jesus Himself could not have made such a statement without believing that

[49] Jeremias, *Proclamation of Jesus*, p. 65.
[50] See ibid., p. 64.

He could do what is reserved to *God* alone,[51] implying that He shared in His Father's divinity.

When the above Q logion is combined with Jesus' claim to bring the Kingdom of God in His own Person, to have divine power within Himself (in the way He performed miracles), and His claim to be the beloved Son of the Father (see Lk 20:13), we can see why the apostles became convinced that Jesus claimed to be not only the Messiah but also divine. Jesus' claim became known beyond His close company of apostles and apparently had reached the Sanhedrin, which is reflected in the questioning of the High Priest at Jesus' trial, who asked Jesus:

> "Are you the Christ, the Son of the Blessed?" And Jesus said, "I am; and you will see the Son of man sitting at the right hand of Power, and coming with the clouds of heaven." (Mk 14:61–62)

Wright believes that this unambiguous declaration is *essentially historical*. Jesus' response to the questions of the interrogators led them to believe that He had not only messianic but also *divine* "pretensions"—worthy of a charge of blasphemy. This charge in Jewish legal proceedings seems to have been wholly original and was apparently formulated specifically to address Jesus' response to His interrogators. According to Wright:

> Since we have no evidence of anyone before or after Jesus ever saying such a thing of himself, it is not surprising that we have no evidence of anyone framing a blasphemy law to prevent them doing so.[52]

In view of the above, it is reasonable and responsible to affirm that Jesus claimed to be not only the Messiah but also the exclusive Son of His Father—sharing in His divine status.

Let us now return to our main point. When the early Church leaders combined Jesus' proclamation of His exclusive divine Sonship with their experience of His glorious Resurrection, His gift of the Spirit of God ("the power of God", which can be given only by God Himself), His miraculous power by His own authority, His

---

[51] See Wright, *Victory of God*, pp. 649–50.
[52] Ibid., p. 551.

unconditional love (manifest by His self-sacrifice on the Cross), and the apostles' power to do miracles in Jesus' name, they had little doubt that He truly is and was the Son of God from all eternity. They proclaimed this truth boldly, though the cost was exceptionally high—the loss of thousands of potential converts, the loss of religious, social, and financial status, and ultimately persecution and expulsion from the synagogue. Nevertheless, they accepted the cost because they were convinced of the truth that Jesus is and was the Lord of Heaven and earth.

In the Letter to the Philippians, we find one among many early New Testament liturgical hymns[53] in which the Church proclaimed Jesus as sharing in the Father's divinity:

[Jesus], though he was in the form of God, did not count *equality* with God a thing to be grasped, but emptied himself, taking the form of a servant, being born in the likeness of men. And being found in human form he humbled himself and became obedient unto death, even death on a cross. Therefore God has highly exalted him and bestowed on him the name which is above every name, that at the name of Jesus every knee should bow, in heaven and on earth and under the earth, and every tongue confess that Jesus Christ is Lord, to the glory of God the Father. (2:6–11)

This was the truth as Jesus revealed it and the apostles and early disciples experienced it. They paid the ultimate price to proclaim and give evidence for it.

VII. Conclusion

The above historical and scriptural evidence for Jesus' ministry, miracles, Resurrection, gift of the Spirit, and self-proclamation as divine

[53] These early liturgical hymns are found in the Letters of Saint Paul and the Gospel of Saint John, and predate those works by many years. Most of them are thought to have been written between A.D. 38 and 48 by well-educated Jewish scribes familiar with both wisdom speculation and Greek philosophical thought (originating from Plato)—for example, Phil 2:6–11; Col 1:15–20; Eph 2:14–16; 1 Tim 3:16; 1 Pet 3:18–22; Heb 1:3; and the Prologue of the Gospel of John. See Jack T. Sanders, *The New Testament Christological Hymns: Their Historical Religious Background* (Cambridge: Cambridge University Press, 1971).

does not stand solely on the methods, credentials, and evidence of great scholars and historians. It stands fundamentally within the undiminished authority of the apostolic tradition and the living tradition of the Catholic Church. The evidence given above complements and corroborates this foundation, but the foundation still stands without it. In the next four chapters, we will be examining scientifically discernible evidence for the Passion and Resurrection of Jesus as well as His Real Presence in the Eucharist and His mother's real presence in the work of salvation. As probative as this evidence is, it, too, is not necessary for believing that Jesus Christ is the Emmanuel—the Son of God, who has come to be with us. As with the historical and scriptural evidence, the forthcoming scientific evidence is complementary to and corroborative of the apostolic tradition, the living tradition of the Catholic Church, and the historical and scriptural evidence. It adds a significant piece to the combined evidence for Jesus' Resurrection, presence, and divinity, while allowing the apostolic and living tradition of the Church to be the undiminished foundation upon which the life and proclamation of Jesus will stand until the end of time.

Chapter Three

Science at the Doorstep to Jesus—
The Shroud of Turin

Introduction

The relic known as the Shroud of Turin can amplify our understanding and reinforce the rational validation of Jesus' Crucifixion, death, and Resurrection. It is the most unique image produced in human history. It has the further distinction of being by far the most scientifically investigated historical artifact. These tests have unveiled a preponderance of scientific evidence implying the Shroud's authenticity. Though the evidence is quite probative, it is not definitive, and additional scientific testing is needed to obtain greater certitude. As noted above, our faith does not rest on any artifact (or the scientific investigation of it), but rather on the apostolic testimony, the teaching of the Catholic Church, the New Testament Scriptures, and the confirmation of the Holy Spirit in us and the rest of the faithful.

As we shall see, the authenticity of the Shroud was seriously questioned by a 1988 carbon dating that placed the Shroud's origin between A.D. 1260 and 1390. However, this carbon dating has recently been discredited by further analysis of the raw data from the tests as well as by the heterogeneous composition of fibers taken from the sample site. Though five new dating tests and other extrinsic dating evidence point to a first-century origin of the Shroud in Judea, additional testing must be done. As we shall see below (Section IV.D), we must test for certain cosmogenic isotopes indicative of a nuclear reaction before doing another carbon dating. If the Shroud's image was produced by a low-temperature nuclear reaction coming from the disintegration of all the stable atomic nuclei in the Shroud

97

man's body, then carbon dating cannot be used to determine the date of the cloth.

Additionally, scientists have unveiled a host of mysteries surrounding the Shroud, which have confounded current models of scientific and naturalistic explanation. Regardless of whether we attribute these science-confounding mysteries to supernatural causation, this investigation of the Shroud will show it to be one of the most, if not *the* most, unique and mysterious historical artifacts ever to have come to light. We may now turn to the Shroud itself.

The Shroud of Turin is a linen burial cloth fabricated by a herringbone, three-to-one twill pattern, measuring roughly 14 feet by 3.5 feet. It has a perfect three-dimensional, photographic negative image of a crucified man, an image executed in such accurate anatomical detail that modern medicine can diagnose many of the injuries by analyzing it. Most intriguing of all, scientific tests have revealed the image was not produced by any kind of paint, dye, chemical, vapor, or scorching (see below, Section IV). The Shroud has 372 bloodstains (159 on the front image and 213 on the back image) with AB+ blood type and enzymes indicating polytrauma, which were embedded on the Shroud prior to the creation of the image.[1] It tells the story of a crucifixion that resembles the unique Crucifixion of Jesus of Nazareth by Roman authorities (see below, Section III).

The provenance of the Shroud is limited because its recorded history begins only when it surfaced in Lirey, France, around 1350 in the hands of a French nobleman, Geoffrey de Charny, whose wife's ancestors were linked to one of the leaders of the Fourth Crusade— Othon de la Roche.[2] Since that time, the journey of the Shroud to Turin, Italy (where it remains today), is quite certain. However, prior to 1350, we have only sightings of what appears to be the Shroud— for example, in Palestine (from the New Testament—Jn 20:5), in Edessa, Turkey (from the Roman emperor who laid siege to Edessa to obtain it in 943), and in Constantinople (from one of the leaders of

---

[1] Barbara Faccini and Giulio Fanti, "New Image Processing of the Turin Shroud Scourge Marks" (Proceedings of the International Workshop on the Scientific Approach to the Acheiropoietos Images, ENEA Frascati, Italy, May 4–6, 2010), http://www.acheiropoietos .info/proceedings/FacciniWeb.pdf.

[2] I give a more detailed explanation of the documented history of the Shroud from around 1350 to the present in Robert Spitzer, *God So Loved the World: Clues to Our Transcendent Destiny from the Revelation of Jesus* (San Francisco: Ignatius Press, 2016), Appendix I, Section I.

the Fourth Crusade who saw it displayed in the Church of St. Mary of Blachernae in 1204—Robert de Clari). We can also trace its history by means of the indigenous pollen grains embedded in it from Palestine, Edessa, Constantinople, France, and Italy.[3]

In sum, the Shroud is a unique and mysterious historical artifact that, through scientific testing, appears to be the burial cloth of Christ, revealing details that correspond to the unusual set of injuries described in the Gospels. It also contains evidence of a source of radiation and mechanical transparency suggestive of the Resurrection of Jesus Christ. We will here give a brief survey of some of this evidence.

We will investigate five major dimensions of the Shroud of Turin:

1. The 1988 Carbon Dating Problems (Section I)
2. Eight Alternative Dating Methods (Section I)
3. The Bloodstains on the Shroud (Section II)
4. The Crucifixion and Death of the Man on the Shroud (Section III)
5. The Image on the Shroud and Evidence of the Resurrection (Section IV)

I. The 1988 Carbon Dating Problems and Eight Alternative Dating Methods

The 1988 carbon dating that placed the Shroud's age between A.D. 1260 and 1390 convinced much of the scientific and religious world the Shroud had been produced by a medieval forger; but as we shall see, the sample produced heterogeneity in the raw data of the carbon dating, invalidating it as establishing a medieval timeframe. It also strongly disagrees with five other dating tests and three extrinsic methods of dating. When all of this is considered, the most probable date of the Shroud of Turin is in the first century, around the area of Palestine. We will explain these findings in three sections:

1. Problems with the 1988 Carbon Dating (Section I.A)
2. Five New Scientific Dating Methods (Section I.B)
3. Three Extrinsic Dating Methods (Section I.C)

[3] For a more detailed explanation of the historical journey of the Shroud prior to 1350, see ibid., Appendix I, Section II.E.

A. Problems with the 1988 Carbon Dating

Before explaining the new methods used to date the Shroud, which place it in the time of Christ, we should address the one test that does not seem to match these findings: a radiocarbon-14 test done in 1988 on a sample of the Shroud that placed it between A.D. 1260 and 1390.[4] As a rule, carbon-14 is a reliable way of measuring the age of an artifact, but several factors compromised the results of this particular test.

First, a single sample was drawn from a controversial place on the Shroud that may have had some fibers associated with repairs to the cloth after the fire of Chambery in 1532. The original protocol for carbon dating recommended that seven samples be taken from different parts of the Shroud—parts that were not compromised by the fire of Chambery and other environmental factors. This was precisely what was *not* done. Instead, a single sample was taken from a part of the Shroud that may have been compromised by repairs after the fire of Chambery. Despite these baffling protocol failures, the peer-reviewed journal *Nature* published the results as if they were indisputable: "The results provide conclusive evidence that the linen of the Shroud of Turin is mediaeval."[5]

These findings were not "indisputable". After trying for decades to obtain the raw data from the carbon dating, French researcher Tristan Casabianca and colleagues finally used a Freedom of Information Act request to obtain the raw data from the British Museum, which finally relinquished the data after thirty years. Casabianca and colleagues made a statistical analysis of the data and found significant heterogeneity among the three parts of the samples as well as within each separate sample, meaning that these samples could not be used to date the Shroud accurately to the Middle Ages.[6] In an interview with *L'Homme Nouveau* (Catholic bimonthly), Casabianca explained their findings as follows:

---

[4] P. E. Damon et al., "Radiocarbon Dating of the Shroud of Turin", *Nature* 337 (1989): 611–15, https://www.nature.com/articles/337611a0.

[5] Ibid., Abstract.

[6] Tristan Casabianca et al., "Radiocarbon Dating of the Turin Shroud: New Evidence from Raw Data", *Archaeometry* 61, no. 5 (2019): 1223–31, https://www.researchgate.net/publication/331956466_Radiocarbon_Dating_of_the_Turin_Shroud_New_Evidence_from_Raw_Data.

Our statistical analysis shows that the 1988 carbon 14 dating was unreliable: the tested samples are obviously heterogeneous, and there is no guarantee that all these samples, taken from one end of the sheet, are representative of the whole fabric. It is therefore impossible to conclude [from the 1988 carbon dating] that the shroud of Turin dates from the Middle Ages.[7]

There are two major hypotheses that may explain an errant dating by carbon-14 testing:

1. *The sample had fibers from a later period embedded in it* (which could have come from repairs after the fire of Chambery). Dr. Raymond Rogers identified cotton and a medieval gum dye mordant (for affixing dye) in the sample used for dating. If the sample had large amounts of these sixteenth-century materials, it would have adversely affected the C-14 test. According to Rogers:

> The color and distribution of the coating implies that repairs were made at an unknown time with foreign linen dyed to match the older original material.... The consequence of this conclusion is that the radiocarbon sample was not representative of the original cloth....
>
> A gum/dye/mordant coating is easy to observe on Raes and radiocarbon [sample] yarns. No other part of the shroud shows such a coating....
>
> [This indicates that the] radiocarbon sample had been dyed. Dyeing was probably done intentionally on pristine replacement material to match the color of the older, sepia-colored cloth....
>
> The dye found on the radiocarbon sample was not used in Europe before about A.D. 1291 and was not common until more than 100 years later. The combined evidence from chemical kinetics, analytical chemistry, cotton content, and pyrolysis/ms [mass spectrometry] proves that the material from the radiocarbon area of the shroud is significantly different from that of the main cloth. The radiocarbon sample was thus not part of the original cloth and is invalid for determining the age of the shroud.[8]

---

[7] Quoted in John Burger, "New Data Questions Finding That Shroud of Turin Was Medieval Hoax", *Aleteia*, July 22, 2019, https://aleteia.org/2019/07/22/new-data-questions-finding-that-shroud-of-turin-was-medieval-hoax/.

[8] Raymond N. Rogers, "Studies on the Radiocarbon Sample from the Shroud of Turin", *Thermochimica Acta* 425, nos. 1–2 (2005): 192–93, http://www.shroud.it/ROGERS-3.PDF.

2. *Particle radiation might have significantly increased the C-14 content of the cloth.* In the particle radiation hypothesis (explained below in Section IV.D), there would have been a neutron fluence coming from the nuclear disintegration of the body. This neutron eradiation would have converted N-14 (indigenous to linen) into additional C-14.[9] Dr. Arthur C. Lind (nuclear physicist and expert in plasmas and polymers) and colleagues show that this additional C-14 stays within the cloth over the long term, despite environmental factors, such as warming from a fire.[10] This would create the appearance that the cloth was much younger than it actually is.[11] If in fact the body did disintegrate with a blast of neutrons, protons, alpha particles, and gamma rays, then the significant increases in C-14 content would make any C-14 dating wholly unreliable—not merely the one done in 1988, but all subsequent C-14 tests.

There are several tests that can be done to determine whether C-14 was created through a blast of neutrons and protons during nuclear disintegration of the body (see below, Section IV.D).

B. Five New Scientific Dating Methods

There are five dating methods that place the origins of the Shroud much nearer to the first century:

1. Wide-Angle X-Ray Scattering (Section I.B.1)
2. Vanillin Test (Section I.B.2)

---

[9] "The neutron fluence that would be needed to cause the radiocarbon date to be medieval instead of first century is 8.3×10^{13} n·cm^{-2} if the nitrogen content of the Shroud is about 570 ppm." Mark Antonacci, "Particle Radiation from the Body Could Explain the Shroud's Images and Its Carbon Dating", *Scientific Research and Essays* 7, no. 29 (2012): 2619, https://academicjournals.org/article/article1380798649_Antonacci.pdf, citing data from Arthur C. Lind et al., "Production of Radiocarbon by Neutron Radiation on Linen" (Proceedings of the International Workshop on the Scientific Approach to the Archeiropoietos Images, ENEA Frascati, Italy, May 4–6, 2010), http://www.acheiropoietos.info/proceedings/LindWeb.pdf.

[10] Lind et al., "Production of Radiocarbon".

[11] Antonacci, "Particle Radiation".

3. Fourier Transform Infrared Spectroscopy (Section I.B.3)
4. Raman Spectroscopy (Section I.B.4)
5. Break-Strength Testing (Section I.B.5)

We will discuss each in turn.

1. Wide-Angle X-Ray Scattering

Dr. Liberato De Caro (specialist in coherent diffractive imaging, crystallography, and solid state chemistry at the Institute of Crystallography and the National Research Council of Italy) and his colleagues developed, tested, and peer-reviewed a new technique in 2019 to measure the age of cellulose (intrinsic to linen) of ancient fabrics: wide-angle X-ray scattering. This new test was shown to measure fabrics accurately (dated by C-14 and other dating techniques) from 3,000 B.C. to A.D. 2,000. On the basis of these results, they applied the technique in March 2022 to a sample of the Shroud of Turin that dated the Shroud to A.D. 55–74—slightly after the time of Jesus' Crucifixion. They described the test and results, published in the peer-reviewed journal *Heritage*, as follows:

> On a sample of the Turin Shroud (TS), we applied a new method for dating ancient linen threads by inspecting their structural degradation by means of Wide-Angle X-ray Scattering (WAXS). The X-ray dating method was applied to a sample of the TS consisting of a thread taken in proximity of the 1988/radiocarbon area (corner of the TS corresponding to the feet area of the frontal image, near the so-called Raes sample). The size of the linen sample was about 0.5 mm × 1 mm. We obtained one-dimensional integrated WAXS data profiles for the TS sample, which were fully compatible with the analogous measurements obtained on a linen sample whose dating, according to historical records, is 55–74 AD, Siege of Masada (Israel). The degree of natural aging of the cellulose that constitutes the linen of the investigated sample, obtained by X-ray analysis, showed that the TS fabric is much older than the seven centuries proposed by the 1988 radiocarbon dating. The experimental results are compatible with the hypothesis that the TS is a 2000-year-old relic, as supposed by Christian tradition, under the condition that it was kept at suitable

levels of average secular temperature—20.0–22.5°C—and correlated relative humidity—75–55%—for 13 centuries of unknown history, in addition to the seven centuries of known history in Europe. To make the present result compatible with that of the 1988 radiocarbon test, the TS should have been conserved during its hypothetical seven centuries of life at a secular room temperature very close to the maximum values registered on the earth.[12]

Evidently, the Shroud was not kept at a secular room temperature of 134°F for seven hundred years, because the secular temperature in Turin, Italy, averages between 46°F (in winter) and 83°F (in summer). Inasmuch as the technique has accurately tested multiple linen fabrics over a wide age range (3,000 B.C.–A.D. 2,000), the estimated age of A.D. 55–74 is probably correct. Multiple tests can be done on the same sample, and it is very likely that additional tests will be done on other samples. When this dating test is combined with Fanti's three dating tests given below (yielding an approximate age of A.D. 90), it is very likely that the Shroud's date of origin is in the first century, well within the time of Jesus' Crucifixion and Resurrection—A.D. 33.

2. Vanillin Test

In 2005, Dr. Raymond Rogers devised a vanillin test of the age of the Shroud, publishing his results in the peer-reviewed journal *Thermochimica Acta*.[13] Vanillin is an organic compound that, like C-14, decays with age. Linens from the Middle Ages typically retained 37 percent of their vanillin when tested, while older artifacts like the Dead Sea Scrolls had lost all of theirs. Comparison of the Shroud's results with these other linens established a possible age range of thirteen hundred to three thousand years old (i.e., a date of origin between 985 B.C. and A.D. 715).[14] This is *profoundly* different from the 1988 carbon dating (even when effects of the fire of Chambery

---

[12] Liberato De Caro et al., "X-ray Dating of a Turin Shroud's Linen Sample", *Heritage* 5, no. 2 (2022), Abstract, https://www.mdpi.com/2571-9408/5/2/47/htm.
[13] Rogers, "Studies on Radiocarbon Sample", pp. 190–92.
[14] Ibid., p. 192.

are considered), which moved Dr. Giulio Fanti to devise additional tests based on mechanical compressibility and tension as well as spectroscopic analysis. He devised three different tests—two for opto-chemical (spectroscopic/laser) analysis and one for mechanical compressibility and tension. The results follow.

3. Fourier Transform Infrared Spectroscopy

Fourier transform infrared spectroscopy was used to detect cellulose degradation.[15] Researchers took nine ancient textiles of different ages (from Egypt, Israel, and Peru), as well as two modern fabrics, and tested them to establish the rate at which cellulose (another decaying compound) disappears over time. By applying this test to the Shroud and comparing it with the other known samples, a date of origin for the Shroud was set at 250 B.C. ± 200 years at a 95 percent confidence level.[16]

4. Raman Spectroscopy

Fanti devised a second opto-chemical method to test for cellulose degradation. Using a different method (Raman spectroscopy) allowed him to determine an accurate general age range.[17] This provides corroboration of the results from each method as well as more accurate adjustment for the effects of the fire of Chambery in 1532, in which the Shroud was involved. The result of the tests showed

[15] Giulio Fanti, Pierandrea Malfi, and Fabio Crosilla, "Mechanical and Opto-chemical Dating of the Turin Shroud", *MATEC Web of Conferences*36 (2015), Workshop of Paduan Scientific Analysis on the Shroud, https://www.researchgate.net/publication/287294012 _Mechanical_ond_opto-chemical_dating_of_the_Turin_Shroud. See also Giulio Fanti, "Optical Features of Flax Fibers Coming from the Turin Shroud", *SHS Web of Conferences* 15 (2015), 2014 Workshop on Advances in the Turin Shroud Investigation, https://www.shs-conferences.org/articles/shsconf/abs/2015/02/shsconf_atsi2014_00004/shsconf _atsi2014_00004.html.

[16] Ibid.

[17] Fanti, Malfi, and Crosilla, "Mechanical and Opto-chemical Dating". See also Giulio Fanti et al., "Non-destructive Dating of Ancient Flax Textiles by Means of Vibrational Spectroscopy", *Vibrational Spectroscopy* 67 (2013): 61–70, http://dx.doi.org/10.1016/j.vibspec .2013.04.001.

an age of the Shroud at about A.D. 30 ± 200 years at a 95 percent confidence level.[18]

5. Break-Strength Testing

Break-strength testing was used to compare the tensile strength of ancient fabrics.[19] By correlating the Shroud fibers with other known ancient fabrics, an estimated date of origin was obtained. Fanti used several samples from the Shroud and other ancient fabrics (of various ages), and then applied a least-squares multiple linear regression (MLR) to the measured mechanical data, which shows a date of origin at A.D. 260 ± 200 years at 95 percent confidence level.[20]

Fanti then averaged the weighted results of the above three tests and concluded as follows:

> [The two opto-chemical/spectroscopic] dates combined with the mechanical result, weighted through their estimated square uncertainty inverses, give a final date of the Turin Shroud of 90 AD ±200 years at 95% confidence level.[21]

Fanti recommended that additional samples and tests be used to confirm the probable date of the Shroud determined above—A.D. 90, close to the time when Jesus was crucified (approximately A.D. 33).[22] As noted above, such a test has been developed, tested, and peer-reviewed by Dr. De Caro and colleagues in March 2022, yielding an age of the Shroud between A.D. 55 and 74. Since the latter test has a much smaller margin of error than the opto-chemical/spectroscopic tests, it is more accurate.

At present, all contemporary scientific dating of the Shroud points to an age much earlier than the carbon-14 tests performed in 1988. It

[18] Ibid.

[19] See Fanti, Malfi, and Crosilla, "Mechanical and Opto-chemical Dating". See also Giulio Fanti and Pierandrea Malfi, "Multi-parametric Micro-mechanical Dating of Single Fibers Coming from Ancient Flax Textiles", *Textile Research Journal* 84, no. 7 (2013): 714–27, http://dx.doi.org/10.1177/0040517513507366.

[20] See ibid.

[21] Fanti, Malfi, and Crosilla, "Mechanical and Opto-chemical Dating", Abstract.

[22] Ibid.

is unlikely that De Caro and his colleagues' dating test and all three of Fanti's distinct dating methods would be off by thirteen hundred years. Therefore, it is likely that the 1988 carbon dating is invalid (as shown in Casabianca's 2019 statistical analysis of the raw data). As noted above, this invalid carbon dating may be attributable to sixteenth-century materials embedded in the carbon dating sample[23] and a neutron fluence that converted N-14 to C-14 in the linen cloth.[24] In light of this, we tentatively conclude that the Shroud originated well before the Middle Ages—in the first century A.D. This is further confirmed by extrinsic dating methods, particularly the Sudarium of Oviedo (Jesus' face cloth).

C. Three Extrinsic Dating Methods

In addition to the above dating methods, there are also extrinsic factors pointing to the Shroud's origin prior to the Middle Ages:

1. The Congruences with the Sudarium of Oviedo (Section I.C.1)
2. The Presence of Palestinian Pollen Grains (Section I.C.2)
3. The Possible Imprints of Roman Coins on the Shroud Man's Eyes (Section I.C.3)

1. The Congruences with the Sudarium of Oviedo

The Sudarium of Oviedo[25] is a bloodstained cloth (without an image) that according to Scripture and tradition was wrapped around the face and head of Christ after His death (see Jn 20:7). Use of a face cloth

---

[23] Rogers, "Studies on Radiocarbon Sample".

[24] Lind et al., "Production of Radiocarbon".

[25] See Guillermo Heras Moreno, José-Delfín Villalaín Blanco, and Jorge-Manuel Rodríguez Almenar, "Comparative Study of the Sudarium of Oviedo and the Shroud of Turin", *III Congresso Internazionale di Studi sulla Sindone* Turin (1998): 1–17, translated from the Spanish by Mark Guscin and revised by Guillermo Heras Moreno, https://www.shroud.com/heraseng.pdf. See also Mark Guscin, "The Sudarium of Oviedo: Its History and Relationship to the Shroud of Turin", Shroud of Turin website, citing *Actes Du III—Symposium Scientifique International Du C.I.E.L.T., Nice 1997* (Paris: C.I.E.L.T, 1998), pp. 197–99, https://www.Shroud.com/guscin.htm#top.

was customary to transport a body to a tomb particularly when the face was disfigured from torture or illness. The Sudarium of Oviedo is also embedded with a predominance of pollen grains from Judea, as well as some from Turkey and Spain (where it was deposited in the Cathedral of Oviedo in A.D. 718).[26]

There is a precise provenance (historical record) of the Sudarium's journey from bishop to bishop—region to region—from A.D. 616 in Turkey to A.D. 718, when it was placed in the Cathedral in Oviedo, Spain, where it remains till this day.[27] There is also a probable provenance of the Sudarium before A.D. 616 in Palestine that is confirmed by pollen grains embedded in the Sudarium. As on the Shroud of Turin, there are thirteen pollen grains from Israel, and four of them are unique to that region,[28] meaning that the Sudarium stayed in the Palestinian area for a long time before it was moved toward Europe (see below, Section I.C.2). This coincides with Bishop Pelagius' account of the Sudarium's history. Mark Guscin sums up the historical record as follows:

> According to this history, the sudarium [of Oviedo] was in Palestine until shortly before the year 614, when Jerusalem was attacked and conquered by Chosroes II, who was king of Persia from 590 to 628. It was taken away to avoid destruction in the invasion, first to Alexandria by the presbyter Philip, then across the north of Africa when Chosroes conquered Alexandria in 616. The sudarium entered Spain at Cartagena, along with people who were fleeing from the Persians.[29]

The face cloth used to transport Jesus from the Cross to the tomb is specifically mentioned in the Gospel of John: "[Peter] saw the linen

---

[26] See ibid.

[27] I have outlined the history of the Sudarium of Oviedo in Appendix I of *God So Loved the World*. The Sudarium has been in Oviedo since 718, where it remains to this day. However, its history prior to that time was traced by Bishop Pelagius in his *Book of the Testaments of Oviedo* and the *Chronicon Regum Legionensium*, written around 1121. Pelagius discovered the line of bishops who received the Sudarium when it arrived in Cartagena (from Palestine) in 616 to its arrival in Oviedo in 718. See Guscin, "Sudarium of Oviedo".

[28] Emanuela Marinelli, "The Question of Pollen Grains on the Shroud of Turin and the Sudarium of Oviedo" (paper presented at the 1st International Congress on the Holy Shroud of Turin, Valencia, Spain, April 28–30, 2012), p. 5, https://www.Shroud.com/pdfs/marinelli 2veng.pdf.

[29] Guscin, "Sudarium of Oviedo".

cloths lying there, and the face cloth, which had been on Jesus' head, not lying with the linen cloths but folded up in a place by itself" (20:6–7, ESV). The cloth wrapped around Jesus' head was in a place by itself because it was taken off of Jesus before He was placed in the linen cloth (the Shroud). This procedure corresponds to Jewish custom. The fact that the face cloth was not on Jesus' face at the time of His Resurrection explains why there is no image on the Sudarium. If the image was produced by a strong source of radiation throughout the body (as will be shown below in Section IV), it is likely coincident with Jesus' Resurrection, which the Gospel accounts and Saint Paul indicate has a component of power and light (see Section IV.E).

The sequence of events that took place after Jesus' body arrived at the tomb is clear:

1. When the body arrives at the tomb, his disciples remove the cloth wrapped around his face and lay it in a place by itself. This explains why only the bloodstains, but not the image, are on the Sudarium.

2. His body is then laid on half of the linen Shroud (on His back), and the other half of the Shroud is placed over His head all the way down to His feet, where it is secured. The blood on the body then makes contact with certain places on the Shroud and stains it.

3. After the blood has stained the cloth, an event occurs (which we now believe was a strong source of radiation and light) that causes a perfect photographic negative image (with three-dimensional layering) to be imprinted on a non-photographically sensitive linen cloth.

We now turn to a brief analysis of the bloodstains (which will be discussed in great detail below in Section II). Dr. Alan Whanger used the polarized image overlay technique on photographs of both cloths and discovered 120 points of congruence in the bloodstains on both cloths—seventy points of congruence on the front portion (the face), and fifty points of congruence on the rear side (the back of the head and the nape of the neck). Additionally, there are significant points of congruence in fluid discharges (e.g., plural edema fluid from the nose). There are so many points of congruence between the wounds and fluid markings of both cloths that Guscin notes, "The only possible conclusion is that the Oviedo sudarium covered the same face as the Turin Shroud."[30] The odds against a purely coincidental congruence of this magnitude (without the same face touching both cloths) are astronomically high.

In view of the similarities in blood type and facial features, as well as the 120 points of congruence in the positioning of blood and fluids on the two cloths, it is difficult to avoid Guscin's conclusion—that the two cloths touched the same face of a man crowned with thorns and severely beaten.

Why is this significant for dating the Shroud? The Sudarium of Oviedo has a continuous traceable provenance (recorded history) starting in A.D. 616 (compared to the Shroud's documented history starting in A.D. 1349). Furthermore, the Shroud came from Palestine in A.D. 614, which is confirmed by the preponderance of pollen grains on it from that region. Now, if the two cloths touched the same face after the same event (beating and crowning with thorns), it establishes that the Shroud must be *at least* as old as the Sudarium (prior to A.D. 616). This means that the Shroud must be at least 780 years older than the 1988 carbon dating indicated.

When we combine the evidence of the Sudarium of Oviedo with the above five dating methods of Dr. Liberato De Caro, Dr. Raymond Rogers, and Dr. Giulio Fanti, it gives strong probative evidence that the Shroud originated in first-century Palestine around the time of Jesus' Crucifixion.

We now proceed to two other kinds of extrinsic dating methods that add probative force to the combined evidence—the pollen grains showing that the Shroud originated in Jerusalem/northern Judea prior to Lirey, France (Section I.C.2), and the controverted existence of Roman coins on the Shroud man's eyes (Section I.C.3).

2. The Presence of Palestinian Pollen Grains

Max Frei Sulzer (commonly known as Max Frei) was a Swiss botanist and criminologist who was arguably the foremost expert on pollen grains throughout the world. He took hundreds of samples from the

[30] Ibid.

Shroud and noted that three-fourths of them were likely of Palestinian origin—thirteen were unique to that region.[31] Emanuela Marinelli sums up the extensive work of Frei as follows:

> Three-quarters of the species found on the Shroud grow in Palestine,
> of which 13 species are very characteristic or unique of the Negev
> and the Dead Sea area (halophyte plants). The palynology thus allows
> us to say that during its history (including manufacturing) the Shroud
> resided in Palestine.[32]

Since the largest number of pollen grains (three-fourths) come from
the region of Palestine (and remained exposed to the open air there
longer than anywhere else), it is likely that Palestine was the place of
the Shroud's origin.

Yet the story is even more interesting, because Frei identified
unique pollen grains from the region of Turkey. As Marinelli indicates, quoting Frei,

> This result [the unique grains from Palestine] does not explain the pres-
> ence of pollen of steppe plants that are missing in Palestine or are ex-
> tremely rare there. According to palynology, the Shroud must have
> been exposed to open air in Turkey because 20 of the found species
> are abundant in Anatolia (Urfa, etc.) and four around Constantinople,
> and are completely lacking in the Central and Western Europe.[33]

The presence of the pollen grains complements the above dating
conclusions about the earlier date of the Shroud's origin in two respects:

1. The probable place of origin was Palestine—not Europe.
2. The Shroud did not go directly from Palestine to Europe. It
   very probably traveled first to Anatolia (Turkey) and remained

---

[31] Marinelli, "Question of Pollen Grains", p. 3, citing Max Frei, "Il passato della Sindone
alla luce della palinologia", in *La Sindone e la Scienza: Atti del II Congresso Internazionale di
Sindonologia*, Turin, October 7–8, 1978 (Turin: Edizioni Paoline, 1979), p. 198. See also
Max Frei, "Note a seguito dei primi studi sui prelievi di polvere aderente al lenzuolo della S.
Sindone", *Sindon* 23 (1976): 5, and Max Frei, "Nine Years of Palynological Studies on the
Shroud", *Shroud Spectrum International* 3 (1982): 3.

[32] Marinelli, "Question of Pollen Grains", p. 3, citing Frei, "Il passato della Sindone",
p. 198.

[33] Marinelli, "Question of Pollen Grains", pp. 3–4, quoting Frei, "Il passato della Sindone",
p. 198.

there long enough to accumulate pollen grains from the open air. It then traveled to Constantinople, where it also accumulated pollen grains.

What does this mean? First, the medieval dating is highly unlikely, because we know the Shroud was in Lirey, France, about the same time that the C-14 test dates its origin. The Shroud was revealed to be in Lirey, France, in 1353 by Geoffrey de Charny, and its carbon dating was said to be between A.D. 1260 and 1390. This means that the Shroud could have been in Europe only throughout its lifetime—specifically, France and Italy. However, this cannot be the case, because there are a large number of pollen grains from Palestine (many of which are unique to that region) and a large number of pollen grains from Turkey (and unique to that region). The presence of so many pollen grains from these other regions indicates that the Shroud was in Palestine and Turkey for at least as long as it was in Europe, making a medieval dating highly unlikely. These conclusions fit well with conclusions concerning the Sudarium of Oviedo (which has an established historical record—confirmed by pollen grains— starting with Palestine prior to A.D. 616).

3. The Possible Imprints of Roman Coins on the Shroud Man's Eyes

Some numismatists (those who study coins) have identified partial imprints of coins on the eyes of the man on the Shroud, but this is controverted by some physicists who believe that a Roman lepton-sized coin can be identified, but not the markings on that coin.[34] We will first examine the claims of those who believed the coin images are real, and then examine the criticism of those claims by materials experts.

In 1982, Father Francis Filas believed that he had identified on the eyes of the man in the Shroud images similar to those on Roman leptons specially minted by Pontius Pilate in Jerusalem in A.D. 30 (±1 year).[35] Alan and Mary Whanger believed they had a positive correlation of these images (by polarized overlay photographic

[34] J. F. Thackeray, "Lepton Coin Diameters and a Circular Image on the Shroud of Turin", Shroud of Turin website, May 23, 2019, https://www.Shroud.com/pdfs/thackeray.pdf.

[35] Francis Filas, *The Dating of the Shroud of Turin from Coins of Pontius Pilate* (Youngtown, Ariz.: Cogan Productions, 1982).

analysis) with collectors' leptons minted by Pontius Pilate in A.D. 30 (±1). They describe their conclusion as follows:

> We have done this by means of the polarized image overlay technique that we developed which enables the highly accurate comparison of two different images and the documentation of the various points of congruence.... Using the forensic criteria for matching finger prints, we feel that there is *overwhelming evidence* for the identification of the images and the matches with the coins [the special minting of Roman leptons by Pontius Pilate in A.D. 30 in Judea].[36]

This photo overlay analysis was supported by computer scientist Nello Balossino (of the Turin Faculty of Sciences), who succeeded in bringing out the sacrificial cup through computer filtering on the coin on the Shroud man's right eye.[37]

There are several enigmas on the coins that show them to be part of this special minting. According to T.V. Oomenn:

> Special coverage is given to a novel extraction of the images since 2000 by a Pilate Coins expert. The image used was a high resolution color image of the Shroud face. The right eye image shows an augur staff and the letters OY KAI AROC, and the left eye image shows the augur staff and the letters TIBERIOY. Obviously, both coins were issued by Pontius Pilate AD 30-31 and a good coin should have the complete set of letters, TIBERIOY KAICAPOC.[38]

Oommen's conclusion is further corroborated by numismatist Agostino Sferrazza, who based his conclusion on the computer-enhanced images developed by Nello Balossino.[39]

---

[36] Alan D. Whanger, "A Reply to 'Doubts concerning the Coins over the Eyes'", Shroud of Turin website, August 24, 1997, https://www.Shroud.com/lombatti.htm (emphasis added). See also Alan D. and Mary Whanger, "Polarized Image Overlay Technique: A New Image Comparison Method and Its Applications", *Applied Optics* 24, no. 6 (March 1985): 766–72.

[37] Daniel Esparza, "Shroud of Turin Coins May Finally Have Been Identified", *Aleteia*, April 26, 2017, https://aleteia.org/2017/04/26/Shroud-of-turin-coins-may-finally-have-been-identified/.

[38] T.V. Oommen, "Shroud Coins Dating by Image Extraction" (paper presented at the Shroud Science Group International Conference, "The Shroud of Turin: Perspectives on a Multifaceted Enigma", Columbus, Ohio, August 14-17, 2008), https://www.shroud.com/pdfs/ohiotvoommen.pdf.

[39] See Esparza, "Shroud of Turin Coins".

As noted above, these correlations are controverted by some image and materials specialists who claim that the weave of the Shroud is too coarse to render clear images on a very small Roman coin (like a lepton).[40] Don Lynn indicates that what was thought to be "markings on a lepton was most likely a coincidental result of contrast manipulation of photographs."[41]

Notwithstanding Lynn's view, there is high probability that a coin almost identical in diameter to the special Roman lepton minted in A.D. 30 is present on the eyes of the man in the Shroud. Thackeray concluded:

> There is no significant different [sic] ($p = 0.05$) between the diameter of the Shroud's circular image (D1, measured as 14.5 mm) and the mean diameter of Pilate leptons (15.0 +/− 0.62 mm, n = 22 measurements).[42]

Inasmuch as the placement of coins on the eyes of the deceased was commonly practiced in first-century Jerusalem,[43] and the leptons and the objects on the man's eyes are identical to the mean diameter of Pilate leptons, it is certainly possible that there are leptons of that kind on the eyes of the man in the Shroud. The scholarly community remains divided about whether the purported images belong to such leptons or whether they are imagined by experts seeing similarities in computer-enhanced images. Whatever the case, it is very likely that there is a coin almost identical in size to a Pilate lepton on the man's left eye.[44] As will be explained in Section IV.C, images from the coins could have been imprinted on the Shroud through neutron radiation coming from the body.

D. Conclusion

Tristan Casabianca (who did the statistical analysis of the raw data of the Shroud's C-14 dating, showing its invalidity) presented a paper at

---

[40] In "Lepton Coin Diameters", Thackeray cites Don Lynn, who indicated this in a personal communication to Barrie Schwortz (photographic expert during the STURP investigation).
[41] Ibid.
[42] Ibid.
[43] Ibid.
[44] See Antonacci, "Particle Radiation", p. 2621.

a Shroud conference in Ancaster, Ontario, Canada, in 2019. According to Joe Marino, he disclosed that of the forty-six Shroud papers in peer-reviewed journals, twenty-nine were against the 1988 C-14 medieval dating (i.e., in favor of the Shroud's ancient origin), twelve were neutral, and only five were in favor of the C-14 medieval dating (i.e., against the Shroud's ancient origin).[45] Some of the authors of the papers in favor of the C-14 test were associated with the labs who performed the tests.[46] Thus, the overwhelming number of peer-reviewed articles were against the 1988 C-14 medieval dating (favoring its ancient origin). When this negative assessment of the carbon dating is combined with the five new dating methods (particularly the 2022 wide-angle X-ray scattering method) as well as the external dating evidence, it is very probable that the Shroud origin was in first-century Palestine. This gives us sufficient grounds for continuing the exploration of the scientific evidence for the authenticity of the Shroud and its implications for the Resurrection of Jesus.

II. The Bloodstains on the Shroud

In addition to the image itself—which is anatomically accurate and a perfect photographic negative with three-dimensional layering—there are 372 bloodstains on the Shroud.[47] The image was formed *after* the bloodstains congealed on the cloth. According to Dr. Kitty Little (retired nuclear physicist from the U.K.'s Atomic Energy Research Establishment at Harwell):

> Another observation of considerable significance is that when blood was, or had been, present over the fibres, it also protected the underlying fibres from whatever caused the image. This means that the blood must have been in place before the image was formed.[48]

---

[45] Joseph G. Marino, "The Radiocarbon Dating of the Shroud of Turin in 1988: Prelude and Aftermath—an English-Language Bibliography", Academia.edu, updated December 14, 2023, https://www.academia.edu/48831028/The_Radiocarbon_Dating_of_the_Turin_Shroud_in_1988_and_its_Aftermath_an_English_language_Bibliography.

[46] Ibid.

[47] Kitty Little, "The Formation of the Shroud's Body Image", *British Society for the Turin Shroud Newsletter*, no. 46 (1997): 19, https://www.Shroud.com/bsts4607.htm.

[48] Ibid.

This means a potential forger would need to place all the bloodstains on the cloth before there was an image on which to place them—unlikely, even before we add the question of how the image itself was put on the Shroud without any paints, dyes, chemicals, vapors, or scorching.

The painstaking analysis of Alan Adler, John Heller, P. L. Bollone, and others shows that the bloodstains on the Shroud are genuine, containing real hemoglobin, bilirubin, AB+ blood type, plasma-serum differentiation, human albumin, human whole blood serum, and human immunoglobins.[49] These typical characteristics of blood are not present in paints or dyes or any other nonblood chemical, which ensures that the stains on the Shroud are in fact real blood.

There are two other dimensions of the bloodstains that are undetectable without contemporary scientific technology:

- The microscopically precise, invisible reactions around the more than one hundred scourge marks throughout the body
- The coagulated bloodstains with serum surrounding borders and clot retraction rings that occur with actual wounds and blood flows, found throughout the front and back of the body, and revealed only by modern scientific technology[50]

Additionally, Elvio Carlino and Giulio Fanti used atomic resolution analysis of nano particles in the blood that detected high levels of ferritin and creatinine—two enzymes that become synthesized when someone is undergoing a polytrauma. They concluded the following:

> Here we show how atomic resolution investigations unexpectedly discover a scenario of violence hidden at the nanoscale in the TS [Turin Shroud] fiber and also suggest an explanation for the controversial results so far obtained. Indeed, a high level of creatinine and ferritin is related to patients suffering of strong polytrauma like torture. Hence,

[49] See John Heller and Alan Adler, "Blood on the Shroud of Turin", *Applied Optics* 19, no. 16 (1980): 2742–44; P. L. Baima Bollone, "The Forensic Characteristics of the Blood Marks", in *The Turin Shroud: Past, Present, and Future; International Scientific Symposium, Torino, 2–5 March 2000*, ed. Silvano Scannerini and Piero Savarino (Torino, Italy: Effatà Editrice, 2000), pp. 125–35; Alan Adler, "The Nature of the Body Images on the Shroud of Turin", Shroud of Turin website, 1999, https://www.Shroud.com/pdfs/adler.pdf; and Antonacci, "Particle Radiation".

[50] Antonacci, "Particle Radiation", p. 2614.

the presence of these biological nanoparticles found during our TEM [Transmission Electron Microscopy] experiments point [to] a violent death for the man wrapped in the Turin Shroud.[51]

This bond of ferritin and creatinine cannot be reproduced by any dye, paint, or pigment, which enables Carlino and Fanti to assert that it is almost certainly beyond the capacity of an ancient/medieval forger to produce:

> This result [ferritin-creatinine bonds] cannot be impressed on the TS [Turin Shroud] by using ancient dye pigments, as they have bigger sizes and tend to aggregate, and it is highly unlikely that the eventual ancient artist would have painted a fake by using the hematic serum of someone after a heavy polytrauma.[52]

Inasmuch as an ancient forger would not be inclined to torture a person in order to obtain blood that would contain the synthesis of these enzymes, it seems most likely that the blood came from a man tormented in a way depicted by the image on the Shroud.

After the 1978 Shroud of Turin Research Project (STURP) investigation, some scientists identify what seem to be anomalous features of some bloodstains on the Shroud that could be construed to be "touch-ups" of the bloodstains by a forger. Adrie van der Hoeven has made a comprehensive study of these anomalous features and found that they validate rather than undermine the Shroud's authenticity. She summarizes her conclusions as follows:

> The anomalous features of the Shroud's bloodstains, instead of being evidence against their authenticity, turn out to be very strong evidence for their authenticity, as these anomalies are the consistent specifics of cold acid postmortem blood that formed pinkish red heme-madder lake on a cold-water-resistant madder-dyed cloth such as, most probably, the Shroud.... Microscopic and other observations preclude that a red madder lake or red madder dye was painted on to produce or retouch all bloodstains. Especially the fluorescent serum margins of

[51] Elvio Carlino and Giulio Fanti, "Atomic Resolution Studies Detect New Biologic Evidences on the Turin Shroud", *PLOS One* 12, no. 6 (2017), https://www.ncbi.nlm.nih.gov/pmc/articles/PMC5493404/#:~:text=Indeed%2C%20a%20high%20level%20of,wrapped%20in%20the%20Turin%20Shroud.
[52] Ibid.

some of the stains, the stains' lack of potassium, and the apparent formation of acid heme-madder lake stains in the right anatomical loca-
tions of the body image before this image—consisting of matterless
image fibers—was formed, render it highly unlikely that the stains and
body image were produced by a medieval artist. Shroud stains containing acid heme and lacking potassium, for lack of reasonable alter-
natives, virtually must have been formed by acid postmortem blood
of which some was clotting and exuding its potassium-rich serum on
a relatively cold surface, such as the cold skin of a dead body, possibly that of Jesus Christ.... A few experiments confirmed that much
serum can drain from human blood on a cold surface and that human
blood is able to form pinkish stains on starched and madder-dyed
linen that remain pinkish while simultaneously formed bloodstains on
pure linen turn brown.[53]

From the vantage point of biochemical analysis, then, every aspect of
the bloodstains on the Shroud appears to be authentic.

In 2018, Matteo Borrini and Luigi Garlaschelli wrote an article
claiming that they had used a blood pattern analysis approach to assess
the authenticity of the bloodstains from the chest, arms, and the lance
wound in the side of the man on the Shroud. They found that the
blood patterns on the Shroud were incompatible with blood patterns
they produced on a live volunteer/mannequin, and concluded that
these inconsistencies showed the Shroud to be the probable work of
a forger.[54] In a report to the website Live Science, Borrini noted:

> You realize these cannot be real bloodstains from a person who was
> crucified and then put into a grave, but actually handmade by the artist
> that created the Shroud.[55]

The Borrini and Garlaschelli "study" continues to receive attention
on the web despite responses from scientists who are more qualified
in blood chemistry and pathology. Two problems should be noted.

---

[53] Adrie van der Hoeven, "Cold Acid Postmortem Blood Most Probably Formed Pinkish-
Red Heme-Madder Lake on Madder-Dyed Shroud of Turin", *Open Journal of Applied Sciences*
5 (2015): 727–29. https://www.scirp.org/pdf/OJAppS_2015113010464750.pdf.

[54] Matteo Borrini and Luigi Garlaschelli, "A BPA Approach to the Shroud of Turin", *Journal of Forensic Sciences* 64, no. 1 (2018): 137–43, https://onlinelibrary.wiley.com/doi/10.1111
/1556-4029.13867.

[55] Charles Choi, "Shroud of Turin Is a Fake, Bloodstains Suggest", Live Science (website),
July 18, 2018, https://www.livescience.com/63093-shroud-of-turin-is-fake-bloodstains.html.

First, Borrini is an anthropologist and Garlaschelli a chemist—neither is an expert in blood, blood flow, blood patterns, and pathology. Their results conflict with experts in the field, such as Dr. Frederick Zugibe[56] (who has done multiple studies on blood flows and blood patterns on the Shroud over fifty years[57]) as well as Dr. Robert Bucklin's pathological studies during the 1978 STURP investigation.[58] Bucklin found no inconsistency between the different sets of wounds on the front and back, and he accounted for the direction of the flows that would have occurred before and after death.[59]

So how did Borrini and Garlaschelli come to their "new" findings? They used a questionable procedure (which has been shown to be inaccurate and is disallowed for court use today) called "blood pattern analysis". Essentially, they used a mannequin and placed artificial blood on areas where bloodstains were found on the Shroud to register flow patterns, and for the lance wound they soaked a sponge in artificial blood and pushed it on a wooden stick against the side of the mannequin to register those flows. Needless to say, this is in no way rigorous, and pales in scientific examination methods when compared to the studies of Zugibe, who used real cadavers of men with hemopericardium (blood accumulating in the pericardial sac), positioned vertically. These cadavers were lanced with a scalpel between the fifth and sixth ribs (as indicated on the Shroud).[60] It is inconceivable that anyone could think that Borrini and Garlaschelli's use of a mannequin employing crude techniques to imitate wounds could replace the rigorous scientific examination with real cadavers carried out by Zugibe and others.

The second problem with Borrini and Garlaschelli's "study" is the bold, but quite incorrect statement of Borrini that the bloodstains

---

[56] Frederick Zugibe was the chief medical examiner of Rockland County, New York, from 1969 to 2002. Zugibe was one of the United States' most prominent forensic experts, known for his research and books on forensic medicine.

[57] Frederick Zugibe, *The Crucifixion of Jesus, Completely Revised and Expanded: A Forensic Inquiry* (New York: M. Evans, 2005), pp. 98–196.

[58] Dr. Robert Bucklin was an experienced pathologist who examined over twenty-five thousand bodies for autopsies over fifty years; he subjected the image/bloodstains of the man on the Shroud to contemporary autopsy studies. His studies coincide with those of Frederick Zugibe. See Robert Bucklin, "An Autopsy on the Man of the Shroud", Shroud of Turin website, 1997, https://www.Shroud.com/bucklin.htm.

[59] Ibid.

[60] Zugibe, *Crucifixion of Jesus*, pp. 98–196, 241–58.

were "handmade by the artist that created the Shroud".[61] The pains-taking analysis of Heller,[62] Adler,[63] and Bollone[64] show conclusively that every bloodstain on the Shroud is in fact real blood with AB+ blood type, and van der Hoeven[65] shows that all of the anomalous aspects of the bloodstains are in fact real blood—not produced by a forger. Evidently, Borrini and Garlaschelli ignored the many peer-reviewed articles on the bloodstains by the above four authors, and made a claim that is patently false about the bloodstain evidence to which we have direct investigative access today!

We conclude with the scathing critique of the shotty method and false assumptions of Borrini and Garlaschelli elucidated by Dr. Alfonso Sánchez Hermosilla (forensic medical doctor and foren-sic anthropologist). Dr. Hermosilla shows significant BPA analysis comparison errors between the living subject (used by Borrini and Garlaschelli) and an actual living body with blood proceeding from wounds with a beating heart as well as significant differences between the blood samples used by Borrini and Garlaschelli and the blood of a living subject who later died and was put into a different position. After pointing out six other critical errors, he concludes as follows:

> The experiment does NOT even remotely reproduce the conditions in which the blood stains of the Turin Shroud have occurred. In these circumstances, the conclusions of the article are TOTALLY devoid of scientific value.
>
> The authors of the article, given their inexperience and lack of the minimum necessary knowledge, have committed serious errors in planning and interpreting the results of their "experiment".
>
> The article is not suitable for publication in a specialised scientific journal; it is assumed that people who have assessed the suitability of the article should have the necessary knowledge and experience. In the case in question, either they do not possess it, or have ignored it for unknown reasons.[66]

[61] Choi, "Shroud Is a Fake".

[62] See Heller and Adler, "Blood on the Shroud of Turin".

[63] See Adler, "Nature of the Body Images".

[64] See Bollone, "Forensic Characteristics".

[65] See van der Hoeven, "Cold Acid Postmortem Blood".

[66] Alfonso Sánchez Hermosilla, "Answer to the Article 'A BPA Approach to the Shroud of Turin', by Matteo Borrini and Luigi Garlaschelli", Shroud of Turin website, July 18, 2018, https://www.Shroud.com/pdfs/Hermosilla%20EN.pdf.

Inasmuch as the blood is real, and the image was created after the bloodstains were embedded on the Shroud (which would have been almost impossible for a medieval forger to accomplish), the Shroud seems to have enveloped a real man who was crucified in a way similar to the unique Crucifixion of Jesus of Nazareth. This is our next subject.

III. The Crucifixion and Death of the Man on the Shroud

Though the Shroud appears to date back to first-century Palestine and the blood is authentic, containing enzymes indicative of a poly-trauma, can we be even more certain that the Shroud is the burial cloth of Jesus Christ? We can—from a pathological examination of the bloodstains and image on the Shroud and a scientific examination of how the image on the Shroud was produced (see below, Section IV). The pathological study of Dr. Pierre Barbet[67] and the updated studies of Dr. Frederick Zugibe,[68] Dr. Robert Bucklin,[69] and Drs. Matteo Bevilacqua, Giulio Fanti, Michele D'Arienzo, and Raffaele De Caro[70] show two kinds of evidence linking the image-blood on the Shroud with the Crucifixion of Jesus:

1. Evidence of Roman (rather than medieval) weapons/customs— such as a Roman legionnaires lance, a Roman flagrum (whip), and Roman crucifixion techniques
2. Evidence of the unique features of Jesus' Crucifixion as described in the New Testament—crown of thorns, pierced with a lance, and whipped to the permitted limit

[67] Pierre Barbet, *A Doctor at Calvary: The Passion of Our Lord Jesus Christ as Described by a Surgeon* (P.J. Kenedy & Sons, 1953).

[68] Zugibe, *Crucifixion of Jesus*, pp. 98–196.

[69] Bucklin, "Autopsy on Man of the Shroud".

[70] Matteo Bevilacqua et al., "Do We Really Need New Medical Information about the Turin Shroud?", *Injury* 45 (2014): 460–64, https://www.injuryjournal.com/article/S0020 -1383(14)00115-6/fulltext. See also Mateo Bevilacqua and Michele D'Arienzo, "Medical News from Scientific Analysis of the Turin Shroud", *MATEC Web of Conferences* 36 (2015), https://www.researchgate.net/publication/307773215_Medical_News_From_Scientific _Analysis_of_the_Turin_Shroud.

We will examine these features in five areas of injury and torture manifest on the Shroud—crowning with thorns (Section III.A), pierced with a lance (Section III.B), scourged multiple times (Section III.C), nailed through the hands and feet (Section III.D), and dislocated shoulder from falling with a large blunt object (Section III.E).

A. Crowned with Thorns

The man on the Shroud was severely wounded by a crown made with long thorns (the Syrian Christ thorn[71]) penetrating the scalp and the bone around his head. It was woven so that the crown would penetrate (like a cap) the top of the man's head as well as the forehead, temples, and back of the head. The thorns were long enough to penetrate the nape of his neck.[72] These thorns would have produced excruciating pain because the scalp and skull area is concentrated with nerves and blood vessels, and the crown used on the man on the Shroud was woven to maximize the number of thorns per cubic inch.[73] The Shroud shows the considerable bleeding that would have been produced by this kind of crown. According to Zugibe:

> [The effects of this crowning] is very dramatically depicted in the Turin Shroud, which shows images representing rivulets and seepage points running down the forehead and confirms that the crown of thorns was plaited in the shape of a cap and not a circlet.[74]

This last point brings up a question: Why would a medieval forger depict an image of Jesus with a capped crown of thorns when he and the people for whom he was making the forged image would have had no example or precedent of such a crown from the Gospels, or medieval art or history? This seems contrary to a forger's project of making a *convincing* image of the Gospel accounts.

[71] Frederick Zugibe, *The Cross and the Shroud: A Medical Examiner Investigates the Crucifixion* (Cresskill, N.J.: McDonagh, 1981), p. 25.

[72] See Stephen Jones, "Were Crowned with Thorns #5: Bible and the Shroud: Jesus and the Man on the Shroud: Shroud of Turin Quotes," October 19, 2015, https://theShroudofturin .blogspot.com/2015/10/were-crowned-with-thorns-5-bible-and.html.

[73] See Zugibe, *Cross and Shroud*, p. 33.

[74] Ibid, p. 35.

The crown of thorns is completely unique to Jesus' Crucifixion, and no evidence of such a crime has been found in the considerable literature on crucifixion. As Zugibe indicates:

> The effect of this type of pain was probably not understood by Pilate because parodies of this type [mock crowning with thorns] were not a usual prelude to crucifixion, and none have ever been related in the vast literature on crucifixion.[75]

The reason that the mock crowning uniquely occurred in Jesus' Crucifixion was the charge leveled against Him by the chief priests: He claimed to be king of the Jews. There are no other criminals in the vast literature of Roman crucifixion accused of such a crime. Without this context, the mock crowning would be inexplicable. If the Shroud originated in first-century Palestine (as the above dating evidence suggests), then the uniqueness of this kind of crucifixion implies that the Shroud is very probably the burial cloth of Jesus Christ.

B. Pierced with a Roman Lance

The man on the Shroud was pierced on the right side between the fifth and sixth ribs at an upward angle by a spear resembling the Roman lancea (lance) carried by Roman militia.[76] It left an elliptical wound "corresponding exactly to excavated examples of the leaf-shaped point of the lancea (lance)".[77] Dr. Robert Bucklin (forensic pathologist and autopsy specialist) indicates that the wound exuded both blood and a watery substance (detectable on the Shroud), which might have been produced by the chest cavity filling up with a clear fluid during a time of stress.[78]

In his Gospel, Saint John reports this feature of Jesus' Crucifixion, making special note of the surprising combination of blood and then water flowing from the wound, which would not have been

---

[75] Ibid, p. 24.

[76] See William Meacham et al., "The Authentication of the Turin Shroud: An Issue in Archaeological Epistemology", *Current Anthropology* 24, no. 3 (1983): 283–311, https://www.Shroud.com/meacham2.htm.

[77] Ibid., under the subhead "Anthropological, Archaeological, and Art Historical Considerations", on the Shroud of Turin website.

[78] Bucklin, "Autopsy on Man of the Shroud".

expected by his first-century Jewish audience. The occurrence is so astonishing that Saint John has to insist that it was seen by a reliable eyewitness:

> But one of the soldiers pierced his side with a spear, and at once there came out blood and water. He who saw it has borne witness—his testimony is true, and he knows that he tells the truth—that you also may believe. (19:34–35)

Is it really possible for blood and water to flow from the side of a man who was pierced with a lance on the right side between the fifth and sixth ribs? It seems so prima facie, because blood partially admixed with a watery substance is detectable at the site of the lance wound on the Shroud. Where could this watery substance come from? Frederick Zugibe made an extensive study of eleven theories, using his vast experience in pathology as well as tests on cadavers. He concluded as follows:

> The spear pierced the right atrium of the heart, hence the blood, and the water resulted from the pleural [the cavity next to the lungs] effusion from the brutal scourging and was contributed to by congestive heart failure from the position on the cross. The sudden thrust of the spear with a quick, jerking motion to pull it out would certainly bring blood out first on the spear and would be followed immediately by the pleural effusion from the pleural cavity.[79]

Why is this important for showing the authenticity of the Shroud as the burial cloth of Jesus? First, there is no account of blood and water flowing from the side of a crucified man except in the Gospel of John, and this occurrence was so astonishing that Saint John had to insist on it coming from a reliable eyewitness. The occurrence seems surprising to this very day, so much so that biblical exegetes (unfamiliar with the evidence on the Shroud) have asserted that it is merely symbolic of the waters of Baptism. Thus, the Shroud shows a remarkable coincidence of blood mixed with a watery substance at the exact spot (between the fifth and sixth ribs on the right side) where a spear could be thrust at an upward angle to produce this result.[80] The fact

[79] Zugibe, *Cross and Shroud*, p. 127.
[80] See ibid.

that this coincidence produces the unexpected result reported by the evangelist indicates both the likely authenticity of the Shroud and the accuracy of Saint John's Gospel account of the Crucifixion.

Second, there is no report in the vast literature on Roman crucifixion of a spear being thrust into a crucified man, because the roman legions did not want to put the criminal out of his misery quickly and incisively; rather, they wanted him to suffer a long excruciating execution. However, in the unique case of Jesus, the soldiers were under a time constraint, because the Sabbath was approaching and the chief priests and people wanted the bodies to be removed before sundown (Jn 19:31). When the soldiers came to Jesus, they found that He was already dead, and to save themselves the trouble of breaking His knees, they simply thrust a lance into His side to ensure that He really was dead (19:32–34).

In sum, the lance wound on the right side of the man in the Shroud shows many features unique to Jesus' Crucifixion—the lance wound itself, the effusion of blood and a watery substance detectable on the Shroud, and a positioning on the right side between the fifth and sixth ribs that would have given rise to the exuding of blood followed by a clear "watery" substance.

C. Scourged Multiple Times

The man on the Shroud was scourged by two individuals (one on his left and a slightly taller individual on his right) by a Roman flagrum having three thongs with dumbbell-shaped lead pellets (*plumbatae*) at the end of the thong.[81] This left lash marks that covered the man's back, thighs, and calves, and reached around to his sides and the lateral parts of his chest.[82] This is precisely the instrument and the manner used by Roman legionnaires for the purpose of torture and execution for serious crimes committed by non-Roman citizens.[83] The flagrums

---

[81] Stephen Jones, "The Shroud of Turin: 3.3; The Man on the Shroud and Jesus Were Scourged", *Shroud of Turin* (blog), July 15, 2013, http://theShroudofturin.blogspot.com /2013/07/the-Shroud-of-turin-33-man-on-Shroud.html, citing Ian Wilson, *The Shroud of Turin: The Burial Cloth of Jesus?*, rev. ed. (New York: Image Books, 1979), p. 38.

[82] Ibid.

[83] Jones, "Man on Shroud and Jesus Were Scourged", citing Orazio Petrosillo and Emanuela Marinelli, *The Enigma of the Shroud: A Challenge to Science*, trans. Louis J. Scerri (Malta: Publishers Enterprise Group, 1996), p. 225.

used on the man on the Shroud are almost the same as those found in archaeological digs at Pompeii and Herculaneum.[84]

Three of the Gospels specifically mention scourging: Matthew 27:26, Mark 15:15, and John 19:1. Luke indicates only that Pilate "chastised" Jesus (23:16, 22). Scourging of criminals for serious offenses was common Roman practice, even for those facing crucifixion. The maximum number of lashes allowable by Jewish law was thirty-nine (2 Cor 11:24), one less than that allowed by Mosaic Law.[85] But the Romans would not have believed this limit to be important.[86] So, why would this number be excessive for the Romans? As noted above, the Romans wanted criminals to suffer a long excruciating ordeal, but the 117 stripes (thirty-nine strikes times three thongs per strike) would have created so much blood loss and weakness that it would surely have led to an early "merciful" death on the part of the criminal. As reported in the Gospels, the scourging had precisely this effect—Jesus was already dead after only three hours when most crucifixions took much longer, from ten hours to two days (Mt 27:45–50; Mk 15:33–39; Lk 23:44–47; Jn 19:28–30).

So, why did the Romans scourge Jesus so many times? Giving rise to what was sure to be a quick and more incisive death? The Gospel supplies the answer. Pilate wished to scourge Jesus to the point of death so that the crowds would be satisfied that Jesus had been punished for His blasphemy, but Pilate did not want to kill Him (Lk 23:16; Jn 19:12). Evidently, Pilate was concerned that Jesus had not committed anything like a capital crime (deserving of death) according to Roman law. So, in John's Gospel he has Jesus scourged miserably to satisfy the crowds. However, despite Pilate's two protestations of Jesus' innocence, they demanded His Crucifixion (19:1–16).

We conclude with the question about the supposed medieval forger. How did he know about the precise dimensions of a Roman flagrum and its three thongs? When there was no evidence of such whips in the medieval period? How did he know about the dumbbell-shaped metal pieces at the end of each thong? How did he know about the Roman

[84]Jones, "Man on Shroud and Jesus Were Scourged", citing "Herculaneum", *Wikipedia*, June 10, 2013.

[85]Jones, "Man on Shroud and Jesus Were Scourged", citing Edward A. Wuenschel, *Self-Portrait of Christ: The Holy Shroud of Turin*, 3rd ed. (Esopus, N.Y.: Holy Shroud Guild, 1954), p. 41.

[86]Ibid.

custom of whipping from two sides? And why did he select 117 stripes? He had no reports of any of these things from the Gospels, or medieval art or history.

D. Nailed through the Hands and Feet

According to Frederick Zugibe, the man on the Shroud was fixed to the cross with a nail proceeding from the palm (in the thenar furrow), angled ten to fifteen degrees downward toward the wrist, which naturally guides the nail toward "the area created by the meta-carpal bones of the index and second finger and the capitate and lesser multiangular bones of the carpus (wrist) called the 'Z' area".[87] This is a remarkably sturdy area that can sustain the weight of a human body, and the exit wound is precisely at the place indicated by the hand on the Shroud. Additionally, this position of the nail explains the lengthening of the fingers apparent on the man in the Shroud.[88]

Dr. Pierre Barbet did not think the nail could proceed through the palm because the soft fleshy tissue would not be capable of sustaining the weight of a human body, and so he conjectured that the nail would have to have proceeded from the front of the wrist to the back of the wrist.[89] Zugibe's explanation is preferable, because it explains the lengthening of the fingers as well as the angle of the exit wound in the back of the wrist area. It also happens to agree with the implied position of the nail wounds on the *hands* (rather than the wrists) of the Risen Christ in the Gospel of Luke (24:39) and the Gospel of John (20:20, 25, 27).

This brings up yet another question about our medieval forger. Evidently, he would have known from the Gospels and from the medieval artistic tradition that the nail would have gone into the palm, but how did he know to make the exit wound come out in the wrist area to depict it properly on the Shroud? Did he have special knowledge of human anatomy and Roman crucifixion practices?

The nail moving through this channel would have injured the median nerve and peripheral nerves, which would have caused

---

[87] Zugibe, *Cross and Shroud*, p. 65.
[88] Ibid.
[89] Barbet, *A Doctor at Calvary*, pp. 83–106.

excruciating pain in a condition called "causalgia". Zugibe describes the pain as follows:

> The pain is described as a peculiar burning sensation that is so intense that even gentle contacts like clothing or air draft cause utter torture. The patient becomes completely preoccupied with avoiding any contact and holding the limb in a particular way. This condition can completely destroy the morale of most stoic individuals.[90]

We now proceed to the nail wounds in the feet. It is difficult to determine whether the feet were nailed side-by-side to the stipes or one foot on top of the other by a single nail.[91] Zugibe favors the two-nail, side-by-side hypotheses because it would have been easier for the Roman soldiers to accomplish.[92] The feet were probably strapped flush to the stipes, and then nailed to the upright to the top of the foot through the heel. No support was necessary for criminals to push themselves up—the feet would have felt glued to the upright.[93] Once again, the pain would have been excruciating, Zugibe describes it as follows:

> The pain in Jesus' feet would have been severe with the iron nail pressing against the plantar nerves like "red hot pokers," similar to that suffered by the median nerve injuries during nailing of the hands. Even a slight movement would incite the incessant, burning pains. After a short period of time on the cross, the severe cramps, the numbness, and the coldness in the calves and thighs, caused by the compression by the bent knees, would force him to push up occasionally and attempt to straighten his legs. This would continue periodically throughout the entire period that Jesus was on the cross.[94]

E. Dislocated Shoulder from Falling with a Large Blunt Object

In 2015, Drs. Matteo Bevilacqua and Michele D'Arienzo published a study that brought to light several new aspects of the crucified man

---

[90] Zugibe, *Cross and Shroud*, p. 76.
[91] Ibid.
[92] Ibid., p. 79.
[93] Ibid., p. 85.
[94] Ibid., p. 87.

on the Shroud.[95] Most importantly, they were able to explain the lowering of the right shoulder ten to fifteen degrees, the hyperextension of the arms, the left twist in the neck, and the retraction of the right eye in the orbit.[96] All of these injuries point to the fact that the man on the Shroud had a large blunt object on his right shoulder, and when he fell forward, the blunt object hit him so hard that it caused a dislocated shoulder and paralysis in the upper right side. They summarized their results as follows:

> This Man shows, on the right side, shoulder lowering, flat hand and henophthalmos, revealing a violent blunt trauma, from behind, to neck, chest and shoulder, with the entire brachial plexus injury and muscular damage to the neck bottom with the head bent forward and turned to the left, on the cross, as he had a stiff neck. Most likely, falling the body forward, the chest trauma caused a heart and lung contusion with hemothorax.[97]

This combination of injuries is very well explained by the circumstances surrounding Jesus' Crucifixion. The Gospel of John makes clear that Jesus started the journey by carrying His own Cross (19:17), but the Synoptic Gospels indicate that along the way, the soldiers compelled Simon of Cyrene to carry the Cross for Him (Mt 27:32; Mk 15:21; Lk 23:26). Why would the soldiers have done this? Certainly, they did not want to ease the burden of the man they intended to torture. In 2015, Drs. Matteo Bevilacqua, Giulio Fanti, Michele D'Arienzo, and Raffaele De Caro (three of whom are specialists in injury and pathology) published a study in the prestigious journal *Injury*, which gives a very probable answer.[98] Referencing the *Injury* article, the Italian daily newspaper *La Stampa* reported that

> the person whose figure is imprinted on the Shroud is believed to have collapsed under the weight of the cross, or the "patibulum" as it is referred to in the study, the horizontal part of the cross. The Man of the Shroud, the academics explain, fell "forwards" and suffered a

---

[95] Bevilacqua and D'Arienzo, "Medical News from Scientific Analysis".
[96] Ibid.
[97] Ibid., Abstract.
[98] Bevilacqua et al., "Do We Really Need New Medical Information?"

"violent knock" "while falling to the ground." "Neck and shoulder muscle paralysis" were "caused by a heavy object hitting the back between the neck and shoulder and causing displacement of the head from the side opposite to the shoulder depression. In this case, the nerves of the upper brachial plexus (particularly branches C5 and C6) are violently stretched resulting in an Erb-Duchenne paralysis (as occurs in dystocia) because of loss of motor innervation to the deltoid, supraspinatus, infraspinatus, biceps, supinator, brachioradialis and rhomboid muscles."[99]

The scrapes on the man's knees confirm the fall—lurching forward, which caused the heavy blunt object to give a severe blunt trauma to the neck, shoulder, and back. At that point, the man's upper body (right side) would have been paralyzed, rendering him incapable of carrying the cross. This is why the soldiers pressed Simon of Cyrene into service. They were not easing Jesus' burden; they had to deal with His upper-body paralysis. This injury explains the hyperextension of the arms of the Turin Shroud man, a feature that was formerly thought to be anatomically incorrect—this injury would have exacerbated the pain of the man when he hung from the cross, and it would have made his breathing difficult, thereby shortening his life.

We now return to our hypothesized medieval forger. How did he get all of the anatomical facts about the blunt-force trauma resulting from a forward fall with a heavy object on the back? The Gospels don't mention a fall, but it makes sense in Jesus' case. He had over two-tenths of a mile to carry His Cross, and He had lost a tremendous amount of blood by receiving the maximum number of lashes. How did the medieval forger put all the details together with the accuracy of a contemporary pathologist—the ten- to fifteen-degree lowering of the right shoulder, the hyperextension of the arms, the eye receding in the right orbit, the scraped knees, and the neck twisted to the left? If he was trying to portray Jesus, how did he know that Jesus fell? This hypothesis seems to be very unlikely—beyond belief for anyone with scientific/medical knowledge.

---

[99] Editorial, "New Study Shows Man of the Shroud Had 'Dislocated' Arms", *La Stampa*, May 8, 2014, https://www.lastampa.it/vatican-insider/en/2014/05/08/news/new-study-shows -man-of-the-shroud-had-dislocated-arms-1.35751980.

F. Conclusion

We may infer from the data on the Shroud that Jesus' Crucifixion (virtually identical on the Shroud and in the Gospel accounts) was incredibly painful. The crowning with thorns on the top of the head, where the nerves and blood vessels are densely packed, would have been very painful, leading to excessive bleeding. The scourging would have been excruciating, and the iron dumbbell-shaped fragments would have torn His flesh to pieces. Carrying the Cross from the praetorium to Golgotha led to a fall with severe blunt-force trauma that dislocated His shoulders, hyperextended His arms, and exacerbated His pain on the Cross. The nailing to the Cross would have damaged two major nerves, causing incredible pain every time Jesus tried to lift Himself to breathe. Such torments are barely conceivable to anyone who has not been tortured.

The fact that the blood evidence on the Shroud so closely resembles the unique features of Jesus' Crucifixion in the Gospel accounts gives warrant for its authenticity. When this is combined with exact anatomical accuracy, and Roman weaponry and crucifixion customs with which a medieval forger would not have been acquainted, it is not unreasonable to infer that this cloth could well be the burial cloth of Jesus Christ. This will be further confirmed in our examination of the image and its implications for Jesus' Resurrection.

IV. The Image on the Shroud and Evidence of the Resurrection

The highly precise, detailed, three-dimensional photographic negative image on the Shroud is unique among images in the history of mankind. As we shall see, it is quite doubtful that a medieval forger—or for that matter, any contemporary forger unfamiliar with nuclear or high energy physics—could reproduce its precise photographic negative features on the uppermost surface of the fibrils of the cloth, superimposing it over preexisting bloodstains. Even if one used a recently dead body and placed the precise wounds of Jesus' Crucifixion on it (described above in Section III), the dead body would not have produced the image on the Shroud unless the forger could induce a significant amount of radiation by one of the very unconventional means described below—instantaneous disintegration of the nuclei of the atoms throughout the whole body (producing a shower of particle radiation) or producing significant amounts of directional vacuum ultraviolet radiation of several billion watts through pulsations of less than one-forty billionth of a second (as in ARF excimer lasers). Inasmuch as these two methods of image production from a dead body are beyond not only nature but also current science and technology, it is quite certain that the Shroud's image was not produced by a forger. But I am getting ahead of myself. How do we know that such bursts of either particle radiation or ultraviolet radiation were needed to produce the image on the Shroud?

There are about forty-five aspects of the image on the Shroud and the condition of the cloth and blood that require explanation, because they are either unique to the Shroud, incapable of explanation by traditional physical, chemical, or biological processes, or seem to be in conflict with one another:

- Thirty-two primary image features
- Eight secondary image features
- Seeming imprints of coins and flowers, which are extrinsic to the body
- The bright red color of all blood residues and stains
- The excellent condition of the cloth after more than seven centuries
- The absence of damage to the bloodstains on the cloth, which would have resulted from the removal of the body wrapped by the cloth
- The equal intensity of the images on the frontal surface and the dorsal surface of the Shroud when the body was in a supine position pressuring the dorsal section

The proposed cause of the image on the Shroud will have to explain the thirty-two primary enigmatic image features and the eight secondary enigmatic image features (see below, Section IV.B). If this proposed cause can also explain other enigmas on the Shroud concerned with the blood and the strength and longevity of the cloth, so much the better (Section IV.C). As we shall see, the only explanation that can come close to explaining these unique and enigmatic features is *radiation*.

Since the 1978 Shroud of Turin Research Project (STURP) investigation, radiation was suspected as the most likely explanation for
image formation, because other physical explanations, such as liquids, rubs, vapors, and scorching, could not explain some of the most
obvious characteristics of the Shroud.[100]

For example, the fact that the image is limited to the uppermost
surface fibrils (a few microns) of the frontal and dorsal surfaces of the
cloth (and does not even penetrate into the middle of the fiber) and
that its precision does not "leak" into adjacent fibers precludes liquids, vapors, and rubs, all of which would penetrate into the middle
of fibers, spangle, and diffuse to adjacent fibers.[101] Furthermore, the
cloth did not make contact with every part of the body, meaning
that the source of the image would have to not only act at a distance
but encode information about relative distance in the image on the
Shroud. This is precisely what radiation does. In the words of Dr.
Luigi Gonella (former scientific advisor to the archbishop of Turin):

> An agent acting at a distance with decreasing intensity is, almost by defi-
> nition, radiation. The limitation of the cloth darkening to the outer-
> most surface pointed to a non-penetrating, non-diffusing agent, like
> radiant energy.[102]

[100] In a comprehensive paper that tested all of these possibilities, physicists John Jackson,
Eric Jumper, and William Ercoline tested all eight major possibilities using laboratory conditions to replicate a nonradiation means of duplicating the image on the Shroud. They
compared the results of the above attempts with the macroscopic and microscopic features
of the Shroud image, and argued that none of the techniques tested can simultaneously
reproduce its main features, from the three-dimensional property to the coloration depth,
to the resolution of the spatial details. They concluded from this that it could not be the
work of an artist or a forger. See John P. Jackson, Eric J. Jumper, and William R. Ercoline,
"Correlation of Image Intensity on the Turin Shroud with the 3-D Structure of a Human
Body Shape", *Applied Optics* 23, no. 14 (1984): 2244–70, https://www.osapublishing.org
/ao/abstract.cfm?uri=ao-23-14-2244.

[101] Ibid. See also John P. Jackson et al., "Infrared Laser Heating for Studies of Cellulose
Degradation", *Applied Optics* 27, no. 18 (1988): 3937–43, https://opg.optica.org/ao/abstract
.cfm?uri=ao-27-18-3937, and Eric Jumper et al., "A Comprehensive Examination of the
Various Stains and Images on the Shroud of Turin", in *Archaeological Chemistry III*, Advances
in Chemistry Series, no. 205, ed. Joseph B. Lambert (Washington, D.C.: American Chemical
Society, 1984), pp. 447–76.

[102] Luigi Gonella, "Scientific Investigation of the Shroud of Turin: Problems, Results, and
Methodological Lessons", in *Turin Shroud—Image of Christ? Proceedings of a Symposium Held
in Hong Kong, March 1986*, ed. William Meacham (Hong Kong: Turin Shroud Photographic
Exhibition Organizing Committee, 1987), p. 31.

In two remarkably synthetic and brilliant papers, physicist Dr. John Jackson proposed that all known enigmatic aspects of the Shroud of Turin's image formation could be explained by two unconventional phenomena:[103]

1. The body as a whole (both inside and on the surface) would have to have emitted radiation that would have caused rapid dehydration of the uppermost fibrils of the cloth—the darker impressions coming from places closest to the body and the lighter impressions coming from places farther away. This radiation would allow action at a distance (imaging in places where the cloth made no contact with the body) and a very precise photographic negative image.[104]

2. As the body was emitting radiation, it would have had to become mechanically transparent simultaneously (i.e., having no physically or chemically solid or resistant properties), allowing the cloth to fall into the body at least three-sixteenths of an inch. This would explain the three-dimensional imaging as well as the images of some of the bones on the frontal and dorsal surfaces inside the body relative to the flesh surrounding them.

Since John Jackson published the two conditions necessary to explain the primary image features on the upper surface of the fibrils, the precise photographic negative image, and the three-dimensional imaging of the surface and inside of the body, two major radiation hypotheses have been proposed and experimentally verified: the ultraviolet radiation hypothesis (URH) and the particle radiation hypothesis (PRH).

First, we consider the *ultraviolet radiation hypothesis* by John Jackson and Paolo Di Lazzaro. These two physicists proposed that directional (collimated) ultraviolet (high frequency/intensity) radiation

---

[103] John P. Jackson, "Is the Image on the Shroud Due to a Process Heretofore Unknown to Modern Science?", *Shroud Spectrum International* 34 (1990): 3–29. See also John P. Jackson, "An Unconventional Hypothesis to Explain all Image Characteristics Found on the Shroud Image," in *History, Science, Theology and the Shroud*, ed. Aram Berard (St. Louis: Man in the Shroud Committee of Amarillo, 1991), pp. 325–44.

[104] See Jackson, "Is Image Due to Process Unknown to Modern Science?", and Jackson, "Unconventional Hypothesis".

was responsible for the imaging. Di Lazzaro (chief of research for the Italian National Research Agency for New Technologies—ENEA) delivered this radiation through exceedingly short pulsations (one forty-billionth of a second). In order to produce these very brief pulses, Di Lazzaro and his team[105] used ARF excimer lasers in a laboratory. Di Lazzaro speculated that in order to obtain sufficient radiation from the entire body, it would take about fourteen thousand ARF excimer lasers, which would deliver radiation at the magnitude of several billion watts for one forty-billionth of a second. In 2010, he reproduced the straw-yellow coloration on the uppermost surface of the fibrils in precise images on a cloth with similar spectral reflectants to the Shroud.[106] It also resembled the Shroud's unique image features in many other ways.[107] He summarized his results as follows:

> Our results showed that Jackson was right. The radiation in the far ultraviolet is able to create a Shroud-like coloration on linen fabrics. Jackson was right as well considering this "radiative hypothesis" outside current paradigm and known scientific phenomena, because we measured the amount of radiation energy and the ultra-short duration of laser pulses required to achieve a Shroud-like linen coloration, and these parameters cannot be generated by any natural phenomenon known to date.[108]

In addition to the short burst of vacuum ultraviolet radiation, the body of the man on the Shroud would have to have become mechanically transparent so that the features on the inside of the body could be brought into three-dimensional proportionality with features on the surface. The ultraviolet radiation hypothesis of Jackson and Di

---

[105] Paolo Di Lazzaro, "Shroud-like Coloration of Linen by Ultraviolet Radiation", Shroud of Turin website, May 2, 2015, https://www.Shroud.com/pdfs/duemaggioDiLazzaroENG.pdf, and Paolo Di Lazzaro, "Could a Burst of Radiation Create a Shroud-like Coloration? Summary of 5-Years Experiments at ENEA Frascati" (paper presented at 1st International Congress on the Holy Shroud in Spain, Centro Español de Sindonología [CES], Valencia, Spain, April 28–30, 2012), https://www.academia.edu/4028955/Could_a_burst_of_radiation_create_a_Shroud-like_coloration_Summary_of_5-years_experiments_at_ENEA_Frascati. See also Giuseppe Baldacchini et al., "Coloring Lines with Excimer Lasers to Simulate the Body Image of the Turin Shroud", *Applied Optics* 47, no. 9 (2008): 1278–85, https://opg.optica.org/ao/abstract.cfm?uri=ao-47-9-1278.

[106] Di Lazzaro, "Shroud-like Coloration".

[107] Ibid.

[108] Ibid.

Lazarro does not explain how this would occur, but only that it must occur for the Shroud to have images from inside the body. In contrast to this, the particle radiation hypothesis *does* explain both the source of the radiation and the mechanical transparency of the body (see below). One last point must be made. Jackson and Di Lazzaro admit that there is no known natural way of producing this kind of ultraviolet radiation (at the far end of the spectrum), which opens upon the possibility of a supernatural cause.

We now turn to the second possible explanation of the Shroud's image: the *particle radiation hypothesis* by Dr. Kitty Little, Dr. Jean-Baptiste Rinaudo, and Dr. Arthur Lind. Dr. Jean-Baptiste Rinaudo[109] (biophysicist at the Centre de Recherches Nucléaires Médicales) and Dr. Kitty Little[110] (retired nuclear physicist at the U.K.'s Atomic Energy Research Establishment at Harwell) independently proposed the likely cause of Jackson's double effect—the instantaneous disintegration of the nuclei of the atoms composing the body of the man in the Shroud.

As will be explained below (Section IV.A), such instantaneous atomic disintegration would produce *low-temperature* particle radiation—a shower of protons, alpha particles, deuterons, neutrons, electrons, and gamma rays (all of which play a part in explaining the many enigmas of the Shroud) on both the frontal and dorsal surfaces of the Shroud with equal intensity.

At the moment of disintegration, the body would have become mechanically transparent, allowing the frontal cloth to pass through the body and the dorsal part of the cloth to be drawn up through a created vacuum into the dorsal part of the body. This profusion of nuclear particles/rays and the simultaneous mechanical transparency of the body not only fulfills the two requirements of John Jackson (given above), but also explains *all* unique, enigmatic, and seemingly contradictory features in the Shroud of Turin image. It also explains all nonimage enigmas concerned with the bloodstains, the cloth, and possible images from extrinsic objects such as coins and flowers. As we shall see, the profusion of neutrons would also elevate the C-14 content in a linen cloth significantly, giving the appearance of a considerably younger age in a C-14 test.

[109] John-Baptiste Rinaudo, "Protonic Model of Image Formation on the Shroud of Turin" (paper presented at Third International Congress on the Shroud of Turin, Turin, Italy, June 5–7, 1998).
[110] Little, "Formation of Shroud's Body Image".

Rinaudo and Little have shown that the particle radiation from this kind of nuclear disintegration (at low temperature—170°F) gives rise to straw-yellow coloration and other distinctive image features on linen cloths similar to the Shroud.[111] As in the ultraviolet radiation hypothesis of Jackson and Di Lazzaro, the particle radiation hypothesis has no known natural cause. There is no known natural agent that can induce instantaneous disintegration of stable atomic nuclei in an entire body. Since the particle radiation hypothesis is more comprehensive in its explanatory power—covering not only the primary and secondary enigmatic image features, but also the mechanical transparency of the body and the enigmas in the blood, cloth, and C-14 content—preference is given in this chapter to that hypothesis. We will examine it in detail in the following three subsections:

1. What Is the Particle Radiation Hypothesis (PRH)? (Section IV.A)
2. How Does the PRH Explain the Enigmas of the Shroud's Image? (Section IV.B)
3. How Does the PRH Explain the Enigmas on the Shroud's Bloodstains, Cloth, and Carbon Dating? (Section IV.C)

I am most grateful for the work of Mark Antonacci, who has detailed the research on the particle radiation hypothesis in his article "Particle Radiation from the Body Could Explain the Shroud's Images and its Carbon Dating", and his book *Test the Shroud: At the Atomic and Molecular Levels*.[112] In the remainder of this section, I will be using many of his explanations and references to articulate this remarkably comprehensive hypothesis of the Shroud's enigmas.

A. What Is the Particle Radiation Hypothesis (PRH)?

In this section we will show that particle radiation is the only explanation that accounts for the forty image features (thirty-two primary and eight secondary), as well as enigmas concerned with the cloth's

---

[111] Rinaudo, "Protonic Model of Image", and Little, "Formation of Shroud's Body Image".
[112] Mark Antonacci, *Test the Shroud: At the Atomic and Molecular Levels* (Brentwood, Tenn.: Forefront Publishing, 2016).

longevity, the bloodstains' bright red color, and other enigmas concerned with the cloth. As noted above, Dr. Jean-Baptiste Rinaudo[113] and Dr. Kitty Little[114] independently proposed the likely cause of Jackson's double effect—the instantaneous disintegration of the nuclei of the atoms composing the body of the man in the Shroud. This nuclear disintegration would give rise to a shower of particles at low temperatures (approximately 170°F) that would not destroy the cloth.[115] Electrical energy would be in the range of three million watts, accompanied by a very bright light.[116]

The instantaneous disintegration of the atomic nuclei would have given rise to trillions of neutrons and gamma rays as well as protons and alpha particles. According to Antonacci, since "the cloth fell straight down receiving heavy charged particles [protons and alpha particles] only from the part of the body directly underneath it, a highly detailed negative image would be encoded on the cloth."[117] As the cloth fell into the area of the disintegrating body, the heavy charged particles (protons and alpha particles) would have stopped when they hit the uppermost surface of the frontal and dorsal parts of the cloth, causing dehydration and conjugated carbonyls[118] that would ultimately lead to the straw-yellow coloration of the cloth with the characteristics of a highly detailed three-dimensional photographic image.[119] These effects of heavy charged particles (e.g., protons and alpha particles) have been confirmed in the laboratory by Dr. Kitty Little.[120]

[113] Rinaudo, "Protonic Model of Image".

[114] Little, "Formation of Shroud's Body Image".

[115] Ibid.

[116] Dr. Kitty Little produced the straw-like coloration and other effects similar to the Shroud on cellulose fibers (resembling those on the Shroud) with the reactor at Harwell running at about 3.0–3.5 million watts, at temperatures between 70°C and 90°C (between 150°F and 194°F). We may suppose from this that the disintegration of the nuclei of the atoms of the man in the Shroud could have produced the coloration and precision we find on the image in the Shroud with similar wattage and temperatures, leaving the cloth undamaged. See ibid. See also Kitty Little, "Photographic Studies of Polymeric Materials", in *Photographic Techniques in Scientific Research*, vol. 3 (Cambridge, Mass.: Academic Press, 1978), p. 171.

[117] Antonacci, "Particle Radiation", p. 2620.

[118] Conjugated carbonyls occur when double-bonded carbon atoms form after single-bonded atoms (within the linen fibers) break apart. See ibid., p. 2613.

[119] Ibid., p. 2620.

[120] See ibid., p. 2617. See also Rinaudo, "Protonic Model of Image"; Little, "Formation of Shroud's Body Image"; and Little, "Photographic Studies of Polymeric Materials".

What about the other particles released in the disintegration of atomic nuclei—neutrons (heavy but uncharged), electrons (charged but not heavy), and gamma rays? These particles/radiations are not responsible for the image, because they easily pass through the linen cloth and would not be stopped at the cloth's surface (as protons and alpha particles). Nevertheless, they explain several unusual features of the Shroud, which will be discussed below in Section IV.C:[121]

1. How the Shroud could have been taken off the body without breaking, distorting, or smearing any of bloodstains (Section IV.C.1)
2. The bright red color of the bloodstains on the cloth (Section IV.C.2)
3. Its unusual strength and nonfriability and its longevity (Section IV.C.3)
4. The possibility of Roman coins on the eyes of the man in the Shroud and the presence of flowers (Section IV.C.4)
5. Increases in C-14 on the Shroud, giving the appearance of a much younger date of origin (Section IV.C.5)

As we shall see, the particle radiation explanation (through low-temperature disintegration of atomic nuclei) best explains Dr. John Jackson's two requirements for the cause of the Shroud's image. It also explains all forty unusual and enigmatic features of the image, cloth, and blood. No other explanation does this.

B. How Does the PRH Explain the Enigmas of the Shroud's Image?

It would take an entire volume to give a detailed explanation of all forty unique, enigmatic, and conflictual features on the Shroud's image, blood, and cloth, requiring that we limit ourselves to a brief consideration of only the major enigmatic features. For additional information, interested readers can consult Mark Antonacci's article "Particle Radiation from the Body Could Explain the Shroud's

[121] See Antonacci, "Particle Radiation", pp. 2621–22.

Images and Its Carbon Dating" or his book *Test the Shroud: At the Atomic and Molecular Levels*. We begin with Antonacci's compilation of enigmatic features of the Shroud's *image* (the bloodstains, cloth, and carbon dating will be discussed below in Section IV.C). Antonacci's list of thirty-two enigmatic features follows:

- lack of fading
- lack of foreign materials or particulates
- straw yellow coloration
- only topmost superficial fibers of threads encoded
- individual fibers encoded
- fibers colored 360° around circumference
- only outer layers of individually encoded fibers are colored
- no coloration inside of fiber
- fibers colored with similar intensity
- oxidation and dehydration of fibers
- containing conjugated carbonyls (double-bonded carbon atoms formed after single-bonded atoms within linen fibers broke apart)
- [image was] developed over time
- accelerated aging of the body image
- stability to water and heating
- insolubility to acids, redox and solvents
- gross mechanical properties of linen intact
- microscopically corroded appearance of fibers
- lower tensile strength of fibers
- reduction of the cloth's fluorescence
- lack of residue
- highly attenuating or absorbing agent
- agent operated over skin, hair (coins and flowers)
- non-diffuse image with sharp boundaries
- equal intensity for frontal and dorsal images
- lack of two-dimensional directionality
- negative images with left/right and light/dark reversals which develop into
    - highly resolved, photographic quality images
    - without any magnification
    - with skeletal and dental features
    - three-dimensionality
    - encoded through the space between the body and the cloth
    - in a straight-line vertical direction.[122]

---

[122] Ibid., p. 2613.

Protons, alpha particles, deuterons, and other heavy charged particles (which would be part of low-temperature particle radiation emitted by atomic disintegration) explain all of the above image features concerned with coloration—the straw-yellow color, the encoding of only the uppermost fibrils of the cloth (the nonpenetration of the coloration into the middle of the fibers and the middle of the cloth), coloration 360° of the circumference, and oxidation and dehydration of fibers.[123] Only radiation can explain these features, because liquids, dyes, rubs, and scorching cannot do so. In a comprehensive paper, physicists John Jackson, Eric Jumper, and William Ercoline tested the eight alternative possibilities using laboratory conditions to replicate a nonradiation means of duplicating the image on the Shroud.[124] They compared the results of these tests with the macroscopic and microscopic features of the Shroud image, and showed that none of the techniques tested can simultaneously reproduce its main features, from the 3-D property to the coloration depth, to the resolution of the spatial details. They concluded from this that it could not be the work of an artist or a forger.[125]

At this juncture, we know that radiation was necessary to produce the image, but was it *particle* radiation (Little and Rinaudo) or *ultraviolet* radiation (Jackson and Di Lazzaro)? Matters of coloration and encoding information at a distance (where there is no contact between the body and cloth) favor both hypotheses equally; however, the considerations discussed below are better explained by the PRH.

Let us now consider the enigma that the Shroud's frontal and dorsal sections have encoded information from the body with equal intensity. This is highly unusual because there is pressure from the body lying in a supine position on the dorsal part of the cloth. However, the PRH explains this, because the body loses its mass and a vacuum is created during nuclear disintegration. The radiation moving in both directions would therefore encode the cloth with equal intensity on the top and the bottom.[126]

Furthermore, there is information inside the body that is encoded with three-dimensional layering on both the frontal and dorsal sections of the cloth—particularly skeletal features such as finger bones,

[123] Ibid., p. 2614.
[124] Jackson et al., "Correlation of Image Intensity".
[125] Ibid., Abstract.
[126] Antonacci, "Particle Radiation".

bones extending over the palm, part of the skull at the forehead, the left thumb, parts of the backbone, and teeth. The PRH explains this because in nuclear disintegration, as the body loses its mass, a vacuum is created. This vacuum would draw the frontal section of the cloth into the front part of the body and the dorsal part of the cloth into the back end of the body with equal intensity. As both sections of the cloth are drawn in, the particles emanating from the nuclear disintegration would encode the frontal and dorsal sections in layers as they moved toward the center. This remarkable three-dimensional layering of information on the frontal and dorsal parts of the Shroud have been confirmed through a VP8 analyzer. As Antonacci notes:

> The Shroud's truly proportional, full-length, and three-dimensional frontal body image was first demonstrated with a VP8 image analyzer indicating a direct correlation between the lightness and darkness at each point on the Shroud's body image with their respective distances from the underlying body.[127]

Dr. Kitty Little showed how the heavy charged particles (protons, alpha particles, and deuterons) flowing out of nuclear disintegration would not only form the three-dimensional highly precise image on the cloth but would do so only on the uppermost surface fibrils.[128] Referencing Little's 1997 article "The Formation of the Shroud's Body Image", Antonacci notes that she also showed how these heavy charged particles would

> break many of the bonds of the molecular structure of the cellulose ... thereby causing some of the single-bonded carbon atoms attached to hydrogen or oxygen to, thereafter, re-form with other carbon atoms into double-bonded, or conjugated, carbonyl groups.[129]

Jean-Baptiste Rinaudo also produced conjugated carbonyl groups by using low-temperature proton radiation on linen samples.[130] There are large numbers of these double-bonded conjugated carbonyl groups throughout the image on the Shroud.

[127] Ibid., p. 2616.
[128] See Little, "Formation of Shroud's Body Image".
[129] Antonacci, "Particle Radiation", p. 2617.
[130] See Rinaudo, "Protonic Model of Image".

Another enigmatic dimension of the Shroud's image is that it does not fluoresce under ultraviolet light. Linen naturally fluoresces under ultraviolet light, but this does not occur on the Shroud in the image areas—only the background areas show natural fluorescing under UV light. Dr. Jean-Baptiste Rinaudo showed that linen would lose its natural fluorescing characteristic when irradiated with 1.4 MeV (1.4 million electron volts) or less.[131] This low-temperature proton radiation would be a result of a nuclear disintegration of the kind described by Dr. Kitty Little.

Rinaudo's experiments on proton irradiation of linen also showed several other enigmatic features manifest on the Shroud. As Antonacci describes it:

> The protons produced uniform superficial coloration on cloth whose fibers and threads lacked any cementation or added pigments or materials of any kind. Where body image fibers crossed, underlying fibers were protected and remained white, as found on the Shroud; in addition, the inner part of the straw-yellow image fibers remained white, like the image fibers on the Shroud. The scientists were also able to duplicate the microchemistry results of dehydratively oxidized, degraded cellulose, as is also found with the Shroud's body image.[132]

Once again, particle (i.e., proton) radiation explains the Shroud's unique image.

Antonacci shows that the PRH also explains the secondary image features on the cloth. Recall, the PRH predicts that as the body disintegrates, the frontal part of the cloth would fall flat into the disintegrating body and the dorsal part would be drawn flat by a vacuum into the back of the body. This flat collapse explains eight secondary features in the image—gaps in the sides of the man's face and the upturned beard, likely explained by a chin band to keep the mouth closed; the vertical lines running down from the man's chin; motion blurs coming from the cloth collapsing through the beard area; encoded neck features; displaced hair images; motion blurs that produced a very faint image

---

[131] Ibid.

[132] Antonacci, "Particle Radiation", p. 2618, citing Rinaudo, "Protonic Model of Image", and John Rinaudo, "A Sign for Our Time", *Shroud Sources Newsletter*, May/June 2–4, 1996, pp. 2–4.

of the face on the left side of the main image; the small distortion at the femoral quadriceps; and the length of the fingers, which were curved but encoded on a flattened surface.[133]

In conclusion, particle radiation of heavy charged particles (such as protons, alpha particles, and deuterons) can explain all thirty-two primary image features listed at the beginning of this section. The flat collapse of the burial Shroud into the instantaneously disintegrating body not only explains the three-dimensional interior imaging, but also all secondary images. Therefore, the PRH explains all forty major enigmas of the Shroud's image. As we shall see in the next subsection, the PRH also explains nonimage enigmas concerned with the bloodstains, the cloth, and the carbon dating.

C. How Does the PRH Explain the Enigmas on the Shroud's Bloodstains, Cloth, and Carbon Dating?

Recall from above that the disintegration of the atomic nuclei in the body (as specified in the PRH) not only produces mechanical transparency but also a flow of heavy charged particles (protons, alpha particles, and deuterons) *and* a flow of other particles—neutrons (heavy uncharged particles), electrons (light charged particles), and gamma rays. As noted above, the heavy charged particles (and mechanical transparency) are responsible for all forty enigmatic *image* features. We now turn to the other particles—neutrons, electrons, and gamma rays—in conjunction with mechanical transparency that explain all five nonimage enigmatic features of the Shroud. We will discuss each in turn.

1. The Body's Disappearance and the Unbroken Blood Marks

As noted above, blood from the body was transferred on to the cloth, but somehow the body was removed from the Shroud without breaking or distorting any of the blood marks. We know that the body had to have left the Shroud within two to three days after the death of

---

[133] See Antonacci, "Particle Radiation", p. 2620.

the man because there are no signs of decomposition manifest on the Shroud (signs of decomposition after three days are very evident on burial cloths). Now here is the enigma—"If the cloth had been removed from the body by any human or mechanical means—some, most or all of these intimately encoded blood marks would have been broken or smeared."[134] So, if human and mechanical means were not used to remove the body, how did it become separated from the Shroud? The PRH explains this quite well, because the process of the nuclear disintegration of the body would not have disturbed the bloodstains already transferred to the Shroud.[135] There does not appear to be any other way of explaining this enigma without obvious notable effects on the Shroud's bloodstains.

2. The Bright Red Color of the Bloodstains

The bright red color of the bloodstains on the Shroud are inexplicable because blood turns dark brown and then black in a relatively short time in open air. This red coloration was thought to be explained by the high content of bilirubin in the blood, but the brightness of the blood on the Shroud did not vary according to the bilirubin content. The explanation was discovered by Dr. Carlo Goldoni, who experimented with blood irradiated by neutrons.[136] According to Antonacci:

> [Goldoni] concluded that when blood marks are first exposed to neutron irradiation and then to ultraviolet light (such as the Shroud would naturally receive from sunlight during exhibitions) it resulted in the blood marks having a bright red coloration.... [This coloration] existed regardless of the blood's bilirubin content.[137]

[134] Ibid., p. 2619.

[135] Ibid.

[136] Neutron irradiation causes collision displacement, transmutation, and ionization effects in materials, resulting in changes of material microstructure and properties. It is commonly used to discover properties of materials and the efficacy of devices.

[137] Antonacci, "Particle Radiation", p. 2621, citing Carlo Goldoni, "The Shroud of Turin and the Bilirubin Blood Stains" (paper presented at the Shroud Science Group International Conference, "The Shroud of Turin: Perspectives on a Multifaceted Enigma", Columbus, Ohio, August 14–17, 2008).

Once again the PRH supplies a coherent explanation to a long-standing enigma—the neutron radiation arising out of the nuclear disintegration of the body would irradiate the bloodstains, making the blood very bright when exposed to UV radiation (like the sun).

3. The Shroud's Excellent Condition and Resistance to Degradation

Another long-standing enigma about the Shroud is its excellent condition—nonfriability, pliability, and resistance to oxygenation and chemical reaction. This was noticed by Drs. Roger and Marion Gilbert during the STURP investigation.[138] These features would be enigmatic even if the Shroud were only seven hundred years old (according to the 1988 carbon dating). Once again, the PRH supplies the explanation. Recall that during nuclear disintegration, about half the particles emitted are heavy charged particles (e.g., protons and alpha particles) that do not penetrate the Shroud—or even the fibers of the Shroud. They affect only the uppermost surface of the fibers, making those fibers (in the image) more friable (brittle and crumbly—*less* substantial). The other half of the particles of nuclear disintegration are not heavy and charged—for example, neutrons are heavy but uncharged and electrons are charged, but not heavy, and gamma rays (the highest form of photon energy) are not heavy. These particles easily penetrate the linen cloth and move through it. Some of them hit the long-chain cellulose molecules in noncrystalline (weaker) regions, causing them to break, and then to cross-link with crystalline (stronger) structures, which would strengthen the cloth—making it more substantial and pliable while increasing its resistance to solubility, oxygenation, and chemical reactions.[139] According to Dr. Kitty Little:

> Given a high crystallinity, such as one would expect to find in good quality linen,... this type of cross-linking also reduces solubility and susceptibility to oxygenation and other chemical reactions, which would account for the lack of degradation and 'ageing' that might be

[138] Roger Gilbert, Jr., and Marion Gilbert, "Ultraviolet-Visible Reflectance and Fluorescence Spectra of the Shroud of Turin", *Applied Optics* 19, no. 12 (1980): 1930–36.
[139] See Antonacci, "Particle Radiation", p. 2622.

expected in a material 2,000 years old, and that had been subjected to repeated handling and ill-treatment.[140]

4. Possible Imprints of Coins and Flowers

As noted in Section I.C.3, there appear to be Roman coins on the eyes of the man in the Shroud. It was common practice to use such coins to prevent the eyelids from opening during rigor mortis.[141] It is highly probable that there are coins of the diameter of Roman leptons on the eyes of the man in the Shroud,[142] but the images on these coins have been called into question. Computer-enhanced images of the coin imprints on the Shroud suggest that they are Roman leptons coming from a special minting by Pontius Pilate in A.D. 30. As noted above, some scientists believe that the cloth is too coarse to admit of such fine imaging, but this is not certain, because we have not sufficiently explored the refinement of surface imaging by heavy charged particle radiation. We have seen how fine the images of Jesus' hair, beard, and skeletal structure are, so we might say that the jury is out.

Notwithstanding this, we must still answer the question of how imprints of a coin could make their way onto the cloth, because the coins are extrinsic to the body, and the body is the source of the radiation. In other words, how did the radiation come out of the coins as well as the body? Once again, the PRH supplies the explanation. Recall from above that a neutron shower is one of the results of the disintegration of atomic nuclei in the body. These neutrons would have flowed out of the man's eyes, and some of them would have hit copper nuclei (in the bronze coins), giving off a heavy charged particle (e.g., a proton, alpha particle, or a deuteron), which would have created an image on the cloth in the precise way that the body's heavy charged particles created its image. As Antonacci explains it:

> When a neutron hits the nucleus of copper, the primary component of ancient bronze coins, the nucleus can absorb the neutron and give off either a proton, alpha particle, deuteron, or a low-energy gamma

[140] Little, "Formation of Shroud's Body Image".
[141] See Thackeray, "Lepton Coin Diameters".
[142] See ibid.

ray. Each of these particles ... encodes superficial images on the cloth
and, if they were given off the coin's surface, could encode the coin's
features. Similarly, flowers contain trace amounts of heavier elements
such as iron, calcium, and potassium. When any of the countless neutrons hit these three heavier elements, each could also absorb the
neutrons and give off protons and alpha particles. Any protons or
alpha particles given off the flowers' surfaces would also encode a
superficial image on the Shroud.[143]

If heavy charged particle radiation can make intelligible imprints on
coarse fabric, then the identification of the sacrificial cup, the augur
staff, and the Greek letters OY KAI AROC would indicate coins
minted by Pontius Pilate in A.D. 30 on the Shroud man's eyes.

5. Carbon Dating of the Shroud

The PRH gives a likely cause for an errant C-14 dating of the Shroud.
Recall from above that a shower of countless neutrons would emerge
out of the instantaneous disintegration of all the atomic nuclei in the
Shroud man's body. Dr. Arthur C. Lind and colleagues show that
such neutron irradiation of a linen cloth filled with indigenous N-14
(nitrogen) would have converted the N-14 into C-14. This would
significantly elevate the C-14 content in the cloth, making the age
of the cloth (by C-14 testing) appear much younger.[144] Furthermore,
this additional C-14 would take the place of the converted N-14 and
remain in the cloth for hundreds or thousands of years. It would also
be resistant to high temperatures (as in the fire of Chambery) and all
cleaning methods used to prepare the linen for C-14 dating.[145] A. C.
Lind and colleagues calculated the very small amount of neutron irradiation it would take to lower the age of the Shroud sample by thirteen hundred years. Antonacci summarizes Lind's results as follows:

> The neutron fluence that would be needed to cause the radiocarbon
> date to be medieval instead of first century is 8.3×10^{13} n·cm^{-2} if the

---

[143] Antonacci, "Particle Radiation", p. 2621.
[144] Lind et al., "Production of Radiocarbon".
[145] Ibid.

nitrogen content of the Shroud is about 570 ppm. This neutron flu-
ence over the 4.4×1.1 m^2 area of the Shroud is 4×10^{18} neutrons,
which weigh only 0.67 μg.[146]

This is a very small fraction compared to the weight of the man on
the Shroud, making it likely that the carbon dating was made to
appear younger by a significant factor—even thirteen hundred years.

D. Confirming the Particle Radiation Hypothesis

Currently, there is very good reason to believe that the particle radiation hypothesis is superior to the ultraviolet radiation hypothesis,
because it explains many more of the forty enigmas on the image,
blood, cloth, and carbon dating of the Shroud. Furthermore, the
PRH explains both of Jackson's requirements for a perfect *three-
dimensional* photographic negative image that records the interior of
the body on both the frontal and dorsal sides:

1. Radiation, to explain action at a distance and tri-dimensional
   photographic negative imaging
2. Mechanical transparency of the body, allowing the cloth to penetrate the frontal part and the dorsal part (by vacuum) to encode
   the interior of the body (e.g., the backbone)

Instantaneous disintegration of the body's atomic nuclei in the
PRH would explain *both* the source of radiation and the cause of
mechanical transparency while the ultraviolent radiation hypothesis
explains only the radiation (but not the mechanical transparency).

As noted above, both the PRH and the URH (ultraviolet radiation hypothesis) appear to require supernatural causation, because
there is no natural explanation for either hypothesis:

- The simultaneous disintegration of every atomic nucleus in an
  entire human body producing a low-temperature nuclear reaction (the PRH)

[146] Antonacci, "Particle Radiation", p. 2619.

- The production of six to eight billion watts of ultraviolet radiation from all points of the body (the URH)

The remarkable ability of the PRH to explain *every* known enigma on the image, blood, cloth, and carbon dating is very probative, but we can gain greater certitude by looking for remnant isotopes (beyond C-14) that would be created in a nuclear reaction (such as nuclear disintegration).[147] Two such isotopes are Cl-36 (a radioactive cosmogenic isotope of chlorine created from nuclear reaction) and Ca-41 (a cosmogenic nucleus created in nuclear reaction).[148] If these radioactive/cosmogenic isotopes are found in abundance (above their very insignificant levels in natural linen), they, along with C-14 abundance, would almost certainly indicate a nuclear reaction similar to nuclear disintegration. This would confirm the strong likelihood of the PRH.[149]

In view of the above, it is probable that the PRH will be validated by the discovery of cosmogenic/radioactive isotopes in the Shroud. However, if such validation does not occur, we would still have to find another source of radiation (such as ultraviolet radiation) and another cause of mechanical transparency. This combination of requirements points to the strong likelihood of supernatural causation of the enigmatic image, bloodstains, and cloth.

E. Does the Image Indicate a Resurrection Similar to Jesus'?

There is general agreement among scholars that the apostles of Jesus experienced Him as risen from the dead, supernaturally transformed. Gary Habermas, who made an extensive study of scholars across the board—from conservative to radical—concluded as follows:

> The latest research on Jesus' resurrection appearances reveals several extraordinary developments. As firmly as ever, most contemporary

---

[147] Mark Antonacci, "Can Contamination Be Detected on the Turin Shroud to Explain Its 1988 Dating?" (paper presented at the International Workshop on the Scientific Approach to the Archeiropoietos Images, Frascati, Italy, May 4–6), pp. 239–47.

[148] Ibid.

[149] Ibid.

scholars agree that, after Jesus' death, his early followers had experiences that they at least believed were appearances of their risen Lord.[150]

Habermas goes on to say that most scholars believe for good reason that the apostles were not collectively imagining Jesus, but rather experienced Him as physically mediated in the world and transformed in appearance.[151] Some scholars believe He appeared as light (e.g., Reginald Fuller[152]) but most scholars believe that He appeared as a spiritually transformed and glorified body (e.g., N. T. Wright[153]). Wright indicates that the idea of spiritual transformation of embodiment is a central Christian mutation of Second Temple Judaism. Early Christians adhered very closely to the implicit doctrines of Second Temple Judaism (which were developed throughout the period of the Second Temple from about 530 B.C. to A.D. 70). However, in a few cases, Christians departed radically from that body of doctrine. Several of those Christian mutations concern the Resurrection. Two of them are germane to this study:

1. The Jewish doctrinal consensus held that the risen body would be a continuation of physical embodiment, but Christians changed it to a spiritual and glorified (transformed) embodiment.
2. The Jewish doctrinal consensus held that the Resurrection was a less important doctrine, but the Christians held that it was the central doctrine upon which everything else is grounded (see, for example, 1 Cor 15:16–19).

Wright asks, why would the early Christians make these changes to the Jewish doctrinal consensus when they did not want to break from the synagogue? Why would they have risked a doctrinal rift

---

[150] Gary R. Habermas, "Mapping the Recent Trend toward the Bodily Resurrection Appearances of Jesus in Light of Other Prominent Critical Positions", in *The Resurrection of Jesus: John Dominic Crossan and N. T. Wright in Dialogue*, ed. Robert B. Stewart (Minneapolis: Fortress Press, 2006), p. 79.

[151] Ibid., pp. 87–89.

[152] See Reginald Fuller, *The Formation of the Resurrection Narratives* (New York: Macmillan, 1971).

[153] See N. T. Wright, *The Resurrection of the Son of God* (Minneapolis, Minn.: Fortress Press, 2003).

when they continued to respect the doctrines of their forefathers?[154] His answer is that there had to be a cause that was powerful and influential enough to make them depart from a doctrinal consensus to which they basically adhered. That powerful and influential cause had to be more than an empty tomb. It had to be an experience of a spiritually transformed and glorified Jesus—an appearance that was continuous with His former embodiment, but radically transformed in spirit, power, glory, and light (see, for example, 1 Cor 15:43; Acts 9:3, 22:6).

So how does the origin of the image suggest the Resurrection of Jesus? Let us return to John Jackson's two requirements for image formation—a powerful source of radiation (which would have given off a bright light) and mechanical transparency—something akin to becoming spiritual; Saint Paul calls it a "spiritual body [*pneumatikon soma*]" (1 Cor 15:44). The combination of powerful radiation giving off light—that is, from nuclear disintegration/reaction in the PRH or several billion watts of energy in the URH—with mechanical transparency suggestive of "becoming spiritual" corresponds with the testimony of the Gospels, the Letters of Saint Paul, and the Acts of the Apostles, particularly with respect to the Christian mutations of Second Temple Judaism. Perhaps we might say that the radiation, light, and mechanical transparency that caused the image on the Shroud were the gateway to a risen body transformed in spirit, power, light, and glory.

V. Conclusion

The likelihood of the Shroud being a forgery—medieval or otherwise—is positively miniscule. There are four significant challenges to this hypothesis that are almost insurmountable:

1. *The problem of radiation and mechanical transparency.* Inasmuch as Jackson's two requirements for a very precise three-dimensional photographic negative image with encoding of the interior of the body on both the frontal and dorsal sides are completely

---

[154] Ibid., pp. 200–210.

beyond natural causation and even contemporary human causation, we must wonder how a medieval forger could produce this immense source of radiation from a dead human body while making that body mechanically transparent (explained above in Section IV).

2. *The problem of how the body was removed from the Shroud.* Even if we suppose that an ingenious medieval forger took a live victim and used the Roman tools/weapons to replicate Jesus' Crucifixion perfectly, and then wrapped the dead man's body in a cloth soon after, we confront the problem of how the Shroud could have been taken off that dead body without breaking, distorting, or smearing most or all of the 372 bloodstains on the Shroud.[155] There is currently no known human or mechanical method to do this, implying that the body literally disappeared from within the Shroud without disturbing those bloodstains (explained above in Section IV.C.1).

3. *The bloodstains are prior to the image.* Recall that the bloodstains appeared on the Shroud before the image. Even if we suppose that the medieval forger had contemporary anatomical information so as to produce the positioning of the bloodstains perfectly, we have to wonder just how he placed those bloodstains on the Shroud before there was any image on which to position them with perfect anatomical precision. We also need to explain how he made an anatomically precise human body image to perfectly overlay on these preexisting bloodstains. Again, this seems to be beyond even a contemporary forger's capabilities (explained above in Section II).

4. *Pollen grains and blood characteristics not available to a medieval forger.* The first mystery is how a medieval forger could have obtained the pollen grains indigenous to Jerusalem and northern Judea. Recall that of the pollen grains taken from the Shroud, three-quarters of them were from Israel (specifically from sediment deposits from two thousand years ago near the Sea of Galilee), including thirteen that are unique to that region (explained above in Section I.C.2). Secondly, we must explain how a medieval forger placed the invisible/undetectable blood

---

[155] Little, "The Formation of the Shroud's Body Image", p. 19.

elements on the Shroud—"the microscopically precise, invisible reactions around the more than 100 scourge marks throughout the body [and] the coagulated blood stains with serum surrounding borders and clot retraction rings that occur with actual wounds and blood flows, found throughout the front and back of the body, and revealed only by modern scientific technology".[156] The only way the forger could have done this is to kill a live victim in precisely the way that Romans would have scourged and crucified a man in the unique way described of Jesus in the Gospels. This gives rise to a host of questions raised above in Section III, which will be reviewed immediately below.

How did our hypothesized forger know the following? Aware only of crowns that were circlets, he inexplicably made his crown of thorns with a top that was intended by the Romans to inflict maximum pain. Without knowing what a Roman lance looked like, he replicated it perfectly (with its elliptical leaf-like blade) and, without knowing the anatomical particulars of the right atrium and pleural cavity, he positioned the spear injury at the precise place (between the fifth and sixth ribs) for a spear to exact blood and a watery substance (which he depicted perfectly on the Shroud). Without knowledge of what a Roman flagrum (whip) looked like, he replicated the three thongs with their dumbbell-shaped lead fragments at the end perfectly, and then portrayed the lashing according to Roman custom, moving from right to left to right, etc. Having no Gospel evidence of a fall or the anatomical effects of a fall with a large object on the shoulders, he perfectly portrayed the five elements of the blunt-force trauma that would have occurred in a fall—the lowering of the shoulders by ten to fifteen degrees, the hyperextension of the arms, the recession of the right eye into its orbit, the left twist in the neck, and severe scraping of the knees. Finally, without knowing how the Romans would have attached the hands and feet to a cross (without the body pulling away), he portrayed the exit wound from the nail that went through the palm but exited through the wrist with remarkable precision. Even if we conjecture that the forger used

---

[156]Antonacci, "Particle Radiation", p. 2614.

a live victim, and forced him to go through everything that Jesus underwent, how would he know how to carry this out in a particularly Roman way? Are we really to believe that any forger could have done this before our present age?

In conclusion, all the above scientific tests, studies, and corroborative evidence do not constitute a formal scientific proof for the authenticity of the Shroud as the burial cloth of Jesus Christ or the supernatural origin of its image. However, it does provide significant evidence for an informal inference (along the lines described by John Henry Newman[157]). An informal inference puts together a series of different sets of evidence, each of which is antecedently probable, that complement and corroborate one another toward a single conclusion. Such inferences do not constitute a formal metaphysical (deductive) proof or a formal scientific proof, but a mutually corroborative informal basis for probable truth. This allows for future modification of any of the individual, antecedently probable sets of evidence without the consequence of falsifying the conclusion. The mutually corroborative nature of the different sets of evidence give probative force according to the number and diverse kinds of evidence pointing to the same conclusion.

So what can we say about the Shroud of Turin? To begin with, the 1988 carbon dating is problematic because of heterogeneity detected in the statistical analysis of its raw data.[158] This problem may well be explained by sixteenth-century material in the sample used for C-14 dating and the strong possibility that the cloth was irradiated by neutrons from a nuclear disintegration/reaction that converted the N-14 intrinsic to the cellulose of the cloth into an abundance of C-14 (lowering the apparent age of the Shroud by a significant factor—even by more than one thousand years).[159] Given this, there are six major sets of evidence, mutually corroborating each other, that point to the authenticity of the Shroud as the burial cloth of Jesus:

1. *Four other dating tests (using methods other than C-14) indicate a probable date of origin for the Shroud between A.D. 55 (De Caro et*

[157] John Henry Newman, *An Essay in Aid of a Grammar of Assent* (Notre Dame, Ind.: University of Notre Dame Press, 1992), pp. 259–342.

[158] Casabianca et al., "Radiocarbon Dating".

[159] Lind et al., "Production of Radiocarbon".

al.) and A.D. *90 (Fanti et al.)*: wide-angle X-ray scattering, Fourier transform infrared spectroscopy, Raman spectroscopy, and break-strength testing (see references in Section I.B above).

2. *In addition to the above dating methods, there are also two extrinsic factors pointing to the Shroud's origin prior to the Middle Ages* (see Section I.C above):

   a. The 120 points of congruence between the bloodstains on the Shroud and those on the Sudarium of Oviedo indicate that the two cloths touched the same face. Since the Sudarium of Oviedo has a definite historical record dating back to A.D. 616, it is highly likely that the Shroud must also date back to A.D. 616 or before. (See the references in Section I.C.1.)

   b. The pollen grains do not give a date of origin, but rather a place of origin—Jerusalem/northern Judea because the majority (three-fourths) of pollen grains come from this area, thirteen of which are indigenous/unique to the area. The Sudarium of Oviedo also was in Palestine prior to A.D. 616. The 1988 carbon dating dates the origin of the Shroud to the time when it appeared in Europe, meaning that it originated and remained *only* in Europe. However, this cannot be the case, because as the pollen grains indicate, the Shroud spent a substantial time in Palestine and another substantial time in Turkey, which would have lasted at least as long as its stay in Europe. This means that the Shroud must be much older than seven hundred years old. (See references in Section I.C.2.)

3. *The bloodstains on the cloth are authentic and came from a victim who was tortured.* This is indicated by the high content of bilirubin and the synthesis of ferritin and creatinine. The bloodstains were produced by a victim who was positioned vertically for an extended time (as in a crucifixion), indicated by the vertical axis of the serum separation from the plasma. These bloodstains could not have been applied to the cloth by a forger because the blood preceded the image on the cloth and contained blood elements and characteristics that are revealed only by modern technology. This means the forger would have had to have crucified a live victim with Roman tools and customs in the same way as Jesus (see Section II above).

4. *The Shroud portrays the image and blood of a man who was crucified precisely according to Roman custom.* The custom consisted
   of using a Roman legionnaire's lance (with elliptical, leaf-like
   shape) and two Roman flagrums (according to Roman custom) and nailing to the cross according to Roman custom.
   Though the Gospels do not mention a fall of Jesus, the Shroud
   portrays all five characteristics of blunt-force trauma of a man
   who fell to his knees with a heavy object on his right shoulder.
   This seems to be beyond the knowledge of a medieval forger.
   Furthermore, the image and bloodstains are anatomically perfect in themselves and their relative positions to one another.
   Much of this data would have been beyond a medieval forger
   (see above, Section III).

5. *The cloth was detached from the dead body by an unknown means.* If any
   human or mechanical means had been used to take the Shroud
   off the body, it would have broken, distorted, or smeared most
   or all of the 372 bloodstains on the Shroud.[160] This points to the
   mechanical transparency (produced by nuclear disintegration in
   the PRH—or some other cause of mechanical transparency),
   which apparently requires a supernatural cause (see above, Sections IV.C and IV.D).

6. *The image was produced by a source of significant radiation (e.g., particle radiation or ultraviolet radiation) that would have given off bright
   light. It was also produced by a mechanically transparent body.* These
   two requirements are beyond the capacity of not only a medieval forger but also a contemporary forger, and all known natural
   causation. Particle radiation from the simultaneous disintegration of all atomic nuclei (producing a low-temperature nuclear
   reaction) is the probable source of the radiation and mechanical
   transparency. Even if this is disproved, there would have to be
   another source of radiation and mechanical transparency that
   would be equally beyond a medieval forger, a contemporary
   forger, and purely natural causation. Whatever the case, the
   light, radiation, and transparency of the body are suggestive of
   the Resurrection of Jesus indicated in the Gospels, the Letters
   of Saint Paul, and the Acts of the Apostles (in spirit, power,

[160] Little, "The Formation of the Shroud's Body Image", p. 19.

glory, and light) and seemingly produced by a supernatural cause (see above, Section IV.E).

By now it will be clear that these six distinct sets of data complement and mutually corroborate one another. Some of the data sets may be modified in the future, but the combination and corroboration of evidence manifest in the image, the blood, and the cloth along with the pre-A.D. 616 dating entailed by congruences with the bloodstains on the Sudarium of Oviedo strongly suggest this is the burial cloth of Jesus Christ, whose Resurrection imprinted the image of His crucified body on the Shroud. If this is correct, then the Shroud validates the accuracy of the Gospel accounts of Jesus' Passion and Resurrection, while the Gospel accounts point to the identity of the man in the Shroud.

Chapter Four

Science and Eucharistic Miracles

Introduction

It might seem peculiar to be speaking about the scientific investigation of miracles, because science begins with empirical observations and seeks *naturalistic* explanation. However, miracles are by definition supernatural phenomena. This means there cannot be a *direct* scientific test for a miracle. Nevertheless, there can be an *indirect* validation of the *possibility* or *likelihood* of a miracle based on an exhaustive scientific investigation that excludes all *known* natural explanations. Though this can point to the possibility or likelihood of a miracle, it cannot prove it in a formal scientific way, because we do not yet know all possible natural explanations of or within our universe.

As noted in previous chapters, the frustrating part of science is that we cannot know what we do not know until we have discovered it in observational data. Nevertheless, an exclusion of all known natural explanations, though not a proof, helps us to *approach* the truth about whether a phenomenon *appears* to be naturalistically inexplicable—"a miracle". Inasmuch as this indirect *"via negativa"* approach opens up the possibility of supernatural causation in our universe (a miracle), then readers might find this and the following chapter to be another avenue through which science approaches the doorstep to God and Christ.

Some scientists might deny the possibility of a miracle because they mistakenly believe that miracles require the suspension of inviolable physical laws. Though there is no reason why God—as a supernatural creative being—would not be able to suspend the laws of nature, it is not necessary to explain miracles this way. C. S. Lewis put it quite succinctly:

> The divine art of miracle is not an art of suspending the pattern into which events conform but of feeding new events into that pattern. It does not violate the law's proviso, "If A, then B": it says, "But this time instead of A, A2," and Nature, speaking through all her laws, replies "Then B2" and naturalizes the immigrant, as she well knows how. She is an accomplished hostess.[1]

Though a miracle may seem to require a suspension of the laws of nature, Lewis shows that a supernatural cause (e.g., God) need only add another law or event into the original laws of nature (if only for a moment), which would yield a new kind of effect by the original natural laws' accommodation to it. Thus, a miracle need not change, suspend, or violate the original set of natural laws, but only temporarily add to them, after which the new cause or event could disappear.

Perhaps the greatest miracle is not the manifestation of supernatural power, but the fact that nature itself not only has regularity, but that this regularity is describable by mathematics in a most surprising—indeed completely—unexpected way. The Nobel Prize–winning physicist and mathematician Eugene Wigner recognized this remarkable coincidence of natural laws and mathematics, referring to it as a "miracle" or "the scientist's article of faith". "It is, as [Erwin] Schrödinger has remarked, a miracle that in spite of the baffling complexity of the world, certain regularities in the events could be discovered."[2] Wigner later goes on to describe a fourfold miracle in the connection between classical physics, quantum physics, higher-level mathematics, and the human mind's ability to recognize it:

> It now begins to appear that not only complex numbers but so-called analytic functions are destined to play a decisive role in the formulation of quantum theory. I am referring to the rapidly developing theory of dispersion relations. It is difficult to avoid the impression that a miracle confronts us here, quite comparable in its striking nature to the miracle that the human mind can string a thousand arguments

---

[1] C. S. Lewis, *Miracles: A Preliminary Study* (New York: HarperOne, 1947), p. 95.

[2] Eugene Wigner, "The Unreasonable Effectiveness of Mathematics in the Natural Sciences", *Communications in Pure and Applied Mathematics* 13, no.1 (1960): 4, https://www.maths.ed.ac.uk/~v1ranick/papers/wigner.pdf.

together without getting itself into contradictions, or to the two miracles of the existence of laws of nature and of the human mind's capacity to divine them.[3]

Though Wigner was using the term "miracle" loosely here, this fourfold nonnecessary coincidence of physics, mathematics, aesthetics, and the human mind is completely inexplicable in terms of logic, mathematics, and physics themselves. Wigner and Schrödinger leave us to draw our own conclusions, but people of faith will see rigorous rationality and creative serendipity in this fourfold coincidence which has all the earmarks of creative intellection coursing through nature—intellection pointing toward a supernatural mind.

The possibility and likelihood of miracles has not been lost on contemporary physicians, many of whom claim to have witnessed something at least approaching the supernatural. An international survey of physicians found that a majority of physicians are openminded to the transcendent interpretation of patients' unexplained recoveries.[4] A U.S. national survey indicated that 74 percent of physicians believed in miracles in the past, and 73 percent believe they can occur today.[5]

The Catholic Church acknowledges both healing miracles and nature miracles (e.g., the Miracle of the Sun at Fatima or a Eucharistic miracle). With respect to healing miracles, the Church makes recourse to scientists and physicians (both religious and skeptical) to determine whether a purported miracle meets three criteria:

- It must be instantaneous or very near instantaneous.
- It must be efficacious throughout the rest of the life of the recipient.

[3] Ibid., p. 5.

[4] Jakub Pawilkowski et al., "Doctors' Religiosity and Belief in Miracles" (paper presented at the 4th European Conference on Religion, Spirituality and Health, Malta, May 2014), https://www.researchgate.net/publication/311680999_Doctors'_religiosity_and_belief_in_miracles.

[5] Shoba Sreenlvasan and Linda E. Welnberger, "Do You Believe in Miracles? Turning to Divine Intervention When Facing Serious Medical Illness", Psychology Today, December 15, 2017, https://www.psychologytoday.com/us/blog/emotional-nourishment/201712/do-you-believe-in-miracles#:~:text=Even%20physicians%20believe%20in%20miracles.%20A%20national%20poll,occur%20today%20%28Poll%3A%20Doctors%20Believe%20in%20Miracles%2C%202004%29, citing Bill Freeman, "Science or Miracle?; Holiday Season Survey Reveals Physicians' Views of Faith, Prayer and Miracles", WorldHealth.Net, December 22, 2004, https://www.worldhealth.net/news/science_or_miracle_holiday_season_survey/.

- It must be beyond any known natural explanation as determined by strict scientific criteria.

Since nature miracles can be quite diverse—for example, from the tilma and image of Our Lady of Guadalupe to the miracle of Fatima (or the Eucharistic miracles discussed below)—the Church relies on the scientific judgement of experts in the kind of phenomena under consideration. So, for example, with respect to the tilma and image of Guadalupe, the Church consults experts in chemical composition of fabrics, paints and natural pigments, optical image enhancement, etc. With respect to the Miracle of the Sun at Fatima, the Church consults physicists, astrophysics, meteorologists, and astronomers. With respect to Eucharistic miracles, the Church consults histological experts, tissue experts, pathologists, and electron and microscope analysts. One major question must be answered: Is there any known natural explanation or likely future natural explanation for this phenomenon? If the answer is no, or highly unlikely, then there is the possibility of supernatural causation—a miracle.

With the above considerations in mind, we may now proceed to some well-known and documented possible miracles in two areas:

1. Eucharistic miracles (this chapter)
2. Marian miracles involving both nature and healing (Chapters 5 and 6)

In this chapter, we will briefly review the history of Eucharistic miracles (Section I), followed by a more detailed account of the scientific investigation of three contemporary Eucharistic miracles:

1. The Eucharistic Host of Tixtla, Mexico (2006)—Section II
2. The Eucharistic Host of Sokółka, Poland (2008)—Section III
3. The Eucharistic Host of Buenos Aires, Argentina (1996)—Section IV

I. A Brief Review of Eucharistic Miracles

There is of course a miracle that occurs during every celebration of the Holy Eucharist (Mass)—the substantial change (called "transubstantiation") in which the bread and wine are ontologically transformed into the Body and Blood of Christ, with His soul and divinity being present as well. The new substance is Jesus' divine-human nature, which infuses itself throughout the atomic and molecular structure of bread and wine, bringing it to a higher divine level of being. Though one can see in a consecrated Host the organic compounds of bread and wine through an electron microscope, they are no longer merely material substances but rather a divine substance that elevates the merely material into a divine reality that retains only the *appearance* of materiality. Our purpose here is not to delve into the theology and philosophy of this kind of substantial change, which I have discussed elsewhere;[6] rather, it is to address a certain class of special events in which a consecrated Host (with organic compounds ordinarily constitutive of bread) undergoes transmutation into living human heart tissue with white blood cells and other living blood constituents. This transmutated consecrated Host integrates both the substance of the Host and the substance of living heart tissue in an exceedingly close way, which for several reasons seems to defy natural explanation as well as human agency (discussed below).

Eucharistic miracles in which flesh and blood have appeared on consecrated hosts have occurred throughout history in the Middle East, Western and Eastern Europe, Latin America, and India.[7] The testimonies are edifying, but the chain of custody of the transformed Host cannot be validated, and none of the Hosts prior to 1996 except for the Host of Lanciano has undergone scientific investigation. The Lanciano Host was subjected to scientific investigation in 1970 with Dr. Edoardo Linoli (professor of anatomy, histology, chemistry, and clinical microscopy) overseeing the investigation.[8] Though this investigation was not nearly as thorough as the ones conducted on the Hosts of Tixtla, Sokółka, and Buenos Aires, the

---

[6] Robert Spitzer, *The Sacraments, Part 1—The Sacred Eucharistic Liturgy*, vol. 9 of the *Credible Catholic Big Book* (Garden Grove, Calif.: Magis Center, 2017), https://discover.magiscenter.com/hubfs/CC%20-%20Big%20Book%20Pdfs/BB-Vol-9-Eucharistic-Liturgy.pdf?hsLang=en.

[7] Summary of these miracles is given in *The Eucharistic Miracles of the World: Catalogue Book of the Vatican International Exhibition*, presented by the Real Presence Eucharistic Education and Adoration Association (Bardstown, Ky.: Eternal Life, 2009).

[8] Edoardo Linoli, "Histological, Immunological and Biochemical Studies on the Flesh and Blood of the Eucharistic Miracle of Lanciano (8th Century)", *Quaderni Sclavo di Diagnostica Clinica e di Laboratori* 7, no. 3 (1971): 661–74.

state of preservation of the flesh and blood (given a valid chain of custody) are inexplicable, and the similarities to the three contemporary Eucharistic miracles, remarkable. The following five points summarize Linoli's investigation:

1. The Host is constituted by authentic flesh consisting of muscular striated tissue of the myocardium.
2. The chalice contains authentic blood. The chromatographic analysis indicated this with certainty.
3. The immunological study shows with certitude that the flesh and the blood are human, and the immuno-hematological test allows us to affirm with objectivity and certitude that both belong to the same blood type, AB.
4. The proteins contained in the blood have the normal distribution, in the identical percentage as that of the serous-proteic chart for normal fresh blood.
5. No histological dissection has revealed any trace of salt infiltrations or preservative substances used in antiquity for the purpose of embalming.

If we suppose that the chain of custody was proper, and that the Host and blood of Lanciano are credible, then we might conclude that Eucharistic miracles have occurred throughout the world over the course of Christian history. Why would it be limited to Lanciano alone? As we shall see, today we have much better control over chains of custody and significantly improved techniques to analyze the flesh and blood of miraculous hosts scientifically.

Our faith in the Holy Eucharist as the real Body, Blood, soul, and divinity of Christ does not depend on miraculous phenomena (such as the ones discussed in this chapter), because we have the Eucharistic words of Jesus that strongly favor a literal interpretation of "This is my body" and "This is my blood of the covenant."[9] We also have the clear testimony of the Gospel of John: "I am the living bread which came down from heaven; if any one eats of this bread, he will live for ever; and the bread which I shall give for the life of the world is my flesh" (6:51). This saying was so controversial that many of Jesus' disciples decided to leave Him (6:66). If Jesus meant "body"

[9] I have explained this in detail in the *Credible Catholic Big Book*, Volume 9, Chapters 1–2.

and "blood" only symbolically, why would His disciples have made such an issue about it so as to leave Him? The testimony of Saint Paul is also consistent with and complementary to the four Gospels (1 Cor 10:16; 11:25–28). Furthermore, Eucharistic faith is based on the virtual unanimity of early Church Fathers on the Real Presence of Christ in the Eucharist,[10] as well as the constant teaching of the Church and the saints over two thousand years.

It is helpful to understand the first-century Jewish idea of the collapse of time (in the mind of God) from the future to the present (through the words of Jesus at the Last Supper) and the collapse of time from the past to the present in the reliving of the words of institution by the priest during the Eucharistic celebration.[11] This shows how Jesus intended to collapse His body on the Cross at Calvary into the bread and wine He gave to His disciples at the Last Supper, and how He intended His apostles and future celebrants of the Eucharist to collapse the transformed bread and wine at the Last Supper into the bread and wine at the Eucharistic celebration (thereby transforming them).[12]

---

[10] For example, Ignatius of Antioch (c. A.D. 110): "I have no delight in corruptible food, nor in the pleasures of this life. I desire the bread of God, ... *which is the flesh of Jesus Christ,* ... who became afterwards of the seed of David ...; and *I desire the drink of God, namely His Blood,* which is incorruptible love." Ignatius of Antioch, *Epistle to the Romans* 7, 3, in *Ante-Nicene Fathers*, vol. 1, *The Apostolic Fathers with Justin Martyr and Irenaeus*, ed. Philip Schaff (Grand Rapids, Mich.: Christian Classics Ethereal Library, 1985), p. 205, https://archive.org/details /ante-nicene-fathers-vol-1/page/n247/mode/1up (emphasis added).

Justin Martyr (c. A.D. 140–150): "For not as common bread and common drink do we receive these; but in like manner as Jesus Christ our Saviour, having been made flesh by the Word of God, had both flesh and blood for our salvation, so likewise have we been taught that the food is which blessed by the prayer of His word, and from which our blood and flesh by transmutation are nourished, is the flesh and blood of that Jesus who was made flesh." Justin Martyr, *First Apology* 66, in *Apostolic Fathers with Justin Martyr and Irenaeus*, p. 495.

Irenaeus of Lyon (c. A.D. 180): "But if this [the body] indeed do not attain salvation, then neither did the Lord redeem us with His blood, nor is the cup of the Eucharist the communion of His blood, nor the bread which we break the communion of His body.... He has acknowledged the cup (which is a part of the creation) as His own blood, from which He bedews our blood; and the bread (also a part of the creation) He has established as His own body, from which He gives increase to our bodies." Irenaeus of Lyon, *Against Heresies* 5, 2, 2, in *Apostolic Fathers with Justin Martyr and Irenaeus*, p. 1304, https://archive.org/details/ ante-nicene-fathers-vol-1/page/n1346/mode/1up.

[11] I have explained this in some detail in *God So Loved the World: Clues to Our Transcendent Destiny from the Revelation of Jesus* (San Francisco: Ignatius, 2016), Chapter 3, Section IV.A. This is summarized in the *Credible Catholic Big Book*, Volume 9, Chapter 1.

[12] Ibid.

We now proceed to the results of the scientific investigation of three apparently miraculous hosts—Tixtla, Mexico (2006); Sokółka, Poland (2008); and Buenos Aires (1996).[13]

II. The Eucharistic Host of Tixtla (2006)

The case of the transformed Host in Tixtla, Mexico, started on October 21, 2006, when Father Leopoldo Roque, pastor of the parish church of Saint Martin of Tours, invited Father Raymundo Reyna Esteban to lead a spiritual retreat for his parishioners. As Father Leopoldo and another priest were distributing Communion, assisted by a religious sister who was to the left of Father Raymundo, the sister turned toward him with a pyx containing a Host for the sick. Looking at Father with eyes filled with tears, an incident that immediately attracted the attention of the celebrant, the sister showed him the Host she had taken to give to a lady parishioner: it had begun to effuse a reddish substance.[14]

The Host was secured, and the ecclesiastical authorities commissioned Dr. Ricardo Castañón Gómez (a professed atheist until 1992, when he examined his first transformed consecrated Host in Buenos Aires)[15] to initiate a scientific investigation of the Host.[16] Gómez had

---

[13] There is an excellent resource summarizing the scientific investigation of the hosts at Tixtla, Sokółka, Buenos Aires, and other sites. The author, a cardiologist, is particularly thorough on the analysis of the blood, cardiac tissue, and DNA. See Franco Serafini, *A Cardiologist Examines Jesus: The Stunning Science Behind Eucharistic Miracles*, eds. Umberto Villa and B Purcell (Manchester, NH: Sophia Press, 2021).

[14] Real Presence Eucharistic Education and Adoration Association, "Eucharistic Miracle of Tixtla, Mexico, October 21, 2006", in *Eucharistic Miracles*, https://www.therealpresence.org /eucharst/mir/english_pdf/Tixtla1.pdf.

[15] See "A Matter of Faith, a Matter of Fact", TheDivineMercy.org, December 19, 2012, https://www.thedivinemercy.org/articles/matter-faith-matter-fact.

Note that there were two transformed Hosts in Buenos Aires—one in 1992 (which is not examined here) and one in 1996 (examined below in Section IV). Dr. Castañón Gómez was involved in the investigation of the Host in 1992, which initiated his journey to God and Roman Catholicism.

[16] I am most grateful for the thorough investigation of Dr. Gómez, who apparently paid for most of the scientific tests performed in multiple laboratories in Latin America and the United States. He presented his results in a book currently available only in Spanish, which was translated into English for the discussion of this chapter by English Coaching in Mexico (available through Fiber). There is a Kindle edition of the Spanish book, which is worth purchasing because of the photographs and multiple appendices on the different scientific tests performed on the Host. Ricardo Castañón Gómez, *Crónica de un Milagro Eucarístico: Esplendor en Tixtla Chilpancingo, México* (Mexico: Grupo Internacional Para La Paz [GIPLAP], 2014), Kindle, translated by English Coaching in Mexico for the author.

previously participated in the scientific investigation of the miraculous hosts found at the church in Buenos Aires (see below, Section IV). The scientific commission included several scientists, physicians, and laboratories, including an anatomical pathologist, two experts in surgical histopathology, experts in DNA biotechnology, forensic genetics, biochemistry, and pharmacy, an expert in legal and forensic medicine, and two computer-imaging experts.[17] The six major studies performed were all "blind"—the researchers did not know they were examining a consecrated Host.[18]

There are five aspects of the Tixtla Host that cannot be explained by natural causation according to current scientific knowledge (and are unlikely to be so in the future):

1. Real blood with cells originating from live tissue is being exuded from the consecrated Host.
2. The source of the blood is live cardiac tissue in the center of the Host.
3. The blood is being exuded from the inside out (with the liquid blood being inside and flowing toward the exterior, where it is coagulated).
4. The blood tissue and blood show signs of heart trauma induced by something like a blow to the chest.
5. There is a DNA conundrum—there exists human DNA material that (like the Buenos Aires Host—Section IV below) cannot be amplified by PCR (Polymerase Chain Reaction) for genetic profiling.

Each of these enigmatic features will be discussed in turn. We might state here the consensus of the scientific experts who studied the Tixtla Host, which will become more evident as we explain each factor:

Technical reliefs and scientific studies highlight the authenticity and reality of an event far from any natural cause that science can explain.... Science ensures the reliability of the results by keeping them away from any type of natural artifact or human manipulation....

[17] For a list of all the individual physicians, geneticists, and laboratories, see ibid., Chapter 8.
[18] The precise reports from each of the laboratories, as well as photographs of the Host and samples, are provided in ibid., Appendices I–XIX.

Therefore, the Bishop of Chilpancingo-Chilapa has pronounced [that the event is attributable] to a Supernatural causality.[19]

A. The Presence of Blood with Living Cells

The reddish substance exuding from the center of the Host is human blood with AB blood type.[20] The presence of hemoglobin was identified by using capillary immunochromatographic techniques.[21] Castañón Gómez, summarizing the conclusion of the scientific experts, noted:

> The histological examination in its microscopic description identified cellular material affected by autolysis (cellular deterioration). It also shows a regular amount of clear cytoplasm and small nuclei of fine chromatin displaced to the periphery, compatible with adipose [fat] cells.[22]

Castañón Gómez continues:

> Red blood cells (erythrocytes) and white blood cells (leukocytes) of different types have been identified. Acidophilic formed elements that resemble degenerated erythrocytes (red blood cells) can be seen, among them other basophilic elements (white blood cells) that appear to have lobed nuclei and others with rounded nuclei.[23]

From the above test results, it is clear that the blood coming from the Host contains not only *living* red blood cells, but more astonishingly, *living* white blood cells. It is well-established that when white blood cells are removed from living tissue systems, they will die (and disintegrate) within three minutes to one hour (maximum) of being outside of a living system.[24] In the investigation of Dr. Ricardo Castañón Gómez and his team of researchers, the Host and the blood

[19] Ibid., Chapter 8.
[20] See ibid., Chapter 8, Appendices III.A, III.B, and III.C.
[21] See ibid., Chapter 8, Appendix III.B.
[22] Ibid., Chapter 8, Appendix IV.A.
[23] Ibid., Appendices VI.a, VI.b, and VI.c.
[24] See ibid., Chapter 8.

coming from it were tested over several years, which would necessitate the death of all leukocytes (white blood cells) at the very least. As he notes:

> In my case, taking the evidence to different laboratories in different nations and continents, without the support of any form of preservation, I could never have imagined that the specialists, despite the years that elapsed since the first effusions (4–8 years), identified in the Host of Mexico "intact white blood cells".[25]

Even more fascinating was that one of the researchers examining the blood sample found macrophages (white blood cells) in the action of engulfing and digesting unhealthy elements in the blood. In his words:

> When the PATMED report talks about macrophage phagocytizing lipids, it shows us that these cells are intact and in full action. In other words, [the researcher] is expressing that, while studying the evidence that I had given him minutes before, through his microscope, the tissue remained active.[26]

These findings show not only that the white blood cells are alive, but the tissue with which they are interacting is also alive. But how can this be since the tissue is issuing from the center of a consecrated Host and is not connected to a living human body and circulatory system? Castañón Gómez and his fellow university professors and researchers were completely mystified by this result, and asked a proactive question pointing to supernatural "causation" in light of the absence of any known natural causation:

> Can a scientist explain the reasons that justify how this cellular dynamism remains active outside the body despite the years? All the experts consulted in biochemistry and hematology affirm that it is impossible to find intact white blood cells outside of their physiological environment, let alone in a piece of bread.[27]

[25] Ibid., Chapter 9.
[26] Ibid., Chapter 9.
[27] Ibid.

B. The Presence of Living Cardiac Tissue

As implied above, there is not only blood present in the Host, but also *living heart tissue* in the circular area out of which the blood is effusing. This was ascertained in three studies:

1. Fibrous-like structures, suggestive of muscle fibers, are present on the transformed Host and in the blood.[28] Unfortunately, autolysis (cell deterioration) does not allow positive identification with heart muscles.[29]

2. Doctors Rodas and Pernillo state: "In other areas, other elements are also seen that have elongated morphology, presenting bifurcations in the form of 'trouser sleeves' reminiscent of cardiac striated muscle fibers, however, no intercalary discs or striations were observed."[30]

3. In view of the ambiguity about whether the tissue was muscle or skeletal, and if muscle, whether it was cardiac muscle, Dr. Rodas recommended studies of molecular biology and immunohistochemistry. Dr. Sánchez Lazo affirms, "After immuno-histochemical tests, it has been determined that the tissue under study *corresponds to the heart* due to its macroscopic characteristics, in addition to showing the cytochemical results before referrals."[31]

In light of the above, it was concluded that the tissue connected to the Host is cardiac tissue, and that blood is effusing from the tissue in the interior of the Host to the exterior of the Host (see Section II.C).

When we combine this evidence with the fact that there are active red blood cells and, more importantly, active white blood cells in the blood, it is concluded that this cardiac tissue is in fact alive. This is confirmed by the discovery of macrophages phagocytizing (engulfing and digesting) lipids, showing "dynamic activity in the tissue".[32]

---

[28] Ibid., Appendix IV.
[29] Ibid.
[30] Ibid., Appendix IV.C.
[31] Ibid., Appendix IV (emphasis added).
[32] Ibid., Appendix XIV.A.

Thus, there is not only living heart tissue connected with the Host, but that tissue has been alive for over five years without any preservation or enhancement techniques employed. How can this be explained naturalistically? As Dr. Castañón Gómez indicates, it is naturalistically inexplicable on all levels:

> Let us remember that this evidence comes from an effusion of reddish liquid, which we already know is human blood with the presence of DNA, but it happened on October 22, 2006 and the PATMED doctors carried out the analysis in July 2011. Almost five years later, without biochemical support or any preservation with the specimen deposited in a dry and sterilized test tube, under the influence of chores and travel across many nations. [How is it possible that] a tissue with still internal dynamism, with intact physiological activity in one sector, [is functioning] when cell decomposition and widespread autolysis abound around it?[33]

From a naturalistic point of view, it is not possible. The mystery deepens even more when we consider that fresh liquid blood is exuding from the interior of the Host to the exterior.

C. The Interior to Exterior Flow of Blood

Two expert studies show that the Host is bleeding from the inside out. Though the blood first appeared in 2006, it continued to bleed until the date of testing in February 2010 and may continue to bleed until today.

The first test performed by Dr. Eduardo Sánchez Lazo (director of the Scientific Division of Legal and Forensic Medicine) concluded the following:

> Outflow of blood from [the Host's] interior to its periphery—that is, the blood comes from the interior to the exterior, similar to a large vessel that bleeds when it is lacerated in its anatomical structure.[34]

[33] Ibid., Chapter 9.
[34] Ibid., Appendix XIII.

The expert testimony accompanying Dr. Lazo's report states:

> It can be seen that the center of the Host has a semicircular shape (the place where the sample under study was taken), and it is elongated on the right side, and shows a dispersion throughout its entire structure, with a centric zone that seems to go through the entire contour or thickness of its structure. In this way, it is understood that the blood comes from its interior, managing to be absorbed by its thickness on its surface. In other words, it can be seen that the structure of the Host has the same characteristics because it is a single piece, having a single infiltration in its center throughout its elongated extension as if it were a vessel that bleeds to the outside. For this reason it is inferred, in a very high degree of probability, that the analyzed sample ... has an inside-out direction taking into consideration the thickness, consistency, and macroscopic and microscopic characteristics observed.[35]

A second study using various kinds of bright light and microscopic enhancements confirms the conclusion rendered by Dr. Sánchez Lazo, that the blood originates from the interior. Dr. Castañón Gómez explains the studies as follows:

> The studies carried out by means of microscopic vision, allow a digital magnification through illumination systems with ultra-bright white light, ultraviolet and fluorescence. This technology has allowed us, after verifying that the *upper part of the blood* [on the surface of the Host] is *coagulated*. However, the enlargement [in the center] reveals the presence of *fresh blood*. The result is significant because the original effusion is from October 2006, while the study was conducted in February 2010.[36]

The fact that the blood in the interior of the center enlargement is fresh (as shown by the microscopic analysis under various kinds of light) and the blood on the surface of the Host around the interior enlargement (from October 2006) is coagulated shows that the blood not only is moving from the interior to the exterior, but at the time of testing (February 2010) was continuing to exude fresh blood (presumably from the heart tissue in the interior of the Host).

[35] Ibid., Chapter 8, citing the expert study, p. 10.
[36] Ibid., Chapter 9 and Appendix XV.

The dynamics of the blood flow (from the first expert study) show that the blood could not have been placed in the interior of the Host from the exterior. If the blood had only been placed in the interior from the exterior, there would be no consistent dynamic flow in the opposite direction over time.[37] How can a consecrated Host give rise to a continuous flow of blood from its interior to its exterior over the course of four years—and perhaps longer? This is yet another facet of the Tixtla Host that appears to have no known natural explanation.

D. Trauma to the Heart

We have already seen that there are macrophages phagocytizing lipids in the blood of the Tixtla Host. This action normally occurs when these white blood cells are in the process of repairing inflamed tissue.[38] We have also seen that the tissue in question is cardiac tissue (see above, Section II.B). In addition to macrophages (in the process of repairing tissue), there are mesenchymal stem cells (pluripotent stem cells that help to repair damaged tissue) present in the blood of the Tixtla Host. These pluripotent stems cells are produced by bone marrow and are discharged into the blood stream when reparative action is needed.[39] This gives rise to two questions:

1. How are these reparative blood cells being produced without bone marrow in a living circulatory system?
2. Given that they are discharged for the purposes of tissue repair and the tissue in question is from the heart, why are they repairing damaged heart tissue?

The first question was answered in Section II.A above—there is no known natural way of producing functioning macrophages (white blood cells) without bone marrow, and no way that white blood cells can live more than one hour outside of a living circulatory

[37] Ibid., Chapter 9 and Appendix XIII.
[38] Ibid., Chapter 8.
[39] Xiaorong Fu et al., "Mesenchymal Stem Cell Migration and Tissue Repair", *Cells* 8, no. 8 (2019), https://www.ncbi.nlm.nih.gov/pmc/articles/PMC6721499/.

system. The same holds true for mesenchymal cells—they are naturalistically inexplicable without bone marrow and a living circulatory system.

The second question is answered by Dr. Marco Blanquicett Anaya (internist and cardiologist) as follows:

> The repair action observed in the Host of Chilpancingo [Tixtla], through the macrophage in question, could very well be located in cardiac tissue, since, according to their investigations, it is the place where most macrophages phagocytize lipids to avoid the formation of atherosclerotic plaques, which in a young subject, rather than being formed by a bad diet, could derive from situations of extreme traumatic stress, blows to the chest, falls and the like.[40]

Inasmuch as Jesus was a young subject in excellent physical shape without a rich (bad) diet, the macrophages may well be phagocytizing lipids produced by severe trauma to the heart and blows to the chest. This indicates that there is not only living cardiac tissue as well as living white blood cells and mesenchymal stem cells in the blood of the Host, but also likely trauma or damage to the heart of the person who produced this blood.

Interestingly, this corresponds to the findings of the cardiac tissue growing out of the Host in Sokółka, Poland (see below, Section III), in which segmentation and fragmentation (manifesting damage to the heart tissue) were found. This also corresponds to the blood on the Shroud of Turin, where elevated levels of bilirubin as well as the synthesis of ferritin and creatinine (manifesting a polytrauma) were found.

E. *The DNA Conundrum*

In the blood of the Tixtla Host, there is evident genetic material. However, researchers were unable to identify polymer chains to amplify the markers through a process called Polymerase Chain Reaction (PCR).[41] According to Dr. Sánchez Lazo:

---

[40] Gómez, *Crónica de un Milagro Eucarístico*, Chapter 9.
[41] Ibid., Chapter 8 and Appendix XI.

In relation to the genetic studies, it was not possible to extract polymer chains from the sample under study even though an attempt was made to amplify the markers, with the result that only genetic material has been identified without the possibility that it could encode a protein.[42]

There are three possible reasons why the genetic material would not yield a genetic sequence/profile:

1. An insufficient amount of DNA from the sample—or an insufficient sample.
2. The DNA was degraded through autolysis.
3. There are no polymer chains with identifiable markers in the genetic material in the blood of the Tixtla Host.

Since the third option is nonnatural, we will investigate the first two more natural causes before examining the nonnatural one.

First, with respect to the quantity of DNA material, there was sufficient material to carry out multiple tests. Dr. Castañón Gómez noted in this regard:

Despite obtaining adequate samples with the presence of DNA, it was not possible to obtain the corresponding profile. This result is a constant in our research from 1995 to the present.[43]

In light of the above, it is unlikely that insufficient DNA or samples was the cause for the absence of a profile. So, how about degradation of the DNA material in the blood? As Castañón Gómez states, "The autolysis [destruction of cells or tissue] artifact is patent";[44] so there is evident degradation taking place in the blood cells and tissues in the Tixtla Host. However, degradation is unlikely to be the whole explanation for the absence of a DNA profile in the Tixtla Host. As Castañón Gómez asks:

If [researchers] find intact white blood cells, macrophages phagocytizing lipids, and tissues, how deteriorated is the sample?[45]

[42] Ibid., Appendix VII, Section VI.
[43] Ibid., Chapter 8.
[44] Ibid., Chapter 8.
[45] Ibid., Chapter 9.

How could the DNA samples be completely degraded when there are nondegraded tissues as well as live blood cells (some of which are in the process of treating inflamed tissue)? This question provokes another—in other consecrated hosts with living tissue and blood has there ever been an amplifiable polymer chain from which to obtain a DNA profile? Castañón Gómez had personal access to the blood of another Eucharistic Host (the Host of Buenos Aires in 1996—see below, Section IV) that he had analyzed by two other genetic laboratories: Trinity DNA Solutions in Florida and Beta Genetics DNA Laboratory. Interestingly, they obtained the same result, procuring DNA material but no DNA profile (from polymer chains).[46]

Dr. Angelo Fiori (Department of Legal Medicine, Gemelli Hospital in Rome, Italy) studied blood from the Eucharistic Host in Argentina as well as the image of Christ of Cochabamba. He wrote on June 24, 1996:

> Surprisingly, the new DNA analysis was completely negative again, for example, the PCR amplification could not be obtained, even though the sample to be analyzed was abundant.... I have no explanation for this unusual phenomenon.[47]

Furthermore, in the case of the Tixtla Host, there was almost no delay between procuring the DNA material from the blood and testing for a DNA profile. As Dr. Castañón Gómez notes:

> When the laboratory studies arrived at the moment in which they rescued the presence of DNA from our sample, ... they informed me that in the next step known as PCR amplification, they would obtain the corresponding profile. I was present in the laboratories at the moments in which the procedure was being carried out. However, after the time elapsed for the operation, the response was always unique and repetitive: "We can't get the sequence, we can't get the profile."[48]

Unfortunately, the DNA material in the blood on the Shroud of Turin appears to be so degraded (after seven hundred to two thousand years) that it does not allow for profiling.

---

[46] Ibid., Chapter 9.
[47] Ibid., Appendix XI.
[48] Ibid., Chapter 9.

So where does this leave us? The DNA material from the blood of two Eucharistic hosts from Argentina, the Eucharistic Host of Tixtla, and the blood from the image of Christ of Cochabamba all give completely negative profiles with adequate test samples. In the case of the Tixtla Host and the Buenos Aires Host (1996), there were living white blood cells in the samples (which ordinarily die within one hour of being outside a living circulatory system). This indicates the distinct possibility that the DNA materials were not completely degraded, and that there could be at least a few small polymer chains allowing for amplification. By way of summary, Dr. Castañón Gómez asks the same question:

> This is one of the great challenges for our research work: Why is it that despite having specimens in good condition, the genetic profile cannot be obtained despite having obtained DNA?[49]

Though we cannot be certain that the absence of a DNA profile is attributable to either degradation or the absence of a profile in these transformed consecrated hosts, we are left with the distinct possibility that a DNA profile is absent in every example of tissue or blood associated with Jesus Christ in the Eucharist (and in the image of the face of Christ of Cochabamba). If this finding is confirmed with future Hosts, we might ask the further question of why the Lord has kept His DNA profile hidden.

F. Conclusion

The transformed consecrated Host of Tixtla appears to be naturalistically inexplicable in three ways:

1. Blood with living cells is issuing from the Host—some of which are white blood cells, which die within one hour of being removed from a living circulatory system. These white blood cells are engaged in healing activities—phagocytizing lipids (engulfing and digesting harmful fat cells). Inasmuch as the bleeding from the Host began in 2006 and the activity

---

[49] Ibid., Chapter 9.

of the white blood cells was taking place during testing in 2010, we must ask how this is naturalistically explicable when the blood of the Host is not connected to a living circulatory system.

2. There is living cardiac tissue in the center of the Host from which the blood is issuing, provoking the question of how living cardiac tissue is integrated with the substance of a consecrated Host, which is molecularly and structurally distinct from that Host. There appears to be no naturalistic explanation for this.

3. The blood is moving from the inside center of the Host to the outside (surface) periphery of the Host, implying that new blood is being created in the center of the Host pushing outward (excluding the possibility that it was inserted into the Host from the outside). How can a Eucharistic Host (even with heart tissue) produce new blood on an ongoing basis (between 2006 and 2010—perhaps longer)? This appears to have no naturalistic explanation.

In addition to these three nonnatural features, we have two additional features associated with the presence of Jesus Himself. First, the macrophages (white blood cells), which are in the process of phagocytizing lipids, are engaged in an action generally associated with the healing of heart tissue (e.g., after a heart attack or blow to the chest). Thus, it seems that we are dealing not only with living heart tissue but living and *wounded* heart tissue. Is it merely a coincidence that the Host is, according to Catholic doctrine, the Real Presence of Jesus' crucified (and risen) body and blood on the Cross? Second, we have the absence of a DNA profile regardless of the fact that there is living tissue and living blood cells in the Host. This is also the case with other Hosts and examples of the purported blood of Jesus. If this phenomenon is confirmed in other examples of future transformed consecrated hosts, it would be not only distinct from any other known human phenomena, but perhaps a sign that there is more than humanity in the flesh and blood of these Hosts and the person with whom they are identified—Jesus Christ. We conclude by again noting the consensus of scientific experts who have examined the Host:

Technical reliefs and scientific studies highlight the authenticity and reality of an event far from any natural cause that science can explain.... Science ensures the reliability of the results by keeping them away from any type of natural artifact or human manipulation.... Therefore, the Bishop of Chilpancingo-Chilapa has pronounced [that the event is attributable] to a Supernatural causality.[50]

III. The Eucharistic Host of Sokółka, Poland (2008)

On Sunday, October 12, 2008, during a Mass at St. Anthony parish church in Sokółka, Poland, a Host was apparently dropped by a vicar, Father Jacek Ingielewicz, who was distributing Communion. He picked up the Host, and after examining it, decided it was too dirty to consume, so he placed it into a glass of water so it would dissolve in a few days.[51] After Mass, the sacristan, Sister Julia Dubowska, placed the Host and container of water in a safe known only to her and the pastor. One week later, the pastor asked her to check the state of the Host in the safe; when she opened it, she smelled the aroma of bread, and then noticed that in the middle of the Host was a curved bright red stain. The water in the container was unaltered by the transformation of the Host. She told the pastor, who then showed the altered Host to two other priests who decided to maintain discretion. The pastor then informed the metropolitan archbishop who also witnessed the transformed Host along with chancery officials.[52]

On November 30, the archbishop asked that the Host be removed from the container of water and placed on a corporal and put in a separate tabernacle in the rectory. In mid-January of 2009, the Host had completely dried. The Host was left in that tabernacle for two and a half years, and was then transferred back to the church in October 2011.

The archbishop then requested that histopathological studies be done. On March 30, the archbishop created an ecclesial commission to study the phenomenon. A piece of the altered Host was taken

[50] Ibid., Chapter 8.
[51] Adam Białous, *Hostia: Cud eucharystyczny w Sokółce*, trans. Jakub Juszczyk (Częstochowa: Edycja Świętego Pawła, 2015), pp. 45–54.
[52] Ibid.

and analyzed independently by two experts, Professor Maria Elżbieta Sobaniec-Łotowska and Professor Stanisław Sulkowski. Independence was maintained in order to ensure the credibility of the results.
Both are histopathologists at the Medical University of Bialystok.
The studies were carried out at the university's Department of Path-
omorphology. The specialists' work was governed by the scientific
norms and obligations for analyzing any scientific problem in accor-
dance with the directives of the Scientific Ethics Committee of the
Polish Academy of Sciences.[53]

There are four dimensions of the histopathological examination
that cannot be naturalistically explained:

1. The Host did not dissolve in water after forty-eight days.
2. The substance of the Host and the substance of the heart tissue
   are so closely intermingled as to be virtually inseparable, and
   unproducible by current technologies.
3. The tissue part of the Sokółka Host is cardiac tissue.
4. The cardiac tissue is that of a dying man—still living.

With respect to the first observation, the Host's resistance to solubility, Dr. Stanisław Sulkowski noted:

> If we submerge the Host in water, normally, it should naturally dis-
> solve in a short period of time. Yet in this case, the part of the Host,
> for reasons beyond our understanding, did not disintegrate [after
> forty-eight days—from October 12, 2009, to November 30, 2009].[54]

Additionally, the tissue (integrated with the Host) also did not suffer
degradation, which is highly unusual.

With respect to the second observation, the inseparability and
intermingling of the substance of the Host and substance of the tissue,
Dr. Stanisław Sulkowski said:

> Even more inconceivably, the tissue that appeared within the Eucha-
> rist was connected to it inextricably—penetrating even the very
> base on which it came into existence. Believe me, even if someone

53 Ibid.
54 Ibid.

attempted a manipulation, he or she would not be able to connect these two materials as inseparably.[55]

The stained part of the Host was found to be a piece of cardiac tissue, which was completely integrated into the bread part of the Host. The completely seamless integration of the tissue part with the bread part is so refined that it appears that the tissue is growing out of the bread part of the Host. During the analysis with an electron microscope, the outlines of the communicating junctions and the thin filaments of the myofibrils were visible, showing that the cardiac tissue was joined to the consecrated Host in an inseparable manner.[56] Indeed, the analysis showed that the integration of the two substances could not be produced by any known human technology. According to Dr. Sobaniec-Łotowska:

> The Host was connected to heart muscle tissues, as observed both by a light microscope and a transmission electron microscope. To me, this extraordinary fact is evidence that no human interference took place.[57]

She concluded, "Even NASA scientists, who have at their disposal the most modern analytical techniques, would not be able to artificially recreate such a thing."[58]

With respect to the third observation, the presence of human heart tissue, Dr. Sobaniec-Łotowska reported:

> We want to emphasize that the piece of the Host that we tested, was such a small sample, we could only observe as many features indicating heart tissue muscle, in terms of morphologic indicators. One such indicator is the phenomenon of segmentation, which is the damage of muscle fibers due to inserts [structures typical of heart

---

[55] Ibid., pp. 48–49.

[56] Real Presence Eucharistic Education and Adoration Association, "Eucharistic Miracle of Sokółka Poland, October 12, 2008", in *Eucharistic Miracles*, http://www.therealpresence.org /eucharst/mir/english_pdf/Sokółka2.pdf.

[57] Maria Sobaniec-Łotkowska, interview by Białous, *Hostia*, p. 49 (emphasis mine).

[58] Quoted in Jeanette Williams, "The Amazing Science of Recent Eucharistic Miracles: A Message from Heaven?", *Ascension Press*, November 3, 2021, https://media.ascensionpress .com/2021/11/03/the-amazing-science-of-recent-eucharistic-miracles-a-message-from -heaven-%EF%BB%BF/.

muscles], and fragmentation. These damages look like minor cracks. Such changes take place only in *non-necrosis fibers* and reflect quick heart muscle contractions *prior to one's death*.... Another important evidence that the material is human heart muscle was the central formation of cellular nuclei in the fibers observed, which is typical for that muscle. Within some fibers, we also found what could look like contraction nodes. Under the electron microscope, outlines of inserts and bunches of delicate myofibrils were visible. In conclusion ... the material sent for evaluation, in the opinion of two independent pathologists, is positive; it indicates heart muscle tissue, or at least, resembles it the most out of all living organism tissues. And, what we believe is vital, the analyzed material characterized that tissue entirely.[59]

As Drs. Sobaniec-Łotowska and Sulkowski independently showed, there are five main indicators of heart tissue: (1) segmentation and (2) fragmentation, which look like minor cracks coming from quick heart muscle contractions; (3) the central formation of cellular nuclei in the tissue fibers (typical of cardiac tissue); (4) the apparent presence of contraction nodes (typical of cardiac tissue) within some tissue fibers; and (5) outlines of delicate myofibrils (typical of cardiac tissue). This enabled Sobaniec-Łotowska and Sulkowski to conclude independently that the tissue is either cardiac tissue or resembles heart tissue more than any other human tissue.

With respect to the fourth observation, the tissue belongs to a person who is alive but is in the process of dying. The analysis of the doctors asserts that the observed segmentation and fragmentation is of a kind that occurs only in living ("non-necrosis") fibers and is typical of rapid contractions that occur prior to death. Evidently, it is remarkable that heart tissue should be growing out of the substance of a consecrated Host, but it is even more remarkable that it is *living* heart tissue undergoing the process of dying. How does one fake living tissue growing out of a consecrated Host?

The findings of Sobaniec-Łotowska and Sulkowski were challenged by Dr. Paweł Grzesiowski (from the Department of Infection Prevention of the National Medicines Institute), who asserted that the red stain on the Host was made by a bacterium that creates a

[59] Białous, *Hostia*, p. 49.

similar pigment. The fact is that Dr. Grzesiowski never observed or tested the Host. It was a completely unsubstantiated conjecture. Dr. Sulkowski responded:

> There indeed exist such bacteria [that produces red pigment] and it is no discovery. However, the doctor's confidence that the bacteria was alien to the material that he has not seen is puzzling. We are deeply disturbed by the fact that such unverified theories are voiced by a representative of an institute that also employs many renowned, prominent scientists, and was founded to diligently inform the public about existing threats and counter them.[60]

Dr. Sobaniec–Lotkowska added:

> I stress that no bacteria known to science can create a tissue with heart muscle characteristics which we found in the Host.[61]

What might we conclude? There are two dimensions of the Sokółka Host that are inexplicable by known natural explanation. First, the heart tissue is inseparable from, inextricably intermingled with, and appears to be growing out of the substance of the Host. Currently, there is no way of replicating this by any known technological means, which seems to put fakery out of the question. Second, the heart tissue at the time of testing was alive and in the process of dying. It is quite beyond known technology to create an inextricable unity between the substance of the Host and *living* tissue in which organic processes are taking place. It is not reasonable to believe that such a Host could be faked, and the reputation and credentials of the histopathologists are unquestionable.

For this reason, Archbishop Edward Ozorowski convened a special committee to confirm the medical-scientific findings as well as the chain of individuals who had custody of the Host from October 8, 2008 (the origin of the Host), to October 14, 2009 (the beginning of the investigation). They concluded that "no signs of hoax were detected, and no interference of outside parties was noted."[62]

[60] Ibid., p. 51.
[61] Ibid.
[62] Ibid., p. 52.

IV. The Eucharistic Host of Buenos Aires (1996)

On August 18, 1996, during Mass at the church parish of Santa María y Caballito Almagro in Buenos Aires, Argentina, a consecrated Host was discarded during the distribution of Communion.[63] It was given to the pastor, Father Alejandro Pezet, who was unable to consume the Host; therefore he placed it in a receptacle of water so that it would dissolve in the fashion prescribed by the Catholic Church. The receptacle was then placed in the tabernacle of the church.

On August 26, Fernandez checked the tabernacle and found that instead of the Host dissolving, it had transformed into what appeared to be a piece of flesh.[64] Father Pezet informed Archbishop Bergoglio of the occurrence, who then asked Father Pezet to have the Host professionally photographed. This occurred on September 6, 1996.[65] It was decided to keep the Host in the tabernacle without publicizing it or its origin. In 1999 (three years later), the tissue had still not decomposed, and so Archbishop Bergoglio invited Dr. Castañón Gómez to undertake the investigation of the Host.[66] On October 5, 1999, Dr. Castañón Gómez went to Buenos Aires to interview the five priests who had been associated with the transformed Host.[67]

On October 21, 1999, Gómez went to the Forensic Analytic genetics laboratory in San Francisco, and as noted above (Section II.E), the researchers there determined that the tissue had some genetic material but they were unable to obtain a profile.[68] In 2001, he took his samples to Dr. Edoardo Linoli at Gemelli Hospital in Rome. Linoli determined that the tissue corresponded most closely

---

[63] Mieczysław Piotrowski, "Eucharistic Miracle in Buenos Aires", *Love One Another!* 17 (2010), https://www.truechristianity.info/en/articles/article_en_0414.php. For a summary of this article, see "Eucharistic Miracle Beheld by Pope Francis?", *Aleteia*, April 22, 2016, https://aleteia.org/2016/04/22/eucharistic-miracle-beheld-by-pope-francis/.

[64] Ibid. See Ary Waldir Ramos Diaz, "The Future Pope Francis Was in Charge of Dealing with This Reported Eucharistic Miracle", *Aleteia*, June 13, 2020, https://aleteia.org/2020/06/13/the-future-pope-francis-was-in-charge-of-dealing-with-this-reported-eucharistic-miracle/.

[65] Piotrowski, "Eucharistic Miracle in Buenos Aires".

[66] Ibid.

[67] Real Presence Eucharistic Education and Adoration Association, "Eucharistic Miracles of Buenos Aires, Argentina, 1992–1994–1996", in *Eucharistic Miracles*, http://www.therealpresence.org/eucharst/mir/english_pdf/BuenosAires1.pdf.

[68] Gómez, *Cronica de un Milagro Eucaristico*, Chapter 9.

to heart tissue and that it contained intact white blood cells (indicating that the tissue was alive at the time of testing—see below).[69] In 2002, Castañón Gómez sent the sample to Professor John Walker (University of Sydney Australia), who determined the presence of muscle cells and intact white blood cells in the tissue.[70] In 2003, Castañón Gómez went to Dr. Robert Lawrence (in San Francisco), who indicated that the tissue "could correspond to the tissue of an inflamed heart".[71]

On March 2, 2004, Castañón Gómez went to Dr. Frederick Zugibe, one of the world's foremost pathologists and experts in diagnostic histochemistry at Columbia University, New York. He did not tell Zugibe that the tissue sample had come from a consecrated Host.[72] Zugibe informed him that the tissue had come from the left ventricle of the myocardium and the presence of intact white blood cells indicated that the tissue was alive when it was presented to him.[73] In his book on the Tixtla Host, Castañón Gómez noted:

> [Dr. Zugibe said,] "The moment the sample was taken, the tissue was alive.... We have found intact white blood cells, and these can only lead to this place (the myocardium—heart muscle), only by blood circulation." In other words, like the Host of Mexico, the tissue was still active, as if it were found in its own natural biological organism.[74]

Later, on April 20, 2004, Dr. Zugibe reported the following to Mike Willesee (a journalist) and Ron Tesoriero (an attorney) in a letter:

> The slides consists of cardiac (heart) tissue that displays degenerative changes of the myocardial tissue (cardiac muscular tissue) with loss of striations, nuclear pyknosis, aggregates of mixed inflammatory cells consisting of chronic inflammatory cells (macrophages) which are predominant and smaller numbers of acute inflammatory cells

[69] Real Presence Eucharistic Education and Adoration Association, "Eucharistic Miracles of Buenos Aires".

[70] Ibid.

[71] Ibid.

[72] Ibid.

[73] Ibid.

[74] Gómez, *Crónica de un Milagro Eucarístico*, Appendix XVII.

(white blood cells primarily polymorphonuclear leukocytes). The directionality of the myocardial fibers indicates that the site of these changes is relatively close to a valvular region in the ventricular area of the heart.... When I was later told that the heart tissue was kept in tap water for about a month and transferred to sterile, distilled water for three years, I indicated that it would be impossible to see white blood cells or macrophages in the sample. Moreover, it would be impossible to identify the tissue *per se* as there would be no morphological characteristics.[75]

In addition to the above comments about living white blood cells, living heart tissue, and absence of decomposition, Dr. Zugibe disclosed to Dr. Castañón Gómez that the person from whom this tissue was taken suffered greatly. In response to Castañón Gómez's question about why his patient had suffered a lot, Zugibe replied:

Because your patient has some thrombi, at certain moments he could not breathe, oxygen did not reach him, he labored and suffered much because every breath was painful. Probably they gave him a blow at the level of the chest.[76]

This feature of suffering pain closely resembles the Host of Tixtla (see above, Section II.D) and Sokółka (see above, Section III). When Dr. Zugibe was later told that the tissue had come from a consecrated Host, he did not believe it. But upon subsequent reflection, he was greatly moved.[77]

So what might we conclude from the scientific investigation of the transformed Host of Buenos Aires? The chain of custody was reasonably secure (in the hands of Church officials, physicians, and scientific experts), but the scientific investigation was not as systematic as the investigation of the Tixtla Host and the Sokółka Host. Dr. Castañón Gómez initiated the investigation in 1999 (after the transformed Host had been in water for three years). He was able to obtain expert scientific opinions from two genetics laboratories, and several experts in

[75] Frederick T. Zugibe, "Testimonial Given in an Interview with Mike Willesee and Ron Tesoriero on March 15, 2005", in e-mail report; available upon request.

[76] Quoted in Real Presence Eucharistic Education and Adoration Association, "Eucharistic Miracles of Buenos Aires".

[77] Ibid.

histopathology, including the well-known diagnostic histopathologist Frederick Zugibe. Four important results were obtained.

First, the host/tissue did not decompose after three years of being immersed in water, for which there is no natural explanation (see Zugibe's testimony above). Second, the tissue was identified by Drs. Linoli and Zugibe as being heart tissue from the myocardium (left ventricle) near the valve area. This is further corroborated by Dr. Walker, who identified muscular cells in the blood, and Dr. Lawrence, who indicated that the tissue could be inflamed heart tissue.

Third, according to Drs. Linoli, Walker, and Zugibe, there are intact living white blood cells in the heart tissue, indicating that the heart tissue *was alive* at the time of testing. Thus, the Host was transformed into *living* heart tissue, for which there is no known natural explanation.

It may be objected that someone could have fraudulently substituted a piece of heart tissue for the Host between August 18 and September 5, but it would have been exceedingly difficult to make such a fraudulent substitution after September 6, when the transformed Host was professionally photographed. This objection cannot be raised against the Sokółka Host, because the substance of the Host is completely intermingled with the substance of the heart tissue as if it were caught in the process of transmutation. Nevertheless, the Buenos Aires Host did not change over the three years it was immersed in water (since the time the photographs were taken), and the samples used for testing were taken from this tissue. Inasmuch as the samples tested by Drs. Walker, Linoli, and Zugibe originated on or before September 6, 1996, and there were intact white blood cells, indicating that the heart tissue was alive at the time of testing (between 2001 to 2004), it must be concluded that this tissue is beyond natural explanation; for how can heart tissue remain alive separate from a human body for eight years? Barring the possibility that the Church and the examining doctors were killing multiple victims to obtain heart tissue samples with intact white blood cells, it seems indisputable that the living heart tissue that was purported to be transmutated out of the consecrated Host is apparently supernaturally caused.

Fourth, how do we explain the results of the genetic investigation? Recall that genetic material was discovered without an amplifiable sequence or profile. As noted in Section II.E above, this unusual occurrence was also found to be the case with the Tixtla Host and the blood from the Christ image of Cochabamba. Although all three instances of the absence of genetic profile might be explained by degradation of the tissues, the existence of intact white blood cells partially mitigates this conclusion (particularly in the case of Tixtla, where the white blood cells are performing actions to heal inflamed heart tissue). As Castañón Gómez asks, with white blood cells in action, how degraded could the samples be?[78] Given that there are two indications of supernatural causation in the Buenos Aires Host (an absence of macroscopic decomposition and remaining alive for eight years despite separation from a living human body), there is the possibility that the absence of a DNA profile may be attributable to this supernatural cause. Perhaps the supernatural cause of the transmutation of the consecrated Host into living heart tissue did not want a profile to be encoded in the genetic material. There will no doubt be other instances of transmutated consecrated hosts in the future, which will provide the opportunity to confirm the absence of DNA profile with very rapid genetic testing prior to the onset of significant autolysis.

V. Conclusion

There are five common features among the three transformed consecrated hosts (Tixtla, Sokółka, and Buenos Aires):

1. The absence of notable macroscopic decomposition in the tissue (though there is autolysis in the cells and fine tissue areas)
2. The presence of living white blood cells in human blood
3. The presence of living cardiac tissue growing out of (or having grown out of) the substance of a consecrated Host, which is molecularly distinct from it
4. Indications of thrombi (Buenos Aires), fragmentation, segmentation (Sokółka), and macrophages phagocytizing lipids (Tixtla), all of which indicate distress to the heart—perhaps from trouble breathing, a blow to the heart, or a heart attack

[78] Ibid.

5. The presence of genetic material without a DNA profile on amplifiable polymer chains

Ideally, improvements could have been made to the scientific investigation of all three transformed consecrated hosts—particularly that of Buenos Aires. Nevertheless, the protocols used for the Tixtla Host and the Sokółka Host were sufficient to establish the above five conclusions, and in the case of the Buenos Aires Host, to point strongly to those conclusions (given the trustworthiness of the clergy and scientists involved in the chain of custody). All three Hosts are naturalistically inexplicable (pointing to supernatural causation) in several areas:

- The absence of macroscopic decomposition after several years
- Living white blood cells independent of a human body after several years
- Living heart tissue independent of a human body after several years
- The transmutation of the substance of a Host into the substance of living heart tissue (which is explicit in the inseparable intermingling of both substances in the Sokółka Host)
- The continuous production and issuing of fresh blood from the Tixtla Host

The absence of a genetic profile might be thought to suggest unreliability of the evidence, but the validity of the histopathological examination and the electron microscope examination stand independently of the genetic testing (and the absence of profiling). Thus, the genetic profiling is not integral to the five conclusions concerning the natural inexplicability of the three transformed Hosts; it might be argued that the humanity of the tissue is questionable in view of the absence of a genetic profile. However, this can be established by other means, such as the biochemical makeup of the blood,[79] the presence of human hemoglobin,[80] and "the central formation of cellular nuclei in the [heart] fibers observed".[81] In view of the presence

---

[79] Gómez, *Crónica de un Milagro Eucarístico*, Appendix III.A.
[80] Ibid., Appendix III.B.
[81] Białous, *Hostia*, pp. 51–52.

of genetic material without a genetic profile in *all three* transformed Hosts as well as that of the image of Cochabamba, Castañón Gómez indicated that it became a control factor:

> In any case, for us, and given the constancy of this variable, the non-result [of a genetic profile] has become a control variable.[82]

As noted above, this consistency of the absence of a profile must be established by the testing of future transformed consecrated hosts using the quickest possible means to bring a sample with minimum autolysis to a DNA laboratory. If the absence of profile is again manifest in one or two "rapid tests", we might interpret the finding to be a strengthening rather than a weakening of the evidence for transmutation of the Host; for in light of the likely supernatural causation of these Hosts, and the constancy of an absence of a genetic profile, we might reasonably infer that this non-result is the intention of the supernatural causative agent.

Finally, with respect to the evidence for heart damage and trauma in all three Hosts, it is important to note that these Hosts point not only to real living flesh and real living blood, but also to the flesh and blood of a person who has undergone considerable trauma and pain. It is not a far stretch to associate this pain with the Crucifixion of the historical Jesus writ large in the Gospel accounts and the Shroud of Turin. When we consider that the blood of the Shroud has AB blood type (the same as all three Hosts) and that the image seems to be naturalistically inexplicable (because of the need for a blast of radiation—ultraviolet or particle radiation—throughout the whole body), the parallels point to the torture and excruciating pain of a common victim—Jesus Christ.

---

[82] Gómez, *Crónica de un Milagro Eucarístico*, Chapter 8.

Chapter Five

Science at the Doorstep to Mary— Guadalupe and Fatima

Introduction

The Church is careful about approving Marian apparitions as valid because a validation that is subsequently falsified would undermine the Church's credibility. The Church's long-standing positive criteria (administered in 1978 by the Sacred Congregation for Propagation of the Doctrine of the Faith, currently, the Dicastery for the Doctrine of the Faith) are as follows:

1. There must be moral certainty, or at least great probability, that something miraculous has occurred, something that cannot be explained by natural causes or by deliberate fakery, as a result of a serious investigation.
2. The person or persons who claim to have had the private revelation must be mentally sound, honest, sincere, of upright conduct, and obedient to ecclesiastical authority.
3. The content of the revelation or message must be theologically acceptable, morally sound, and free of error.
4. The apparition must yield positive and continuing spiritual assets—for example, prayer, conversion, and increase of charity.[1]

[1] Sacred Congregation for the Doctrine of the Faith, *Norms regarding the Manner of Proceeding in the Discernment of Presumed Apparitions or Revelations* (February 25, 1978), I, https://www.vatican.va/roman_curia/congregations/cfaith/documents/rc_con_cfaith_doc_19780225_norme-apparizioni_en.html.

Over the last five centuries, there have been nine Marian apparitions approved by the Church. We will discuss three of them that have undergone particular historical and scientific scrutiny:

1. The apparition of Our Lady of Guadalupe (Section I)
2. The apparition of Our Lady of Fatima (Section III)
3. The apparition of Our Lady of Lourdes (Chapter 6)

I. The Image on the Tilma of Guadalupe

According to several well-attested accounts, the Blessed Virgin Mary appeared to a native Aztec, Juan Diego, on December 9, 1531. She asked him to ask his bishop—Juan de Zumárraga—to build a church atop Tepeyac Hill (now within the confines of Mexico City). Juan Diego did as he was instructed, but after relating his story to Zumárraga, the bishop did not believe him. The Blessed Virgin appeared again to Juan Diego that same day (December 9) and asked him to return to the bishop. On December 10, Juan Diego returned to Zumárraga, but he still had doubts, and asked Juan Diego to return to the hill and ask the Virgin for a miraculous sign. He did as he was instructed, and the Lady promised a sign the next day (December 11). However, before Juan Diego could return to the hill on December 11, his uncle Juan Bernardino became quite ill and Juan Diego stayed with him to find medical assistance and a priest. On December 12, when Juan Diego left his uncle to find a priest, the Virgin met him on the road and assured him that his uncle would be cured and told him to proceed to the hill, where he would find the sign required by Bishop Zumárraga. He went to Tepeyac Hill and found Castilian roses growing there (not native to Mexico); he gathered them and put them in his tilma. When he returned with the roses to Bishop Zumárraga and opened his cloak to allow the roses to fall, the picture of the Lady of Guadalupe appeared on the tilma. Apparently, the roses and the image were sufficient to convince Bishop Zumárraga to build the first church (and sanctuary for the image) atop of Tepeyac Hill.

Some scholars have challenged the veracity of this story because it was not found either in the writings of Bishop Zumárraga or in an ecclesiastical report about the image. However, in 1995, Jesuit historian Xavier Escalada discovered a hitherto unknown sheet of parchment made of tanned animal skin, which depicts an illustration of the Lady of Guadalupe and Juan Diego on Tepeyac Hill, as well as the dates 1531 and 1548.[2] Escalada then published a four-volume encyclopedia on the image and history of Our Lady of Guadalupe, in which he reports his analysis of the parchment, referring to it as Codex 1548, which later came to be called "Codex Escalada".[3] The Codex 1548 was signed by Bernardino de Sahagún (Franciscan friar, ethnologist, and the major translator of Nahuatl—Juan Diego's native language).[4] Sahagún's signature was authenticated by Banco de Mexico and Charles E. Dibble.[5] Additionally, the Institute of Physics of the National Autonomous University of Mexico determined that the Codex 1548 originated in the sixteenth century.[6] The authentication of the signature—along with the parchment's dating, illustrations, language, and style—validates the parchment as well as the existence and vision of Juan Diego. On the codex there is a glyph of Antonio Valeriano, who was a governor as well as professor at the College of the Holy Cross; however, it is considered to be a later addition because his name is misspelled and the glyph is quite crude in comparison to the illustrations on the codex.[7]

The image itself has many extraordinary attributes that seem to be naturalistically inexplicable. Four attributes have been scientifically tested in the twentieth and twenty-first centuries:

[2] Daniel J. Castellano, "The Codex Escalada", Part XI in *Historiography of the Apparition of Guadalupe*, ArcaneKnowledge (website), 2013, https://www.arcaneknowledge.org/catholic /guadalupe11.htm.

[3] "Códice 1548 o 'Escalada'" ["Codex 1548 or 'Escalation'"], Insigne y Nacional Basílica de Santa María de Guadalupe (website), 2013, https://web.archive.org/web/20150108213938 /http://basilica.mxv.mx/web1/-apariciones/Documentos_Historicos/Mestizos/Codice _1548.html.

[4] Castellano, "Codex Escalada", 17.2.

[5] Ibid.; Alberto Peralta, "El Códice 1548: Crítica a una Supuesta Fuente Guadalupana del Siglo XVI", Proyecto Guadalupe (website), 2003, https://web.archive.org/web/2007 0209082837/http://www.proyectoguadalupe.com/apl_1548.html; Stafford Poole, "History versus Juan Diego", *The Americas* 62, no. 1 (July 2005), https://doi:10.1353/tam.2005.0133; and Stafford Poole, *The Guadalupan Controversies in Mexico* (Stanford, Calif.: Stanford University Press, 2006).

[6] "Códice 1548 o 'Escalada'"; see Castellano, "Codex Escalada".

[7] "Códice 1548 o 'Escalada'" and Castellano, "Codex Escalada", 17.3.

1. The tilma's agave fiber material has extraordinary longevity (Section I.A).
2. There are several enigmas concerned with the pigment and production of the original figure in the image (Section I.A).
3. The image's cornea shows the Purkinje-Sanson triple reflection and the natural curvature of the cornea (Section I.B.1).
4. The corneas reveal at least two figures—possibly as many as thirteen (Section I.B.2).

A. The Material, the Pigment, and the Production

The first enigma concerns the material of the tilma (handwoven from the coarse fibers of the *maguey/agave* cactus[8]); it has maintained its biochemical and structural integrity for almost five hundred years, which is inexplicable. All known replicas of tilmas with the same chemical and structural composition last only *twenty to thirty years* before analyzable decomposition occurs.[9] This means the tilma has lasted seventeen times (470 years) longer than normal without significant decomposition. Furthermore, the tilma was displayed without protective glass for its first 115 years and was subjected to soot, candlewax, incense, and touching throughout its history.

---

[8] In his *Historiography of the Apparition of Guadalupe*, Dr. Daniel Castellano provides an exceedingly comprehensive scholarly study of virtually every aspect of the image of Our Lady of Guadalupe—from the historiography to the fabric to the pigments and painting to the technique and beyond. He indicates that the histochemical studies performed by Isaac Ochoterena show that the material is definitely from the genus *Agave* (*maguey*), but of indeterminate species. Perhaps it is *Agave americana*. Castellano, "The Image in the Twentieth Century", Part XIII in *Historiography of the Apparition*, 19.4, revised 2018, https://www.arcaneknowledge.org/catholic/guadalupe13.htm.

[9] Philip S. Callahan, *The Tilma under Infra-Red Radiation*, CARA Studies on Popular Devotion, vol. 2, Guadalupan Studies, no. 3 (Washington, D.C.: Center for Applied Research in the Apostolate, 1981), p. 5. See also the comments by engineer Dr. José Aste Tonsmann of the Mexican Center of Guadalupan Studies during a conference at the Pontifical Regina Apostolorum Athenaeum in 2001, cited in ZENIT, "Science Sees What Mary Saw from Juan Diego's Tilma", 2001, https://web.archive.org/web/20100620110845/http://www.catholiceducation.org/articles/religion/re0447.html, and Giulio Dante Guerra, "La Madonna di Guadalupe: Un Caso di 'Inculturazione' Miracolosa", *Christianita*, nos. 205–6, June 7, 1992, https://alleanzacattolica.org/la-madonna-di-guadalupe-un-caso-di-inculturazione-miracolosa/. (The article is in Italian. There is an option to translate to English on AlleanzaCattolica.org. Citations refer to the translation available on AlleanzaCattolica.org.)

There is currently no scientific explanation for its physical and biochemical longevity.[10]

The second enigma concerns the tilma's pigment. According to Nobel Prize–winning biochemist Richard Kuhn, who seemingly analyzed a sample of the fabric (1936), the pigments used were from no known natural source, whether animal, mineral, or vegetable. Given that there were no synthetic pigments in 1531, this discovery would be naturalistically inexplicable.[11] Daniel Castellano indicates that Ernesto Sodi Pallares (a well-known forensic scientist) testified that fibers from the tilma were given to Dr. Fritz Hahn, who in turn conveyed them to Richard Kuhn (at the Kaiser Wilhelm Institute in Germany), who gave the above test results. Unfortunately, no record was made about whether Kuhn did the tests himself or conveyed them to an assistant, and there is no record that the threads given to Fritz Hahn were taken directly from the tilma; hence, Kuhn's certification is not absolutely certain.[12] Nevertheless, analysis of the pigments and paints by artists is consistent with Kuhn's analysis and shows the truly remarkable nature of the image. Castellano examined the testimony of several artists/historians, and the consensus is summed up by Francisco Camps Ribera (internationally known painter), who examined the tilma in 1954 and 1963:[13]

1. "The cloth had not been treated with any plaster, and [he] judged that no human could have painted on the cloth without preparing it",[14] giving rise to the question of how the image could have been placed on the cloth without preparation or sizing—an enigma in itself.
2. He could not find any evidence of a brushstroke under strong magnification.[15]
3. He was not able to explain how the image was made, though he excluded oil, tempera, and watercolor.[16]

---

[10] Callahan, *Tilma under Infra-Red Radiation*; see also comments by Aste Tonsmann at the Pontifical Regina Apostolorum Athenaeum, and Guerra, "La Madonna di Guadalupe".
[11] Castellano, "Image in the Twentieth Century", 19.4.
[12] Ibid.
[13] Ibid., 19.5.9.
[14] Ibid.
[15] Ibid.
[16] Ibid.

4. The pigment/coloration of the original figure of the Madonna (excluding the sun rays, moon, angel, and a few other details) has several inexplicable features—no cracking, no flaking, and no degradation of coloration after four hundred years. As Castellano says of the report of Camps Ribera:

> The Image is over 400 years old. Any other painting that age is cracked, has lifted paint, and is darkened with a tobacco color. The Virgin of Guadalupe still has brilliant colors, "no major signs of age," and still gives a "sensation of a fresh and eternal youth."[17]

All of the above enigmatic features were confirmed by the infrared study performed by Dr. Philip Serna Callahan (biophysicist at University of Florida) in 1979, who presented his findings in a monograph, *The Tilma under Infrared Radiation*. His general assessment follows:

> The original figure, including the rose robe, blue mantle, hands and face ... is inexplicable. In terms of this infrared study, there is no way to explain either the kind of color luminosity and brightness of pigments over the centuries. Furthermore, when consideration is given to the fact that there is no under drawing, sizing, or overvarnish, and the weave of the fabric is itself utilized to give portrait depth, no explanation of the portrait is possible by infrared techniques. It is remarkable that after more than four centuries there is no fading or cracking of the original figure on any portion of the agave tilma, which—being unsized—should have deteriorated centuries ago.[18]

With respect to this last point, Callahan emphasizes:

> One of the really strange aspects of this painting is that not only is the tilma not sized, but there is absolutely no protective coating of varnish. Despite this unusual total lack of any protective overcoating, the robe and mantel are as bright and colored as if the paint were newly laid.[19]

[17] Ibid.
[18] Callahan, *Tilma under Infrared Radiation*, p. 15.
[19] Ibid., p. 11.

Though paint touch-ups and enhancements were made to the tilma between 1531 and 1606 (the date of the Echave copy[20]), to the moon, tassel, angel, sunburst, gold trim/black outline, and stars (in that order), the original image of the Blessed Mother appears to have been accomplished without sizing, known paint type (oil, tempera, watercolor), and without undersketching and brushstrokes. All of this implies that the image was fashioned in a single step without preparation of any kind. In addition to this, as noted above, the pigments of the original image have not cracked, flaked, or degraded in color in over four hundred years—quite inexplicable from a naturalistic point of view.

Callahan observed yet another enigma—the image makes artistic use of the lines in the coarse agave fabric to bring out the beauty of the Madonna. As Castellano notes: "There is one place where a coarse fiber is raised above the rest of the weave, following perfectly along the ridge of the top of the lip. Roughness in the cloth creates similar effects below the highlighted area of the left cheek, and to the lower right of the right eye."[21] After careful examination of several such instances, Callahan concluded:

I would consider it *impossible* that any human painter could select a tilma with imperfections of weave positioned so as to accentuate the shadows and highlights in order to impart such realism.[22]

One final enigma was observed by Callahan: the use of the coarse fabric in giving the impression of depth:

Infrared photography shows that eyes and shadows around the nose have no underdrawing, but are part of the face pigment. [Callahan noted,] "Close to the painting, the highlights of the eyes are subdued

[20] The Spanish painter Baltasar de Echave Orio produced the earliest extant painting of the image of Our Lady of Guadalupe on the tilma, signing and dating it in 1606. Therefore, if the paint on the tilma is on the Echave copy, it would have been added prior to 1606. See Clara Bargellini, "Echave Orio, Baltasar de", Encyclopedia.com (website), March 14, 2024, https://www.encyclopedia.com/humanities/encyclopedias-almanacs-transcripts-and-maps/echave-orio-baltasar-de-c-1558-c-1623.

[21] Castellano, "Historiography of the Apparition", Part XIII.

[22] Callahan, *Tilma under Infrared Radiation*, p. 14 (emphasis added).

to the extent that they appear nonexistent."[23] The face and hands have a phenomenal tonal quality resulting from the diffraction of light off the coarse fabric. [Callahan] judges it *impossible* to achieve this marvelous effect by human design.[24]

Inasmuch as Callahan is not only a biophysicist and infrared expert, but also an artist in his own right, the word "impossible" in the above two quotes should be taken seriously. For Callahan, the idea of a sixteenth-century artist integrating the irregular lines of course fabric into the image before any pigment was placed on the cloth and using the coarse fabric to communicate depth in the eyes and other facial features of the Madonna is inconceivable. Whoever designed the image on the cloth had judgment far beyond human artistic capability. This judgment is shared by Camps Ribera and other contemporary artists:

> Drawing upon his knowledge, Camps Ribera deduces that no Spanish, Flemish, or Italian painter of the sixteenth century could have produced the image. No foreign painter residing in Mexico demonstrated sufficient sensibility or technique. He finds it incredible that any of three Indian painters who worked for the Franciscans—Marcos Cipac, Pedro Chachalaca or Francisco Xinamamal—could have represented the Virgin in such an authentic Christian spirit, as they were all recent converts from a very different religion.[25]

So, what might we conclude about the pigment and production of the tilma and its image? First, the coarse fabric of the tilma (without sizing or overvarnish) should have disintegrated nearly five hundred years ago. There is no natural explanation for how it retains its biochemical structure with relatively little decomposition.

Second, there is no natural explanation for how the pigment used to portray the original image of the Madonna (the rose robe, the blue mantel, face and hands) has no cracking, flaking, decay of pigment, or degradation in the brightness of color after nearly five hundred years. All the features that were painted or touched up with paint manifest all these aspects of degradation.

---

[23] Ibid., p. 15, quoted in Castellano, "Image in the Twentieth Century", 19.7.
[24] Castellano, "Image in the Twentieth Century", 19.7 (emphasis added).
[25] Ibid., 19.5.9.

Third, the pigment used is not anything that would have been known by a sixteenth-century artist (e.g., oil, tempera, or water-color), and if the Richard Kuhn statement is correctly attributed, it is from no known natural source—animal, mineral, or vegetable.

Fourth, the production and artistic technique seem to be beyond human capability in several respects—the canvas was not sized or pre-pared for painting, there are no brushstrokes and no undersketching (as if it were produced in one step), and there is integration of the lines of the coarse fabric into the facial features of the image, as well as use of the coarse fabric to communicate depth in the eyes and facial features. All of these production/technique features are considered by many contemporary artists to be beyond human artistic capability.

B. The Eyes of the Madonna

In 1929, the photographer Alfonso Marcué González, examining some negatives of the image of Our Lady of Guadalupe, identified what appeared to be a human half-length figure in the cornea of the Madonna's right eye. This was kept secret until it could be confirmed by additional tests.[26]

In 1951, the official photographer for the sanctuary, Carlos Salinas Chávez, validated this finding in a written public declaration, noting, "[There is] the head of Juan Diego reflected in the pupil of the right side of the Virgin of Guadalupe,... and on the left side as well."[27]

Between 1951 and 1980, over twenty well-known ophthalmolo-gists verified the presence of what seemed to be Juan Diego's head as well as the distinctive triple reflection[28] in the cornea called "the Purkinje-Sanson reflection" (named after Jan Purkinje and Louis Sanson; see below, Section I.B.1).

Dr. José Aste Tonsmann (Ph.D. in engineering from Cornell University and an expert in electronic image processin) validated the Purkinje-Sanson reflection and much more. Using computer-enhancing techniques (explained below, Section I.B.1), he enlarged

---

[26] Guerra, "La Madonna di Guadalupe".

[27] Ibid.

[28] Carlos Salinas Chavez and Manuel de la Mora compile and document these expert testi-monies in an important volume: *Descubrimiento de un Busto Humano en los Ojos de la Virgen de Guadalupe* (Mexico: Editorial Tradición, 1980).

the image reflected on the cornea/iris up to twenty-five hundred times its original size,[29] which revealed not only what seemed to be the head of Juan Diego, but a scene involving all the characters who witnessed the moment at which the image appeared on his tilma.[30] This led to two additional enigmas that cannot be naturalistically explained:

1. The Purkinje-Sanson triple reflection and the proper corneal curvature on the Madonna's eyes (see below, Section I.B.1)
2. The images on the cornea appearing to resemble Juan Diego and other figures present when Juan Diego revealed the tilma (see below, Section I.B.2)

*1. The Purkinje-Sanson Triple Reflection and the
Natural Curvature of the Cornea*

The first ophthalmologist to identify both the Purkinje-Sanson effect and the precise corneal curvature in the images in both of the Virgin's eyes was Dr. Javier Torroella-Bueno in 1956. Jody Brant Smith explains the Purkinje-Sanson triple reflection as follows:

> In its simplest form, the Purkinje-Sanson law states that whenever we see any object, the object is reflected in each eye, not once but in three different places. This threefold reflection is caused by the curvature of the eye's cornea. Two of the reflections are always right side up and one is always upside down. Depending on the angle at which the object is seen, the three reflections occur on different parts of the eye, because of the differing angles of curvature of the cornea. The curvature also causes the reflected images to be distorted in varying degrees.[31]

On May 26, 1956, Dr. Torroella-Bueno sent the following letter to Carlos Salinas:

[29] Ibid.
[30] Ibid.
[31] Donald DeMarest and Coley Taylor, *The Dark Virgin: The Book of Our Lady of Guadalupe—A Document Anthology* (Fresno, Calif.: Academy Guild Press, 1957), p. 109.

If we take a light source and put it in front of the eye ... we see the cornea, the only part of the eye which can reflect an image in three places [images of Purkinje-Sanson]: the front surface of the cornea, and both front and rear of the lens's surfaces, immediately behind.... The image of the Virgin of Guadalupe, which has been given to me for study, contains in the cornea these [three] reflections.... In the images in question, there is a perfect correspondence (in agreement with this principle), with the distortion of the figures concurring with the predicted curvature of the cornea.[32]

The cornea of the Blessed Mother is simply extraordinary—not only does it contain the half-image of a man thought to be Juan Diego, but also the triple reflection of that image—two right side up and one upside down—according to the precise expected measurements and corneal curvature originally recorded by Jan Purkinje and Louis Sanson. It would be very challenging to imitate this with the best technology we have today. How could a supposed sixteenth-century artist (who would not have known about this triple reflection) have imitated this perfectly along with the corneal curvature to replicate it? This is apparently beyond naturalistic and human explanation.

The mystery is even more inexplicable because the Purkinje-Sanson reflections on the Madonna's corneas act as if they have depth of field corresponding to a real human eye. But how can an image on a flat surface have depth of field? This enigmatic quality of the Purkinje-Sanson reflections was validated in an experiment by Dr. Rafael Torrija Lavoignet. Jody Brant Smith gives Brother Eymard's report on Dr. Lavoignet's examination of the eyes of the Virgin on the tilma as follows:

It is impossible to attribute to chance, to a textile accident, or to pictorial matter this extraordinary coincidence between the localization of the reflections in the Virgin's eyes and the most elaborate and up-to-date laws of optical physiology, especially as it seems that these three reflections code a different focal distance. It is their most amazing property, revealed by an experiment made by Doctor Lavoignet. If

---

[32] DeMarest and Taylor, Dark Virgin, p. 109, citing Dr. Javier Torroella-Bueno, "Letter Examining the Purkinje-Sanson Triple Reflection in the Guadalupe Tilma", May 26, 1956.

the light of an ophthalmoscope, set with a suitable lens, is directed onto the reflection corresponding to Purkinje No. 2, the reflection fills with light and "shines like a little diamond." Now the same result is achieved with the third reflection provided the lens is changed.... Each of these reflections, therefore, has recorded the [proper] focal distances of the two faces of the crystalline [transparent] "painted" eye on a flat opaque surface which reacts in the presence of light as though it were a living eye. Mysteriously, the light enters into a "depth [of field]" which explains the phenomenon mentioned above. When Lavoignet aims the light of the [ophthalmoscope at the Madonna's eye],—as though to inspect the back of the eye—[her] eye lights up and the iris becomes brilliant.[33]

There are three inexplicable characteristics of the Madonna's eyes. They manifest not only all three Purkinje-Sanson reflections, but also depth of field (though painted on a flat surface). Furthermore, this depth of field is manifest in the same way as an ophthalmologist would see it in a living human eye. In a living eye, pointing an ophthalmoscope with a lens of proper focal length at one of the reflections will fill that reflection with brilliant light. However, pointing an ophthalmoscope with a lens of wrong focal length at one of the reflections will produce no effect. Now here is the amazing finding of Dr. Lavoignet. When he pointed an ophthalmoscope light with a lens of proper focal length at one of the reflections in the *Madonna's* eyes (painted on a flat surface), it filled the reflection with shining light. However, pointing an ophthalmoscope lens with a wrong focal length at the Madonna's eyes produced no effect. It is simply astonishing that the reflections not only have the proper corneal curvature and depth of field (though painted on a flat surface), but also the ability to become luminescent when a light from an ophthalmoscope having a lens with the proper focal length is used. This is naturalistically and humanly inexplicable. How does a flat image contain depth of field and encode the proper focal length to obtain luminosity?[34]

[33] Jody Brant Smith, *The Image of Guadalupe: Myth or Miracle?* (Garden City, N.Y.: Doubleday, 1983), pp. 79–80.

[34] Recall that the Shroud of Turin has depth of field encoded on it from both the surface and interior of the body. See Chapter 3.

2. The Images on the Cornea of the Blessed Mother

As noted above, Dr. José Aste Tonsmann enlarged the area of the Madonna's corneas by up to twenty-five hundred times and identified thirteen figures—six in the more peripheral iris area under the cornea (including the man thought to be Juan Diego), and seven in the central pupil area under the cornea. These additional figures (beyond the man thought to be Juan Diego) are thought by some scholars to be too imprecise to render a probative conclusion (e.g., Daniel Castellano[35]). Though I do not agree with this negative judgment (because of the credibility of the technique for enhanced imaging used by Aste Tonsmann), it should be noted that the supernatural origin of the original figure of the Madonna on the enigmatically long-lasting agave tilma does not rise or fall on certainty about the additional figures on the eyes of the Madonna. This case can be made with considerable probative force on the basis of the evidence given above—for example, the longevity of the agave tilma (almost five hundred years), the absence of significant image and pigment degradation on the original figure of the Madonna, the absence of brushstrokes, undersketching, and sizing and the production of the original figure, the Purkinje-Sanson triple reflection of the image of the man thought to be Juan Diego, and the appearance of depth in the eyes of the Madonna (painted on a flat surface), among other enigmas. Inasmuch as a chain is only as strong as its weakest link, I do not want to make the Aste Tonsmann evidence a vital link in the "credibility chain" of the tilma's supernatural origin. Nevertheless, I think there is good reason to study Aste Tonsmann's enhancement procedure and to explore the additional reflected figures (beyond the man thought to be Juan Diego) because there are parallels between the enhanced images in the right and left eyes, which are unlikely if produced by mere "cloud-reading"[36] and pure chance. If Aste Tonsmann is correct, it adds to the mysterious and supernatural character of the image, though it is not necessary to establish it.

So what did Aste Tonsmann do? In 1979, when he heard about the discovery made by Carlos Salinas and others concerning the

---

[35] See Castellano, "Image in the Twentieth Century".
[36] Ibid., 19.6.

Purkinje-Sanson reflections of the man thought to be Juan Diego in the eyes of the Madonna, Aste Tonsmann procured very high-quality photographs of the image to apply a technique used by NASA in space photography to the Madonna's eyes. Essentially, Aste Tonsmann broke down the Madonna's corneas into thousands of little points, and then measured the luminosity of each of those points, which he translated into a binary computer code. He then "reassembled" the corneas by translating the binary code back to corneal images. He then used a set of algorithms to eliminate the "noise" of the rough fabric and weave as much as possible. This process revealed twelve additional figures (beyond that of the man thought to be Juan Diego, which is visible with almost no enhancement). If these additional figures were identifiable in only one eye, they might easily be relegated to the realm of "cloud-reading", but consistency between these enhanced figures in both the right and left eyes makes it much more difficult to ignore them. Aste Tonsmann not only identified parallels between additional figures in the right and left eyes; he used a precise mathematical translation table to calculate the expected differences between the right and left figures on the basis of the curvature of the corneas and the position of the people (being observed by the Virgin) relative to the position of her eyes. In my view, the congruences between the figures in the right and left eyes are too numerous and improbable to be ignored (even though they are enhanced by up to twenty-five hundred times).

So what did Aste Tonsmann see in the enhanced images? As noted above, there are two distinct sets of figures in both eyes of the Madonna—one in the iris area under the cornea, and another in the central pupil area under the cornea. We will begin with the figures in the iris area under the cornea because they are more identifiable than those in the central pupil area. Furthermore, they can be related to specific characters that might have been present when Juan Diego came into the presence of the bishop. Photographs of both sets of figures are available on the internet.[37]

Giulio Dante Guerra (part of the Italian National Counsel of Research) sums up Aste Tonsmann's discovery as follows:

---

[37] "The Mystery in Our Lady's Eyes", Our Lady of Guadalupe (website), accessed March 15, 2024, http://sancta.org/eyes.html.

In the eyes of Our Lady of Guadalupe is reflected the entire scene of Juan Diego opening his *tilma* in front of Bishop Juan de Zumárraga O.F.M. and the other witnesses of the miracle. In this scene it is possible to identify, from left to right looking at the eye: a *seated Indian*, looking up; the profile of an elderly man, with a white beard and a head marked by advanced baldness and something similar to the cleric of the friars, very similar to the figure of Bishop Juan de Zumárraga O.F.M. as he appears in Miguel Cabrera's painting depicting the miracle of the *tilma;* a younger man, almost certainly the interpreter Juan González; an *Indian* with marked features, with a beard and mustache, certainly Juan Diego, who opens his cloak, still without the image, in front of the bishop; a dark-faced woman, perhaps a black slave; a man with Spanish features—the one already identified by the ophthalmoscopic examinations on the *tilma* and initially mistaken for Juan Diego—who looks thoughtfully at the *tilma*, stroking his beard with his hand. All these characters are looking towards the *tilma*, except for the first one, the seated *Indian*, who seems to be looking rather at Juan Diego's face. In short, in the eyes of the image of Our Lady of Guadalupe there is a "snapshot" of what happened in the bishopric of Mexico City at the time in which the image itself was formed on the tilma.[38]

Three items should be noted. First, the figure first seen through an ophthalmoscope and identified with Juan Diego can now be identified as a Spaniard stroking his beard—not Juan Diego. Second, Juan Diego is clearly the Indian figure with a mustache and beard who is opening his tilma before the other figures. Third, the principle figure in the middle with the beard and balding head is very likely Bishop Juan de Zumárraga. He is wearing Franciscan robes and indeed was a Franciscan (O.F.M.—the Order of Friars Minor), and he has a beard, balding head, and a strong jawline similar to almost every extant painting of him.[39] As we shall see, this scene is almost identical in both eyes, differentiated only by the different angles and curvature of the two corneas and the different position of the eyes relative to the

---

[38] Guerra, "La Madonna di Guadalupe" (emphasis in original).

[39] There are a variety of paintings of Bishop Zumárraga that may have been painted during his lifetime—or taken from portraits painted during his lifetime. The consistency of appearance seems to indicate a source portrait during his lifetime. These paintings are readily available online.

scene being observed by the Blessed Mother. (The observation point of the Blessed Mother will be discussed below.)

Before further discussing this image and the image in the area of the pupil, it is essential to explain Aste Tonsmann's method for validating the presence of human figures and the correspondence between the left and right eyes. He describes the four steps taken after enhancing the Madonna's eyes by up to twenty-five hundred times:

(1) Human proportions were taken into account and measured, (2) Spots and shadows were identified separated and cuts were made in order to verify that the image did and truly exists, (3) I called in the help of independent observers for verification. (4) I also used a methodology that manages the overlapping of photographs and images through a series of transitions, and contrasting filters to obtain the final results.[40]

The fourth step is of particular importance, because Aste Tonsmann not only enlarged the images and removed the noise, but also used a program to manage the overlapping and comparison of images (in the two eyes, which are in different positions relative to the scene observed) through transition algorithms. The correspondence between the figures in the right eye and left eye are so precise as to be beyond the domain of cloud-reading and pure chance. Should readers want to validate this for themselves, they can do so by procuring Aste Tonsmann's eBook entitled *Our Lady of Guadalupe's Eyes*.[41] This book is only available for smartphones (but not in print) because Aste Tonsmann made the painstaking effort to allow the reader to observe the superimposition of the figures in the right and left eyes through a series of small videos that coordinate the figures in the right and left eyes.[42] The result is beyond coincidence, wishful thinking, and cloud-reading—which is quite remarkable. Additionally, Aste Tonsmann provides a graph showing the precise mathematical correlation of the images, which he describes as follows:

[40] José Aste Tonsmann, *Our Lady of Guadalupe's Eyes* (Digital Discoveries, 2012), p. 3, Apple Books. This book can be downloaded on to a smartphone or tablet by clicking on this link: https://books.apple.com/us/book/our-lady-of-guadalupes-eyes/id548965040.
[41] Ibid.
[42] Ibid., pp. 34–50.

A graph shows the values of the right horizontal coordinates (shown in black) and the same coordinates calculated using a mathematical function for its result (shown in red). You can see the high value of the correlated coefficient of the mathematical function relation to the coordinates conversion that represent the shift, the scale, and its rotation.[43]

The results of Aste Tonsmann's careful image enhancement and coordination of images are so precise that I find it difficult to believe that some scholars have relegated his findings to "cloud-reading". It would be important for these scholars to read his eBook, closely examine the videos coordinating the right and left eyes, and check his coordination functions and coefficients on the provided graph so that they can knowledgably and intelligently justify their skeptical position. For those who believe that the correlation of the figures in the left and right eyes is sufficiently probative, it will be clear that it could not have been produced by a human artist or by pure chance. In the words of Guerra:

> It is materially impossible to paint all these figures in circles of about 8 millimeters in diameter, such as the irises of Our Lady of Guadalupe, and moreover in absolute compliance with optical laws totally unknown in the sixteenth century.[44]

Additional validation of Aste Tonsmann's imaging enhancement and coordination can be obtained by comparing the image of Bishop Zumárraga to early portraits of him. Recall that early paintings of Bishop Zumárraga resemble the figure of him in his Franciscan robes, balding head, and beard in the Blessed Mother's eyes. The figure of Juan Diego (showing the tilma) has a beard and mustache that seem similar to early portraits of him, but the image in the corneas is too inexact to have certitude of identification.

So what was the vantage point of the Blessed Mother as she looked on the scene of Juan Diego revealing the tilma to Bishop Zumárraga and his retinue? From the vantage point of Juan Diego facing Bishop Zumárraga, the Blessed Virgin would have been on Juan Diego's left

---

[43] Ibid., p. 35.
[44] Guerra, "La Madonna di Guadalupe".

and Bishop Zumárraga's right, slightly closer to Bishop Zumárraga.[45] She would have been elevated two or more feet above the ground with her eyes inclined toward the left (as you face her). Given this position, the differences in the figures in the Madonna's right and left eyes can be coordinated. Evidently, the scene was not observed from the vantage point of the tilma (which is blank), but rather from behind and above the gathering—to the right of Bishop Zumárraga and the left of Juan Diego.[46]

We now turn to the second set of images in the center pupil under the cornea. These figures are more difficult to identify than the ones in the iris area under the cornea. The scale is much smaller, but under considerable enlargement several figures were identified by Aste Tonsmann in the pupil areas under the right and left corneas. Since the scale is even smaller than the scene in the iris area, the identification of the individual figures is more speculative. Nevertheless, these small (highly enlarged) figures can be coordinated in both the right and left eyes, and so they are worth noting—not so much to ascertain the supernatural origin of the tilma (which is already validated by the evidence given above) but for the message that it seems to contain. Aste Tonsmann notes in this regard:

> It is extraordinary that in the center of the cornea there are images that conform to what I believe to be a family unit. It is made up of a mother (in the center), a father (wearing a point hat—also known as "a sombrero"), who appears a little bit lower, and between them two small images of a boy and a girl. Behind the mother there is another woman, and later on when we observe the right cornea, we will see a man standing next to this woman. I believe these two images represent the grandparents. Something worth mentioning is that if you look very closely, the mother is carrying a small child on her back (something indigenous woman do using what is called "a rebozo").[47]

The message is clear. In the late twentieth century—the only time since the origin of the image that the figures in the center (pupil area) of the Madonna's eyes could be revealed—the Blessed Virgin is, as

45 Aste Tonsmann, *Our Lady of Guadalupe's Eyes*, p. 21.
46 Ibid., p. 36.
47 Ibid., p. 22.

it were, pointing to the most significant building block of society, the culture, and the Church—the family. Though the identity of the family is unknown, three generations present together in the cohesive community through which children are created, cared for, educated, and formed in faith, and through which the older generation contributes their love and wisdom while finding support, is made manifest by the mother of families. It is almost as if the Blessed Mother anticipated the decline of the family in our era, and she gave a message through her eyes about what really is important in life, love, and faith—a message that would come to light only with today's science and technology.

As noted many times above, the supernatural origin of the original figure of the Madonna and the long-lasting tilma itself does not require validation from these additional figures seen in the iris area and the pupil area under the corneas of both eyes. Nevertheless, the shapes, positions, and scale of these figures are inexplicably coordinated in the left and right eyes of the Blessed Mother, and this coordination is mathematically justified by the transition functions and coefficients identified by Aste Tonsmann. If this analysis is correct, it is quite beyond human artistry or natural occurrence. This should give us pause because it seems that the Blessed Virgin is speaking specifically to us in our scientific-technological culture through these images in her eyes—relating not only events of long ago, but also to the remedy for our societal and cultural malaise: her Son,[48] herself, and the family.

C. Conclusion

The evidence of the supernatural origin of the tilma and its original image of the Madonna has significant probative force:

- The longevity of the agave tilma, which should have decomposed over four hundred years ago

---

[48] The Blessed Mother is pregnant in the image, which can be seen not only by her shape, but by the dark purple belt that signified pregnancy to the Aztecs. The image would have conveyed to them that the Blessed Mother was ready to give the world her Son—the Savior. See Guerra, "La Madonna di Guadalupe".

- The absence of degradation in the pigment and brightness of color on the original figure of the Madonna, which should have suffered degradation and color change over four hundred years ago
- The absence of sizing, brushstrokes, and undersketching in the production of the original image of the Madonna, which many artists consider beyond human capability
- The use of the coarse fabric and weave to present depth (e.g., in the eyes of the Madonna) or add artistic enhancement (e.g., on the lips of the Madonna), which appears to be beyond human capability—at least certainly in the sixteenth century
- The appearance of the Purkinje-Sanson triple reflection on the corneas of the Madonna, which is beyond human capability
- The appearance of depth in the eyes of the Madonna (detected through ophthalmoscope), though they are painted on a flat surface
- The likely appearance of multiple figures in a scene of Juan Diego showing his tilma to Bishop Zumárraga and others that is coordinated between the right and left eyes through expected transition functions and coefficients calculated by Aste Tonsmann, which is beyond human capability or natural causation

There is another remarkable characteristic on the image of the Madonna that cannot be ascertained with certainty though it should be briefly mentioned: the stars on the Madonna's mantle. These stars have a remarkable characteristic that heightens the Madonna's message—not only to sixteenth-century culture but also to our contemporary scientific culture. Dr. Philip Serna Callahan asserted that the stars were later additions to the tilma based on his detection of paint on the stars in his infrared examination, but this conclusion was probably too hasty. There is indeed paint on the stars, but this paint/gold overlay would have to have been added before 1606 (the date of the Echave copy, which shows stars), or more likely, in the mid-sixteenth century,[49] or most likely, they were part of the original image (not an addition).[50] Daniel Castellano implies that the evidence favors the view that the stars were part of the original image. He notes in this regard:

---

[49] See Castellano, "Image in the Twentieth Century", 19.4.
[50] See ibid.

If this part of the codex [Escalada] is authentic, then it follows either that (a) all embellishments were completed by the mid-sixteenth century, or (b) the stars were not late embellishments. In support of the latter consideration [the stars were not added, but on the original image], we note that the stars had been retouched in many places, which could account for their overlay of other elements.[51]

If we accept that the stars are part of the original image (and the paint was added to make them "more golden"), then it will not be surprising to learn what Father Mario Rojas Sánchez discovered about their placement:

[Father Rojas Sánchez] was able to ascertain, thanks to the collaboration of some astronomers and the Laplace Observatory in Mexico City, that [the stars] correspond to the constellations present over Mexico City at the winter solstice of 1531—solstice which, given the Julian calendar in force at the time, fell on December 12—seen, however, not according the normal "geocentric" perspective, but according to a "cosmocentric" perspective, that is, as an observer placed "above the vault of heaven" would see them.[52]

There are maps available on the internet of constellations that would have been present in the sky above Mexico City on December 12, 1531 (the winter solstice of 1531) that convert the geocentric reference frame (an observer looking at the stars from the earth) into a cosmocentric reference frame (an observer looking at the stars from beyond those stars toward the earth).[53] The cosmocentric reference frame requires rotating the cardinal axis ninety degrees counterclockwise.[54] This reference frame corresponds to the placement of the stars on the Madonna's mantle.[55] This is truly extraordinary, because no sixteenth-century Indian or European would have such knowledge that would be available beginning only in the twentieth century.

[51] Ibid., 19.10.

[52] Guerra, "La Madonna di Guadalupe".

[53] Peter Darcy, "Why the Stars on Guadalupe's Mantle are Miraculous", SacredWindows .com, https://sacredwindows.com/why-the-stars-on-guadalupes-mantle-are-miraculous/.

[54] Janet Barber, *A Handbook on Guadalupe* (New Bedford, Mass.: Franciscan Friars of the Immaculate, 2001). For an online explanation see Anonymous, "A Map of the Constellations?", https://www.myguadalupe.com/blog/a-map-of-the-constellations.

[55] Darcy, "Why the Stars on Guadalupe's Mantle are Miraculous".

Though we do not need this evidence to confirm the miraculous origin of the tilma and the original figure of the Madonna, it does provide an interpretive key to accompany the message given by the Blessed Mother to the world about her Son, herself, and the family (see above). To a contemporary scientifically sophisticated audience, the implication seems clear—that the Blessed Mother is not only beyond (and therefore greater than) the earth, but also beyond (and greater than) the stars. She is not only Mother of the world, but Mother of the heavens—or as we might say today, Mother of the universe. The child that she evidently bears would therefore be not only the Savior of the world, but also the Savior of the heavens—and the universe. In view of this, we can rely on Him and her not only for help but for the teaching that transcends all time and universes—the very pathway to eternity.

II. The Miracle of the Sun at Fatima

I am not using the term "miracle" here with reckless abandon, but the scientifically inexplicable phenomenon of the sun at Fatima seems to be the most evident and undeniable instance of a miracle in the last two scientifically oriented centuries. The following analysis should substantiate this claim. Before we can address whether the phenomenon of the sun at Fatima was a miracle, we must first examine the eyewitnesses, the precise events, and any possible naturalistic/scientific explanation of the phenomenon. This will be taken up in three subsections:

1. Eyewitnesses and Professional Testimonies (Section II.A)
2. The Sequence of Events at Midday, October 13, 1917 (Section II.B)
3. Explaining the Miracle of the Sun (Section II.C)

A. Eyewitnesses and Professional Testimonies

Father Stanley Jaki (Ph.D. in physics and theology and major contributor to contemporary philosophy and history of science) has researched and synthesized the vast majority of eyewitness testimonies and resources concerned with the Fatima phenomenon.[56] I have no intention of repeating this careful and rigorous work, but I rely heavily upon it, particularly regarding the testimonies of Avelino de Almeida,[57] Dr. Gonçalo Xavier de Almeida Garrett,[58] Dr. José Maria Pereira Gens,[59] Leopoldo Nuñes,[60] and Father Pio Scatizzi.[61] It should be noted that Father Pio Scatizzi (professor of astronomy at the Gregorian University of Rome) was the first to give a scientific account of the "dancing of the sun" at Fatima (thirty years after the event).

After Father Pio Scatizzi's scientific examination, he declared this event to be the "most obvious and colossal miracle of history".[62] Scatizzi constructed his scientific analysis on the basis of Dr. Almeida Garrett's observations as well as those of other scientists and professionals who were present. Additional smaller scientific reflections were made after the time of Scatizzi's analysis (1947), but the next comprehensive, systematic, scientific account of the Fatima miracle was done by Father Stanley Jaki in 1999. Since that time, Dr. Ignacio Ferrín (Ph.D. in astro-geophysics and full professor, Institute of Physics, University of Antioquia) published another scientific explanation of the miracle in 2021[63] (see below, Section II.C).

Avelino de Almeida was the editor in chief of a secular masonic daily—O Seculo (The Century) in Lisbon. His perspective was hostile to miracles and the Church, and he went to the Cova da Iria (the site of the prospective "miracle") in the hopes of disproving it to

---

[56] Stanley L. Jaki, *God and the Sun at Fatima* (Royal Oak, Mich.: Real View Books, 1999).

[57] Avelino de Almeida, "The Dance of the Sun", *O Seculo*, October 18, 1917, p. 2, section "Ecos".

[58] Gonçalo Almeida Garrett, "Letter to Cannon Formigao concerning Observations at Fatima on October 13th", trans. Antonio Maria Martins, in *Documents on Fatima and Memoirs of Sister Lucia* (Alexandra, S. Dak.: Fatima Family Apostolate, 1992), pp. 173–75.

[59] José Maria Pereira Gens, *Fátima: Como Eu A Vi e Como A Sinto (Memória dum Médico)*, vol. 1 (Leiria, Portugal: Oficinas da "Graffica de Leiria", 1967).

[60] Leopoldo Nuñes, *Fátima—História das Aparições de Nossa Senhora do e aos Pastorinhos na Cova de Iria* (Lisbon: Tipografia Luzitania, 1927).

[61] Pio Scatizzi, S.J., *Fátima alla Luce della Fede e della Scienza* (Rome: Coletti, 1947). For English translations for much of Scatizzi's documented testimonies, see Father C.C. Martindale's *The Meaning of Fatima* (New York: P.J. Kenedy & Sons, 1950). Citations refer to Father Martindale's translation.

[62] Francis Johnston, *Fatima: The Great Sign* (Rockford, Ill.: Tan Books, 2010), p. 66.

[63] Ignacio Ferrín, *An Astronomical Explanation of the Fatima Miracle* (Amazon Kindle Edition, 2021).

embarrass the Church.[64] Indeed, in his October 13 morning article in
O Seculo, he not only casts doubt that the miracle would occur, but
suggests how people would be duped when it did not go according
to plan.[65]

Evidently, Almeida was shocked to find that the phenomenon of
the sun really did occur. In his article in *O Seculo* the next day,[66] he
risked the ruin of his secular journalistic career, endured a host of criticism, and reported what he called the "dancing of the sun" (a phrase
that was immediately picked up by the press and historians throughout Portugal). When other secular newspapers (written by people
who had not been at Fatima) attacked him, he reasserted his claims
by writing another article in *O Seculo* pointing to the truth that tens
of thousands of other people witnessed as well. The fact that a hostile
source reported several dimensions of the dancing sun at Fatima, estimated the crowd to be at least thirty thousand people, and endured
significant attacks and professional threats to maintain his position,
speaks very strongly to the fact that the dancing sun occurred and
was witnessed by tens of thousands of others who were photographed
kneeling in the mud, staring at the sun, and crying out in exaspera-
tion and fear on the date, time, and place that Lucia dos Santos (ten-
year-old peasant girl) predicted on three separate occasions—July 13,
August 19, and September 13. When this testimony is combined with
that of many other secular reporters, the chief of police who was
present to prevent the crowd from gathering (and converted on the
spot after the miracle), several professors, medical doctors, attorneys,
and other credible and hostile witnesses[67] (who agreed on much of
the sequence of events described below), it is quite difficult to deny
that the unique, astonishing, and inexplicable "dancing of the sun"
really did occur. This collective evidence is further confirmed by
the similar amazement and reaction of the crowd, the almost immediate drying of the ground and people's clothes (after a morning of
rain), and the fact that people over ten miles away also witnessed the
phenomenon. In sum, it cannot be credibly denied that something
truly unique, extraordinary, amazing, and inexplicable occurred at

[64] See Jaki, *God and the Sun at Fatima*, Chapter 1.
[65] Ibid., p. 24.
[66] Almeida, "Dance of the Sun".
[67] Nuñes, *Fátima*, p. 86.

the Cova da Iria at midday on October 13, 1917. In order to obtain a clear and confirmed set of observational data for scientific analysis, we must first speak about three other witnesses and researchers.

Dr. Gonçalo Xavier de Almeida Garrett, professor of mathematics and science at the University of Coimbra (and well acquainted with scientific observation), personally observed (with his wife and son—an attorney) six distinct aspects of the phenomenon of the sun (and crowd reaction) that were validated by other scientific colleagues as well as large numbers of nonscientific witnesses.[68] Almeida Garrett's testimony is complemented and supplemented by that of Dr. José Maria Pereira Gens (a medical doctor who published his own account and gave a public interview[69]). As will be seen, Dr. Pereira Gens' testimony gives scientifically important details regarding the fall of the sun and the remarkably quick drying of the ground and people's clothes after a morning of rain. Leopoldo Nuñes (writer), who was present at the scene, indicated that witnesses included the minister of education and the Masonic government, as well as illustrious men of letters, arts, and sciences, who came attracted by the prediction.[70] Finally, Father Pio Scatizzi's compilation of testimonies from scientific and lay witnesses corroborates the events of the day as seen by people throughout the Cova as well as up to ten miles away.[71]

Before giving a general account of the events at midday on October 17, we should take up the matter of how many witnesses there were and whether the phenomenon could have been an activity of the sun itself. First, with respect to the number of witnesses, Avelino de Almeida (the secular masonic newspaper editor) estimated the crowd to be about thirty thousand people, while Almeida Garrett estimated the crowd to be between seventy thousand and one hundred thousand people.[72] Though Avelino de Almeida may have been

---

[68] In his letter to Canon Formigao, Almeida Garrett details the claims that will be discussed below. See Garrett, "Letter to Cannon Formigao", pp. 173–75. Stanley Jaki has assembled Almeida Garrett's testimony, along with that of scientific and other professional witnesses, in *God and the Sun at Fatima*. Father Pio Scatizzi, S.J., made a significant compilation of testimonies to the phenomenon in his *Fátima alla Luce*.

[69] See Gens, *Fátima*.

[70] See Nuñes, *Fátima*.

[71] See Scatizzi, *Fátima*.

[72] John de Marchi, *The True Story of Fatima* (St. Paul, Minn: Catechetical Guild Educational Society, 1956), https://www.ewtn.com/catholicism/library/true-story-of-fatima-5915.

motivated to underestimate the crowd because of his secular view-point, it is more likely that in his area and vantage point the Cova was less densely packed than the area and vantage point from which Almeida Garrett made his estimates. In order to account for these imprecisions, it is probably best to estimate the crowd to have been around fifty thousand people, as well as many others who were up to twenty miles distant from the Cova. It is significant that virtually no one who was present at the Cova da Iria denied the events that took place there at midday on October 13, 1917. Indeed, John de Marchi and his team of interviewers never encountered a single "denier" in seven years:

> Believe only that we, who are reporting it here, lived for more than seven years within sight of the Cova da Iria, and have yet found no one to confound or deny with just reason, the events of this memorable day.[73]

There are five major collections of written testimonies and interviews of those who witnessed the events:

- Stanley Jaki, *God and the Sun at Fatima* (1999)
- John Haffert, *Meet the Witnesses of the Miracle of the Sun*—transcriptions of television interviews (1961)[74]
- John de Marchi, *The True Story of Fatima* (1956)
- Pio Scatizzi, *Fatima all Luce della Fede e della Scienza* (1947)
- Chanoine Barthas, *Fatima: Merveille inouïe; Les Apparitions; Le Pèlerinage; Les Voyants; Des Miracles; Des Documents* (1943)[75]

Can the phenomenon be explained as an activity of the sun itself? The answer is most certainly no. Almeida Garrett made various inquiries to scientific institutions and observatories that were later verified by more distant observatories showing that the sun itself did nothing out of the ordinary on that day. There were not even

---

[73] Ibid.

[74] John M. Haffert, *Meet the Witnesses of the Miracle of the Sun* (Spring Grove, Penn.: American Society for the Defense of Tradition, Family, and Property—TFP), 1961.

[75] Chanoine Barthas, *Fatima: Merveille inouïe; Les Apparitions; Le Pèlerinage; Les Voyants; Des Miracles; Des Documents* (Toulouse, France: Fatima Editions, 1943).

strong solar flares or other rare observable solar phenomena.[76] So this leaves us with only two naturalistic explanations[77] that will be explained in detail below (Section II.C)—a meteorological phenomenon (described by Stanley Jaki[78]) or a nonsolar astronomical phenomenon, a small comet entering the earth's atmosphere (explained by Ignacio Ferrín[79]). Though these explanations are naturalistic, the odds of them occurring are exceedingly rare under any circumstances, but the odds of their occurring at the precise date, place, and time predicted by Lucia on three separate occasions (July 13, August 19, and September 13) are infinitesimally small, making them virtually impossible by any naturalistic standard. This is why Stanley Jaki, Ignacio Ferrín, Pio Scatizzi, and other scientists acquainted with this phenomenon consider it to be a miracle.

B. The Sequence of Events at Midday on October 13, 1917

After having seen visions of an angel in 1916, Lucia dos Santos (a ten-year-old shepherd girl, as mentioned above) and her two cousins Jacinta and Francisco Marto (also shepherd children) witnessed a vision of the Blessed Virgin Mary on May 13, 1917. She was brighter than the sun and had a mantle edged with gold. She was holding a rosary and asked the children to pray the Rosary daily for peace and an end to the war (World War I). Lucia's parents did not believe her, but undaunted, Lucia, Jacinta, and Francisco returned to the Cova on June 13 as the Blessed Virgin had asked them. On June 13 the Blessed Mother returned and revealed that Jacinta and Francisco would go to Heaven soon, but Lucia would live longer to spread the message of

[76] Stanley Jaki, *A Mind's Matter: An Intellectual Autobiography* (Grand Rapids, Mich: Eerdmans, 2002), pp. 232–35.

[77] Several individuals have speculated that the phenomenon was mass hypnosis or some other auto-suggestive subjective explanation. As will be explained below, this does not withstand the scrutiny of critical explanation because the phenomenon had real effects—such as drying the ground and people's clothes—and was witnessed by many people up to ten miles away, well beyond the influence of collective auto-suggestion. It should be noted that there has been no instance of mass hypnosis with any group, even remotely as large as fifty thousand people.

[78] Jaki, *God and the Sun at Fatima*, Chapter 8.

[79] Ferrín, *Astronomical Explanation*, Chapter 6.

the Rosary and the Immaculate Heart of Mary. She allowed them to see a vision of Hell (to warn the world about Satan) and then revealed some secrets concerning the Church and the world.[80] The children returned on July 13, and the Blessed Virgin revealed that she would perform a miracle at midday on October 13, 1917. The children attempted to return on August 13 but were prevented from doing so by the provincial administrator, Artur Santos, who arrested and jailed them to prevent them from being "politically disruptive". When they were released, they had an apparition of the Blessed Virgin on Sunday, August 19, at the town of Valinhos. The Blessed Virgin again indicated that she would perform a miracle for all to see to overcome doubts at midday on October 13, 1917. On September 13, the children returned to the Cova and the Blessed Virgin repeated her promise to manifest a miracle that would convince everyone.

In view of Lucia's insistence about the date, time, and place of the miracle, and the publicity that had been given to it throughout Portugal, we should not be surprised that at least fifty thousand people made their way over to the Cova da Iria that day. The crowd included skeptics as well as believers, scientists, and professionals. The crowd also included the uneducated faithful, wealthy as well as poor; many of the poor came by foot and donkey, while the wealthy by horse, carriage, and motorcar. Throughout the night it had been raining; the crowd approached the Cova amid a steady drizzle. At ten o'clock, the sun was completely hidden by the clouds, and it began to rain in earnest.[81] Recall that Lucia reported that the Blessed Virgin Mary had told her the miracle would take place at midday on October 17. If by "midday" is meant the high point of the sun, this would have occurred at twelve noon before World War I. However, the Portuguese government, in sympathy with their troops who were fighting in France, moved their clocks ahead two hours to correspond to the time in France. This meant that the high point of the sun ("midday") occurred by this revised time at 2:00 P.M.

---

[80] Some people wonder why the Blessed Virgin would have allowed children to see a vision of Hell. The reason seems to be that she wanted the children (who had no political, ecclesial, or cultural agenda) to give a warning to an increasingly secularized world about Satan and the eternal domain into which he is trying to seduce us. Note that she mitigated the children's personal tribulation by telling them they would be going to Heaven.

[81] See de Marchi, *True Story of Fatima*.

We may now proceed to the Miracle of the Sun itself, beginning with a summary statement from Dr. Pereira Gens (who tried to maintain a stance of scientific objectivity throughout the event):

> Then Lucia cried: Look at the sun! All that crowd, which now hoped that the sun itself would become the miracle that had been announced, turned their gaze anxiously towards the royal star [sun], which, visibly decreasing the brilliance of its rays, allowed itself to be looked at by the astonished eyes of all.... Hardly two or three persons shaded their eyes with their hands.... However, the sun, at the zenith cleared of clouds, stands out like a colossal host [i.e., a Eucharist host]. I still have before my eyes that strange and unforgettable scene. The midday sun, suspended in front of us, a sun without any defense, neutral, without any of the aggressive rays that characterize it. Suddenly it begins to turn on itself with a vertiginous speed, and, at certain moment, seems to approach and threatens to fall on us. Then something unforgettable happens. The crowd, deeply impressed, kneels in the mud, praying and crying. All across the Cova there spreads a huge sound, a true tidal wave of enthusiasm and [uninhibitedness]. However, the sun stops, in order to recommence, after a brief pause, its course, its strange dance, turning on itself, giving us the sensation of moving away from us and of approaching us. A light whose color varies from one moment to the next is reflected on the people and the things, and if it is true that the brightness of the sun diminished, its heat did not lose anything of its force.[82]

We can divide this ten-minute event into six stages, examining other testimonies from scientists, journalists, and other participants, principal among whom are Avelino de Almeida and Dr. Gonçalo Xavier de Almeida Garrett:

1. The stopping of the rain and the emergence of the sun as Lucia predicts
2. The reappearance of the sun with a pearl/silver rim in a fashion that could be gazed upon without harm
3. The dancing of the sun

---

[82] The English translation of this quotation was given in Jaki, *God and the Sun at Fatima*, p. 327. It is taken from an interview with Chanoine Barthas as reported in his book *Fatima*, p. 358.

4. The segmentation and different colors of the rays coming from
   the sun
5. The falling of the sun
6. The drying of the ground and clothing

Let us proceed to the first stage.

1. The stopping of the rain and emergence of the sun as Lucia predicts. At
about 1:30 P.M. (about thirty minutes before midday), the two girls
were brought to the place where the altar was erected. Lucia saw
a flash in the sky (which occurred before other appearances of the
Blessed Virgin[83]) and said, "Jacinta, Jacinta, here comes Our Lady. I
just saw the flash."[84] Lucia had everyone close their umbrellas, despite
the pouring rain, because, she said, "Our Lady is coming." The
crowd closed their umbrellas and waited. The clouds suddenly dispersed, and the sun came out—then moved by an interior impulse,[85]
Lucia cried out, "Look at the sun." As we shall see, this was not
a command to do something self-destructive, but rather to witness a
spectacular phenomenon.

*2. The reappearance of the sun with a pearl/silver rim in a fashion that could
be gazed upon without harm.* At the moment Lucia cried out, "Look at
the sun", virtually the whole crowd turned toward it. The sun was
not opaque, but the entire crowd could gaze upon it for about ten
minutes (with only two slight interruptions) without pain, having to
look aside, or incurring retinal damage. From a physiological point
of view, this is inexplicable. How can fifty thousand people gaze at
the sun for ten minutes without retinal damage—or any other physiological side effect?

The rim of the sun was covered in a silver or pearl-like radiance
quite distinct from the brightness of the sun itself. Avelino de Almeida
(the secular masonic editor and chief of O Seculo) noted:

[The crowd] saw a unique spectacle, [an] unbelievable spectacle for
anyone who did not witness it. From the road ... one could see the

[83] Ferrín, *Astronomical Explanation*, Chapter 5.
[84] Ibid.
[85] Ibid.

immense multitude turn towards the sun, which appeared free from clouds and in its zenith. It resembles a dull silver disc, and it is possible to look at it without the least discomfort.[86]

A more detailed and precise description of the sun at this stage of the miracle was given by Dr. Almeida Garrett:

It must have been nearly two o'clock by the legal time, and about midday by the sun. The sun, a few moments before, had broken through the thick layer of clouds which hid it, and shone clearly and intensely. I veered to the magnet which seemed to be drawing all eyes, and saw it as a disc with a clean-cut rim, luminous and shining, but which did not hurt the eyes. I do not agree with the comparison which I have heard made in Fatima—that of a dull silver disc. It was a clearer, richer, brighter colour, having something of the luster of a pearl. It did not in the least resemble the moon on a clear night because one saw it and felt it to be a living body. It was not spheric like the moon, nor did it have the same colour, tone, or shading. It looked like a glazed wheel made of mother-of-pearl. It could not be confused, either, with the sun seen through fog (for there was no fog at the time), because it was not opaque, diffused or veiled. In Fatima it gave light and heat and appeared clear-cut with a well-defined rim.[87]

In Section II.C, we will discuss two potential explanations for how the sun could maintain its brightness and heat without causing damage or discomfort to the eyes, as well as what might have caused the great pearl-light rim. For the moment, suffice it to say that these naturalistic explanations are so exceedingly improbable at the precise date, time, and place indicated by Lucia that supernatural intervention seems virtually unavoidable (see below, Section II.C).

3. *The dancing of the sun.* As noted above, this metaphorical phrase is attributable to Avelino de Almeida (the editor in chief of *O Seculo*). It refers to a series of events leading up to the apparent falling of the sun from the sky. We begin with Almeida's written account in *O Seculo* on October 14, 1917:

[86] Almeida, "Dance of the Sun".
[87] Gonçalo Xavier de Almeida Garrett, "Letter to Cannon Formigao", cited in de Marchi, *True Story of Fatima.*

The sun trembled, the sun made sudden incredible movements outside
all cosmic laws—the sun "danced" according to the typical expression
of the people.[88]

Dr. Almeida Garrett adds,

The sun's disc did not remain immobile. This was not the sparkling of
a heavenly body, for it spun round on itself in a mad whirl.[89]

This trembling, "dancing", and whirling was just the beginning of
the ten-minute episode that led to the apparent falling of the sun.

4. *The segmentation and different colors of the rays coming from the sun.*
As the sun rotated on its own axis, it began to throw out segmented
flares of different colors, which were described by several witnesses.
Dr. Almeida Garrett describes his observations (in a response to Dr.
Formigao) as follows:

The impression was not that of an eclipse, and while looking at the
sun I noticed that the atmosphere had cleared. Soon after I heard a
peasant who was near me shout out in tones of astonishment: "Look,
that lady is all yellow!" And in fact everything, both near and far, had
changed, taking on the colour of old yellow damask. People looked as
if they were suffering from jaundice, and I recall a sensation of amuse-
ment at seeing them look so ugly and unattractive. My own hand was
the same colour. All the phenomena which I have described were
observed by me in a calm and serene state of mind, and without any
emotional disturbance. It is for others to interpret and explain them.[90]

The Lisbon daily *O Dia* validated Almeida Garrett's observations,
noting:

The light turned a beautiful blue, as if it had come through the
stained-glass windows of a cathedral, and spread itself over the people
who knelt with outstretched hands. The blue faded slowly, and then

---

[88] Avelino de Almeida, "Dance of the Sun", p. 2. Also cited in de Marchi, *True Story
of Fatima.*
[89] Almeida Garrett, "Response to Dr. Formigao", cited in de Marchi, *True Story of Fatima.*
[90] Ibid.

the light seemed to pass through yellow glass. Yellow stains fell against white handkerchiefs, against the dark skirts of the women. They were repeated on the trees, on the stones and on the serra. People wept and prayed with uncovered heads, in the presence of a miracle they had awaited. The seconds seemed like hours, so vivid were they.[91]

Ti Marto (the father of Jacinta and Francisco) gave additional details on the segmented colors and the dance:

> We looked easily at the sun, which for some reason did not blind us. It seemed to flicker on and off, first one way, then another. It cast its rays in many directions and painted everything in different colours—the trees, the people, the air and the ground. But what was most extraordinary, I thought, was that the sun did not hurt our eyes. Everything was still and quiet, and everyone was looking up. Then at a certain moment, the sun appeared to stop spinning. It then began to move and to dance in the sky.[92]

Father Pio Scatizzi, S.J., analyzed the many accounts of the separated colored flares, noting that they were monochromatic and segmented—that is, each flare had a particular color and was distinct from other flares coming from the rotating sun. This clearly distinguishes the phenomenon from a rainbow or other similar reflective medium:

> In the case of Fatima, it is extremely difficult to place such a phenomenon within a framework when outside the solar disc there was only limpid air without any reflecting agent, as [would be the case] with a rainbow, when along each monochromatic ray numberless drops of water renew the prismatic effect. In Fatima, as seen by motionless observers, the monochromatic sectors appeared to revolve and to subsist without any support. We must conclude that each colored ray was maintained autonomously, with its origin in the solar body, the air providing no means of transmission.[93]

In another article entitled "The Miracle of the Sun: A Critical Note", Scatizzi asserts that the monochromatic, segmented, autonomous flares

---

[91] Editorial, *O Dia*, October 13, 1917, cited in de Marchi, *True Story of Fatima*.
[92] Ti Marto, "Testimony of Events", cited in de Marchi, *True Story of Fatima*.
[93] Pio Scatizzi, "The Miracle of the Sun", reprinted in de Marchi, *True Story of Fatima*.

seemingly dispersed without a medium are unique and apparently not physical in origin, implying a transphysical cause:

> Immediately there began to radiate from [the sun's] centre thousands upon thousands of colored monochromatic lights in sectors, which, in the form of spirals, began to whirl around the centre itself rather like a catherine wheel, while the colored rays spread out in a centrifugal movement covering the sky as far as the curtains of clouds turning everything various colors as if by magic. Such a spectacle of red, yellow, green and violet rays from the sun, spreading and sweeping over the sky, cannot be explained by any known law nor has such a thing been seen before.[94]

Though Scatizzi asserts that this phenomenon cannot be explained by any known physical law, Jaki and Ferrín believe that it might be explained by a coincidence of highly improbable meteorological phenomena (Jaki) or a highly improbable astronomical phenomena, a small comet (Ferrín) (see the discussion below in Section II.C).

Though the precise cause has not yet been fixed, one thing is certain—the cause of the rotating, segmented, monochromatic rays was not the sun itself, because the phenomenon was not manifest outside of a thirty-mile perimeter around the Cova da Iria.[95]

5. *The descending sun.* At this point, the whirling sun seems to detach itself from its position above the earth and rapidly descend toward the Cova, causing the crowds to cry out and protect themselves. Dr. Almeida Garrett describes it as follows:

> Suddenly, one heard a clamour, a cry of anguish breaking from all the people. The sun, whirling wildly, seemed to loosen itself from the firmament and advance threateningly upon the earth as if to crush us with its huge and fiery weight. The sensation during those moments was terrible.[96]

[94] The article of Scatizzi is published in full as an appendix in de Marchi, *True Story of Fatima.*

[95] De Marchi, *True Story of Fatima,* Chapter 1. According to Scatizzi, "The above-mentioned solar phenomena were not noted in any observatory. Impossible that they should escape the notice of so many astronomers and indeed the other inhabitants of the hemisphere." Scatizzi, "Miracle of the Sun", Appendix I. See Jaki, *Mind's Matter,* pp. 232–35.

[96] Garrett, "Response to Dr. Formigao".

The quotation from Dr. Pereira Gens citied above validates Dr. Almeida Garrett's observations. It might prove helpful to repeat part of that quotation for purposes of correlation:

> I still have before my eyes that strange and unforgettable scene. The midday sun, suspended in front of us ... suddenly begins to turn on itself with a vertiginous speed, and, at a certain moment, seems to approach and threatens to fall on us.... The crowd, deeply impressed, kneels in the mud, praying and crying. All across the Cova there spreads a huge sound, a true tidal wave of enthusiasm and [uninhibitedness]. However, the sun stops, in order to recommence, after a brief pause, its course, its strange dance, turning on itself, giving us the sensation of moving away from us and of approaching us.... If it is true that the brightness of the sun diminished, its heat did not lose anything of its force.[97]

The physicist-astronomer Father Pio Scatizzi analyzed the event as follows:

> It now remains to examine ... the movement of the sun, which appeared to detach itself from the sky and to fall on the earth in a zigzag path. It can be affirmed that such a phenomenon is outside and against all natural and astronomical laws. It appears that with this final occurrence, all doubts as to the natural origin of the events, all skepticism on our part, must be laid aside.... Suddenly, without the intervention of any new factor, the multitude is seized with terror as if menaced by a cataclysm. Everyone feels threatened by imminent catastrophe. There is a sensation that the sun is about to fall on the earth; that it is being torn from the cosmic laws of its eternal path. Hence the invocations, the prayers, the cries of affliction, as in a universal cataclysm.[98]

Dr. Pereira Gens (cited above) draws attention to fact that even though the brightness of the sun was reduced, the heat of the sun was still present, and seems to have increased as the sun approached, so much that it dried the ground as well as the people who were

---

[97] Translated text in Jaki, *God and the Sun at Fatima*, p. 327, from an interview with Chanoine Barthas as reported in his book *Fatima*, p. 358.

[98] Scatizzi, "Miracle of the Sun".

soaking wet (see no. 6 below). As an astronomer, Father Scatizzi draws attention to the fact that the phenomenon (which cannot be the sun itself—see below) must be occurring through an agency that is beyond the parameters of known physics and astronomy. We will discuss this explicitly in Section II.C below. We now turn to the testimony of some nonscientific people whose impressions complement the observations of the above scientists. We begin with Alfredo da Silva Santos, who de Marchi identifies only as a professional from Lisbon:

> The sun began to move, and at a certain moment appeared to be detached from the sky and about to hurtle upon us like a wheel of flame. My wife—we had been married only a short time—fainted, and I was too upset to attend to her, and my brother-in-law, Joao Vassalo, supported her on his arm. I fell on my knees, oblivious of everything, and when I got up I don't know what I said. I think I began to cry out like the others.[99]

The next testimony comes from Maria da Capelinha (formerly known as Maria de Carreira), an uneducated domestic who became the caretaker of the chapel on the site of the apparitions (hence her new name):

> [The sun] turned everything to different colours—yellow, blue and white. Then it shook and trembled. It looked like a wheel of fire that was going to fall on the people. They began to cry out, "We shall all be killed!" Others called to our Lady to save them. They recited acts of contrition. One woman began to confess her sins aloud, advertising that she had done this and that.... When at last the sun stopped leaping and moving, we all breathed our relief. We were still alive, and the miracle which the children had foretold, had been seen by everyone.[100]

6. *The drying of the ground and clothing.* Dr. Pereira Gens, in his medical autobiography, reported that the rapid approach of the sun dried his wet clothing:

[99] Alfredo da Silva Santos, "Interview by de Marchi", in de Marchi, *True Story of Fatima*.
[100] Maria da Capelinha, "Interview with de Marchi", in de Marchi, *True Story of Fatima*.

It is a truth that although the sun's rays diminished in intensity, the heat maintained itself and although my clothes were somewhat soaked through, I felt them suddenly almost completely dry.[101]

As we shall see, this fact is important because it gives physical evidence (along with other evidence discussed below in Section II.C) that discredits the hypothesis of auto-suggestion and mass hypnosis as an explanation for the miracle.

We now turn to an American, Dominic Reis, who was seventeen years old and living with his parents in Lisbon at the time of the miracle. He noted not only the drying of people's clothing, but also the ground, which had been drenched by the rain:

As soon as the sun went back in the right place the wind started to blow real hard, but the trees didn't move at all. The wind was blow, blow and in few minutes the ground was as dry as this floor here. Even our clothes had dried. We were walking here and there, and our clothes ... we don't feel at all. The clothes were dry and looked as though they had just come from the laundry. I believed. I thought: Either I'm out of my mind or this was a miracle, a real miracle.[102]

Father Inácio Lourenço (a priest from Alburitel—about eleven miles from Fatima) left the following written report, which de Marchi validated with the teachers and various witnesses in the village. He reported the same phenomenon of the approaching sun rapidly drying the clothes of the people in the Cova:

In that hectic noontime, while the great star [sun] hung in cloudless clarity, the people, who had been drenched and soggy with the pelting, unremitting rain, were suddenly and completely dry—their shoes and stockings, their skin and their clothes, as though the Lady of the Rosary had invoked the power of some new machine.[103]

In his book *Meet the Witnesses*, John Haffert gives a written transcript of television interviews with many witnesses, thirteen of whom

---

[101] Translated texts taken from Jaki, *God and the Sun at Fatima*, p. 326, from the Portuguese text of Gens, *Fátima*.

[102] Dominic Reis, "Transcript of a Television Interview with Him", in Haffert, *Meet the Witnesses*, p. 11.

[103] Inácio Lourenço, "Report", in de Marchi, *True Story of Fatima*.

verify that not only their clothes but also others' clothes (as well as the ground) dried very rapidly with the approach of the sun.[104]

In view of the following witnesses and circumstances, it is virtually undeniable that the above six stages of the Miracle of the Sun took place from a little before two o'clock P.M. (midday—high point of the sun) at the Cova da Iria on October 13, 1917, as predicted by Lucia dos Santos on July 13, August 19, and September 13:

- 50,000 witnesses
- The observations of scientists, professors, and doctors who affirm the unique phenomenon
- The testimony of many journalists and government officials, some of whom were formerly hostile to miracles and the Church
- Photographs of the crowd gazing up at the sun, none of whom suffered pain or eye damage
- The simultaneous action of the crowd in kneeling in the mud and screaming out at the approach of the sun
- De Marchi (who lived for seven years after the apparition within sight of the Cova) and his team of interviewers, who testified that they never heard a single person confound or deny the Miracle of the Sun[105]

Since the above witnesses and circumstances go far beyond any scientific, historical, or courtroom criterion for historical veracity, the above six stages of the Miracle of the Sun are considered reasonably and responsibility established as historical, and therefore are those which must be explained. This is our task in Section II.C.

C. Explaining the Miracle of the Sun

Even if the above facts about the six stages of the phenomenon of the sun at Fatima are completely accurate, we cannot leap to the

[104]Dominic Reis (p. 11); Mrs. Guilhermina Lopes da Silva, speaking of her relative Louis Lopes (p. 41); Maria Teresa of Chainca (p. 45); José Joaquim da Assuncao (p. 53); Mario Godinho (p. 56); Augusto Pereira dos Reis (p. 60); João Carreira, son of Maria da Capelinha (p. 64); Maria do Carmo Menezes (p. 65); Joaquim da Silva Jorge (p. 69); Maria Candida da Silva (p. 89); Manuel Antonio Rainho (p. 90); Maria José Monteiro (p. 90); and Joaquim Vincente (p. 91).

[105]De Marchi, *True Story of Fatima*.

conclusion that it was a miracle. The unique extraordinary character of these events strongly suggests that something supernatural occurred, but it is the job of scientists to try in every way possible to find (or rule out) naturalistic causes before appealing to the supernatural. This has always been the Church's procedure for certifying a miracle.[106] So, what have scientists hypothesized as naturalistic explanations for the six stages of the Miracle of the Sun? We may divide the discussion into four parts:

1. Subjective (or collective subjective) explanations, such as mass hypnosis (Section II.C.1)
2. Simple naturalistic explanations, such as the sun, a rainbow, or the aurora borealis (Section II.C.2)
3. Complex objective naturalistic explanations—meteorological (Section II.C.3)
4. Complex objective naturalistic explanations—astronomical (Section II.C.4)

As we shall see, the two complex objective explanations can explain the six stages of the Miracle of the Sun naturalistically, but this does not mean that supernatural intervention (a miracle) was not necessary, for the odds against the two complex naturalistic explanations occurring by pure chance on the date, time, and place predicted by Lucia on three separate occasions are in the order of one part in fifty thousand trillion—exceedingly unlikely! This will be explained below in Section II.C.4.

1. Subjective (and Collective Subjective) Explanations

Critics of the Miracle of the Sun very rarely choose to deny the event, because the witness testimonies are overwhelming. Instead, they choose to explain the phenomenon of the sun in terms of mass hypnosis or mass hallucination—a religiously zealous crowd of fifty thousand were so expecting a miracle that they all simultaneously hallucinate the same general set of events when Lucia calls out, "Look at the sun." This hypothesis has not been given credibility since it was first formulated by hostile journalists for four major reasons:

[106] See Chapter 4, Introduction.

1. There were real *objective* physical events that occurred that cannot be accounted for by a *subjectivist* explanation:
   a. The approach of the sun rapidly dried the ground and people's clothing. Hallucination cannot account for this extraordinarily rapid objective drying process.
   b. About fifty thousand people were looking at the sun for nearly ten minutes, none of whom experienced pain or retinal damage. Anyone who looks at the sun for ten minutes will under normal *objective* circumstances experience both.

2. Many people up to ten to twenty miles away (who did not hear Lucia say, "Look at the sun") experienced most of the six stages of the Miracle of the Sun. How could they be caught up in mass hypnosis or mass hallucination when many of them were not looking for the miracle, were not in proximity to the crowd, and were unaware of what the crowd was experiencing?

3. The crowd was not entirely religiously oriented. There were several skeptics, such as Avelino de Almeida, who went to the grotto specifically to disprove the "miracle", but reported the same events as those who were looking forward to a miracle.

4. There were six aspects of the phenomenon that were completely unique in the history of solar phenomena: (i) The sun dancing and trembling, (ii) the sun rotating rapidly on its own axis, (iii) the sun shooting out segmented monochromatic flares of different colors, (iv) the different colors projecting themselves onto the people and objects in the Cova (e.g., everyone turning blue, and then everyone turning yellow), (v) the sun falling toward the Earth, and (vi) the sun returning to its normal place in the sky. How could fifty thousand or more people simultaneously hallucinate the same six unique events without any preparation? Recall that they knew only there would be a miracle—they did not even know that it would be a miracle of the *sun*.

In view of the unlikelihood of a subjectivist explanation, we must now examine *objective* explanations.

2. Simple, Objective, Naturalistic Explanations

Throughout the last century, five simple objective explanations have been proposed:

1. Some abnormal activity in the sun itself
2. A rainbow
3. The aurora borealis
4. A cloud of stratospheric dust
5. A cloud of dust from the Sahara

Let's consider each hypothesis to determine whether it explains the six stages of the Miracle of the Sun described in Section II.B. The first hypothesis—the extraordinary Fatima phenomenon was caused by the sun itself—was already discussed above. As noted there, Dr. Almeida Garrett[107] (a witness to the miracle at Fatima) made inquiries at national observatories near the time of the miracle, noting that nothing unusual occurred that day. Independent inquiries were also made by Father Pio Scatizzi (professor of astronomy)[108] and Father Stanley Jaki (professor of physics),[109] confirming that the sun did not demonstrate any unusual activity on October 13, 1917.

The second hypothesis—a rainbow—was also discussed above. As Pio Scatizzi noted, the solar flares were manifest in mostly limpid air without reflective agents (like droplets of water) to transmit the prismatic effects.[110] Furthermore, a rainbow does *not* explain the rotational effect of the sun, the falling of the sun and its return, and the rapid drying effect during the fall of the sun.

The third hypothesis—the aurora borealis—was shown to be inadequate by Scatizzi, who noted three important distinctions between the "miracle of the sun" and the aurora borealis:

▪ In auroras, the sun is always invisible, but at Fatima it was clearly visible.

[107] See de Marchi, *True Story of Fatima*, Chapter 1.

[108] According to Scatizzi, "The above-mentioned solar phenomena were not noted in any observatory. Impossible that they should escape the notice of so many astronomers and indeed the other inhabitants of the hemisphere." Pio Scatizzi, "Miracle of the Sun".

[109] Jaki, *Mind's Matter*, pp. 232–35.

[110] According to Scatizzi, "In the case of Fatima, it is extremely difficult to place such a phenomenon within a framework when outside the solar disc there was only limpid air without any reflecting agent, as [would be the case] with a rainbow, when along each monochromatic ray numberless drops of water renew the prismatic effect. In Fatima, as seen by motionless observers, the monochromatic sectors appeared to revolve and to subsist without any support. We must conclude that each colored ray was maintained autonomously, with its origin in the solar body, the air providing no means to transmission." Scatizzi, "Miracle of the Sun".

- The phenomenon at Fatima was well ordered—segmented, distinct, and at regular intervals; however, auroras are precisely the opposite—chaotic.
- An aurora is visible over a wide area, but no aurora was seen by a European observatory on October 13, 1917.

The fourth hypothesis—a cloud of stratospheric dust—was conjectured by Steuart Campbell, who noted that such a cloud made the sun easy to look at and caused it to appear yellow, blue, and violet, and to spin. In support of his hypothesis, he reported that a blue and reddened sun was registered in China and documented in 1983.[111] In response to his hypothesis, it must be said that the phenomenon registered in China, though it did produce atmospheric coloration, was not anything like the Miracle of the Sun at Fatima. There was no spinning effect, no segmented monochromatic flares, no projection of the coloration to objects on the ground, no falling of the sun, and no rapid drying effects. It is a huge leap to move from stratospheric dust causing atmospheric coloration to the six stages of the Miracle of the Sun. He has not given any grounds for why such a leap is scientifically justified. If he is speculating without empirical evidence that stratospheric dust clouds *could* do such a thing, then his conjecture is almost purely hypothetical. To raise it to the level of a theory, he would have to provide evidence closely resembling the Fatima phenomenon, which the China event significantly lacks.

The fifth hypothesis—a cloud of dust from the Sahara—as conjectured by Paul Simons,[112] is purely speculative for the same reasons. Though Saharan dust clouds can produce a reddish atmospheric coloration as well as "blood rain" (when the red pigment of the dust is mixed with rain), such effects leave 95 percent of the Fatima phenomenon completely unexplained—the sun rotating on its own axis, segmented monochromatic flares, projection of multiple colors onto the ground, the falling of the sun, the return of the sun, and rapid drying effects after prolonged rain. As we saw above, to raise this hypothesis to the level of even a speculative theory, there would

---

[111] Steuart Campbell, "The Miracle of the Sun at Fatima", *Journal of Meteorology* 14, no. 142 (October 1989): 334–38, http://www.ijmet.org/wp-content/uploads/2014/09/142.pdf.

[112] Paul Simons, "Weather Secrets of Miracle at Fatima", *Times*, February 17, 2005.

have to be some empirical indication that Saharan dust clouds can do anything like the activities constituting the Fatima phenomenon.

The problem with simple, objective, naturalistic explanations is that they fall short—very short of being an explanation for the six stages of the Fatima phenomenon described above (Section II.B). Each simple explanation shows basic similarity in one or two respects, but leaves unexplained virtually all of the unique objective dimensions of the Fatima phenomenon—such as the sun's rotation, segmented, mono-chromatic flares, and the falling and return of the sun. As such, we must leave them behind as inadequate hypotheses and move to hypotheses that can explain most or all of the reported objective events that took place at midday on October 13, 1917, over the Cova da Iria.

3. Complex, Objective, Naturalistic Explanation—Meteorological

The best complex meteorological proposal has been given by Father Stanley Jaki, who summarized his conclusions from *God and the Sun at Fatima*[113] in his intellectual autobiography *A Mind's Matter*:[114]

> Enough data are on hand to force one to recognize the meteorological nature of "the miracle of the sun" and to look askance at the phrase, "the sun danced over Fatima." That the miracle was not solar, that it did not imply any "solar activity" in the scientific sense of that term, is indicated by the fact that nothing unusual was registered by observa-tories about the sun at that hour. Prior to that hour, rain was coming down heavily over the area from the late morning hours on, with the clouds being driven fast by a westerly wind across the sky. A cold air mass was obviously moving in from the Atlantic, only at about 40 kms from Fatima, which itself is at about 15 kms to the east from the line where the land begins to form a plateau well over 300 meters above sea level. The hollow field, Cova da Iria, outside Fatima is itself at about 370 meters. An actual view of the geographic situation is a great help for an understanding of the true physical nature of "the miracle of the sun," especially when one takes a close look at cloud patterns typical over the Cova.

[113] Jaki, *God and the Sun at Fatima*, pp. 343–60.
[114] Jaki, *Mind's Matter*, pp. 232–35.

I feel that at this juncture I must summarize my explanation of the miracle. It began at about 12:45 P.M., solar time, after the rain suddenly stopped, and lasted about ten to fifteen minutes. During all that time, the sun, that had not been seen for hours, appeared through thin clouds, which one careful observer described as cirrus clouds. Suddenly the sun's image turned into a wheel of fire which for the people there resembled a "rodo de fuogo" familiar to them in fireworks. The physical core of that wheel was, as we now have to conjecture, an air lens full of ice crystals, as cirrus clouds are. Such crystals can readily refract the sun's rays into various colors of the rainbow.

The references to the strong west-east wind and to the continued drift of clouds may account for the interplay of two streams of air that could give a twist, in a way analogous to the formation of tornadoes, to put that lens-shaped air mass into rotation. Since many present there suddenly felt a marked increase in temperature, it is clear that a sudden temperature inversion must have taken place. The cold and warm air masses could conceivably propel that rotating air lens in an elliptical orbit first toward the earth, and then push it up, as if it were a boomerang, back to its original position. Meanwhile the ice crystals in it acted as so many means of refraction for the sun's rays. Some eyewitnesses claimed that the "wheel of fire" descended and reascended three times; according to others this happened twice. Overwhelmed by an extraordinary sight that prompted most of the crowd to fall on their knees, even "detached" observers could not perform as coolly as they would have wished. Only one observer, a lawyer, stated three decades later that the path of descent and ascent was elliptical with small circles superimposed on it.

Such an observation would make eminent sense to anyone familiar with fluid dynamics or even with the workings of a boomerang. There is indeed plenty of scientific information on hand to approach the miracle of the sun scientifically. This is, however, not to suggest that one could reproduce the event say in a wind tunnel. *The carefully co-ordinated interplay of so many physical factors would by itself be a miracle*, even if one does not wish to see anything more in what actually happened. Clearly, the "miracle" of the sun was not a mere meteorological phenomenon, however rare. Otherwise it would have been observed before and after, regardless of the presence of devout crowds or not. I merely claim, which I did in my other writings on miracles, that in producing miracles God often makes use of a natural substratum by greatly enhancing its physical components and their interactions. One can indeed say, though not in the sense intended by some Fatima writers, that the fingers of the

Mother of God played with the rays of the sun at that extraordinary hour at Fatima.[115]

We can glean three significant explanatory factors from Jaki's complex meteorological explanation:

1. Unlike the simple, objective, naturalistic explanations (given in Section II.C.2), Jaki's explanation explains all the unique phenomena witnessed by thousands over the Cova da Iria at midday on October 13, 1917:
   a. The rotation of the sun produced by a lens-shaped air mass (an air lens) filled with ice crystals taken from the nearby cirrus clouds
   b. The refraction of different colors produced by the refraction activity of the ice crystals
   c. The rotation of the air lens produced by strong winds moving in different directions causing the air lens to rotate like a tornado
   d. The falling and returning of the sun produced by a sudden temperature inversion causing a warm air mass along with the cold air mass. The cold air mass would be pushing the air lens toward the Cova while the warm air mass (nearer to the ground) would push it back into the sky causing a boomerang effect (like that in fluid dynamics).
   e. The rapid drying of the ground and people's clothes produced by the warm air mass from the temperature inversion (nearer to the ground)
   Jaki's explanation truly accounts for all unique aspects of the Fatima phenomenon in a single plausible, complex, naturalistic explanation; therefore, it meets the requirements for an explanatory theory.
2. Though Jaki's complex explanation is entirely naturalistic, it still entails a supernatural force to bring this complex assemblage into unity at a particular point in time. Note that there are four different naturalistic phenomena that have to be very precisely coordinated over the Cova da Iria—the formation of an air lens,

---

[115] Ibid., pp. 232–35 (emphasis added).

the filling of that air lens with ice crystals from nearby cirrus clouds, the positioning of the air lens in the midst of strong winds moving in different directions, and the occurrence of a sudden temperature inversion in such a way that the resultant warm air mass would interact with the cold air mass above it to produce a boomerang effect. What are the odds that these four natural events would coincide to produce the Fatima phenomenon? Given that this kind of event (the Fatima phenomenon) is completely unique in recorded history, we might infer that it is very rare. Indeed if we look at the probability of the precise coordination of these four natural forces in one restricted locale to produce the Fatima phenomenon, it would likely be less than one chance in a trillion, but this is just the beginning of the improbability. We now must explain how this highly improbable coordination of natural forces occurred at the date, time, and place predicted by Lucia dos Santos on July 13, August 19, and September 13. We are now in the range of one part in thousands of trillions—exceedingly, exceedingly unlikely.

3. This lead Father Jaki to conclude that there is a very strong likelihood of supernatural agency to bring this unique diverse array of natural forces into the Fatima phenomenon at the date, time, and place predicted. He describes this as a supernaturally enhanced naturalistic phenomenon: God using a natural substratum by greatly enhancing its physical components and their interactions.[116] In this sense, he can say that "the fingers of the Mother of God played with the rays of the sun at that extraordinary hour at Fatima."[117]

4. Complex, Objective, Naturalistic Explanation—Astronomical

In 2021, Dr. Ignacio Ferrín (Ph.D. in astro-geophysics, University of Colorado—Boulder) proposed another complex, objective, naturalistic explanation based on an astronomical event instead of a complex of meteorological events. He shows that a small comet entering the

[116] Ibid.
[117] Ibid.

earth's atmosphere hurtling toward Fatima that disintegrates completely as it approaches the Cova could explain all six of the unique aspects of the Fatima phenomena. It should be noted that such a small comet that completely disintegrates in the atmosphere immediately before making contact with the earth is an exceedingly rare event.[118] Such a comet oriented to disintegrate precisely above the Cova da Iria at the day and time predicted by Lucia is an exceedingly, exceedingly rare event (explained below).[119] The following is a summary of Ferrín's explanation taken from his book *An Astronomical Explanation of the Fatima Miracle*:[120]

1. *The mitigation of the sun's aggressive rays so that the crowd could gaze upon the sun for over ten minutes without pain or retinal damage.* Ferrín interprets this as

   > at that moment the small comet had entered the atmosphere from the direction of the Sun, and it was covering its disk with dust producing a solar eclipse of a different type, "a dimming by a dust cloud." The surface temperature of the Sun is about 5700° Centigrade, so the surface is too bright to look at. When the comet enters, at the contact with the atmosphere it begins to heat up and evaporates dust and gases. This evaporation creates an atmosphere (a halo), that subdues the Sun's light. The Sun is seen as if through clouds, smoke or snowflakes. From now on the people are looking at the Sun, but through the dusty comet's atmosphere that is growing extraordinarily in opacity.[121]

2. *The trembling or "dancing" of the sun.* Ferrín explains:

   > Since the [comet] may not be spherical, and since it may rotate, and since it is losing mass, the mass may come apart in different alternative sides giving the impression of "dancing [or trembling]."[122]

---

[118] Ferrín, *Astronomical Explanation.*
[119] Ibid.
[120] Ibid.
[121] Ibid.
[122] Ibid.

3. *Dancing continued; the descent and reascent of the rotating sun.* Ferrín interprets this as follows:

> "Whirling upon itself," rotating upon itself. "Giving the impression of approaching and receding." As the comet moves down the atmosphere, the front part is heating more and more as it traverses thicker regions. Due to the heat and pressure interactions, it sheds mass that moves toward the observer. The mass is carried over to the sides by the shock wave. This phenomenon happened several times, giving the impression of approaching (sheds mass) and receding (the mass is carried away to the sides).[123]

4. *The segmented monochromatic flares from the rotating sun, which turned objects on the ground blue then yellow.* Ferrín explains this as follows:

> Comets are composed of many kinds of molecules. The most abundant are water (H_2O), carbon Monoxide (CO), carbon dioxide (CO_2), ammonia (NH_3), methane (NH_4), cyanogen (CN), sodium (Na), nickel (Ni), and magnesium (Mg). When excited, these molecules exhibit spectral lines in different parts of the spectrum. For example iron and CO_2 will shine brightly in the visual region, yellow. CN will produce a line in the blue, and ammonia and methane in the red. Sodium will be yellow-orange, nickel green, and magnesium yellow-green and amethyst. Therefore, the sighting of colors can be explained if different layers of the comet are being exposed and excited to brighten depending on the temperature of the object.
>
> The CN Cyanogen molecule is the most intense in all comets. So, when the comet entered the atmosphere, this molecule overcame all others with a deep blue-violet hue (amethyst would be a good comparison), since the band shines at 3883 Angstrom of wave length in the blue-violet region of the spectrum. Actually it is a beautiful almost monochromatic color. Later on, when the temperature of the bolide increased due to the heat generated in the atmosphere, the hue changed to yellow.[124]

---

[123] Ibid.
[124] Ibid.

5. *The falling of the sun toward the earth.* Ferrín explains this as follows:

> The comet was reaching to the end of its trajectory in the atmosphere. Now it may have been at only 10 to 20 km above, just a few seconds from hitting ground. It was actually falling, and that is what the people saw.
>
> "The Sun began to move as if being detached from the sky and was falling upon us. It was as if a wheel of fire was about to fall on the people."
>
> That was precisely what was happening. Indeed the comet was falling after having traversed most of the atmosphere. So it had to be hot, fiery and falling.[125]

6. *The rapid drying of the ground and the people.* Ferrín interprets this as follows:

> It is well known, that whenever a bolide enters the atmosphere, it will explode under the atmospheric pressure, and will send a sonic wave or air blast in the direction of the impact.... [According to witness Dominic Reis], "the wind started to blow real hard," "and in a few minutes the ground was dry as this floor here. Even our clothes had dried," is the best demonstration that it was an atmospheric impact that produced it, and a good proof that the event was real.[126]

Ferrín's astronomical explanation is not beyond the realm of possibility. Indeed, he provides several pictures and exhibits showing how meteor and comet behavior could produce the unique visual features of the Fatima phenomenon. Though this explanation is naturalistic, it shares the same feature as the meteorological explanation of Stanley Jaki—the need for supernatural intervention. Recall that Jaki calls this, supernaturally enhanced natural causation—"God using a natural substratum by greatly enhancing its physical components and their interactions."[127]

[125] Ibid.
[126] Ibid.
[127] Jaki, *Mind's Matter*, p. 235.

Ferrín addresses the need for "supernatural enhancement of natural causation" as follows:

> The probability of [a small comet] "falling on Fatima on that date and at that time within 1 hour, coming from the Sun's direction," is 1 in 6,540×405,896×365×10 = 9,500,000,000,000 (one result in 9.5 trillion "tries"). But we know the time of the event with an error of ±15 minutes, so the above number has to be multiplied by a factor of 2, or a probability of 1 in 19,378,000,000,000 (one result in 19.4 trillion "tries").
>
> This number is the probability of a stone falling on Fatima. Above we calculated that in a hundred years, the probability of a comet falling is less than one in 3,170. If we multiply the above number by 3,170 we get one chance 61,428,000,000,000,000 (one result in 61,000 trillion "tries"). That is small![128]

When Lucia made her predictions on July 13, August 19, and September 13, she had one chance in sixty-one thousand trillion times of being correct. At this point, her chances of being correct were virtually nonexistent—yet she was. The forces of nature by themselves could not have realistically brought this about. So Ferrín concludes his complex astronomical explanation, as Stanley Jaki did in his complex meteorological explanation, that there has to be an intervening supernatural agency that enables this exceedingly improbable result to occur. He calls the result of this supernatural agency, "a miracle".[129]

D. Conclusion

As noted in Chapter 4, there can be no *direct* scientific test for a miracle, because science is concerned with observable natural facts while miracles are about transphysical/supernatural events. However, we can make a reasonable and responsible inference that supernatural agency is virtually necessary for a particular event to occur—an indirect scientific validation of a miracle. The Miracle of the Sun at Fatima is perhaps the clearest case of such an indirect scientific

---

[128] Ferrín, *Astronomical Explanation*.
[129] Jaki, *Mind's Matter*, p. 235.

validation as we will ever get, because we can estimate or calculate the odds against its occurrence through a complex of natural causes.

Throughout Section II, we first showed that the number and professional status of the witnesses, the affirmation of the events by hostile witnesses, and the physical effects (rapid drying of the ground and clothes as well as no retinal damage or pain after gazing at the sun for ten minutes) reasonably substantiate the veracity of the six stages of the Miracle of the Sun at Fatima (Sections II.A and II.B). We then turned to science to explain this phenomenon—the Catholic Church's procedure for determining a miracle (Section II.C). This critical analysis showed that collective subjectivistic explanations (such as mass hypnosis and mass hallucination) did not stand up to the facts (Section II.C.1). We then assessed five simple, naturalistic, objective explanations—the sun itself, a rainbow, the aurora borealis, a cloud of stratospheric dust, and a Saharan dust cloud. All of these hypotheses fell far short of explaining most of the unique features of the Fatima phenomenon, and so they were judged to be highly unlikely as valid scientific explanations (see Section II.C.2).

We then turned to two complex, objective, naturalistic explanations— a meteorological one (proposed by Stanley Jaki—Section II.C.3) and an astronomical one (proposed by Ignacio Ferrín—see Section II.C.4). Both of these explanations fulfilled the requirements for a valid scientific explanation because they naturalistically explain all the unique features in the six stages of the Fatima phenomenon. However, they both fall short of a *complete* explanation in one respect—the odds against them are in the order of one result in tens of thousands of trillions of "tries". Since such complexes of rare natural causes are so exceedingly improbable to occur naturally, both Jaki and Ferrín infer the presence of a *supernatural* agency to bring it about.

Perhaps another scientist will provide yet another complex, naturalistic explanation for this event, but it too will likely be exceedingly improbable, which could explain why the Fatima miracle is unique in recorded scientific history. In view of this, we have very strong indirect inferential grounds that a true miracle of grand proportions took place on October 13, 1917, at the time and place predicted over three months by Lucia Dos Santos.

Did the supernatural causative agent (e.g., God or the Blessed Virgin) have to use a complex of rare natural causes to create the six stages of the Fatima phenomenon? Such a supernatural agent certainly could have used these natural events (as described by Jaki and Ferrín), but this agent did not *have to* do so. The supernatural agent could have created a supernatural lens in the atmosphere above the Cova da Iria with all the properties needed to produce the unique features of the Fatima phenomenon. Of course this would be just as miraculous as bringing together the complexes of rare natural events over the Cova da Iria at the date and time predicted by Lucia. Whatever the case, we have good scientific grounds for establishing that supernatural agency (a miracle) is virtually unavoidable.

Who or what was the source of this supernatural agency? According to Lucia, it was the Blessed Virgin Mary—the Mother of our Lord, Jesus Christ—who came with a warning about the evil spirit and his domain, as well as a message of hope founded in repentance and prayer—especially the Rosary. If Lucia was correct about her triple prediction, why would we not believe her about the Blessed Virgin and her message of warning, repentance, hope, and prayer?

Chapter Six

Science at the Doorstep to Mary—Lourdes

Introduction

The earlier discussion of miracles (Chapter 4) explained that science cannot *directly* validate a miracle, because science begins with observable data and seeks a *naturalistic* explanation, while miracles are by definition *beyond* nature. Nevertheless, science can give an *indirect* account of the *possibility* or even the likelihood of a miracle by showing that the phenomenon in question cannot be explained by any known natural cause after exhaustive examination.

Thus, even with the strictest reasonable criteria for ascertaining a transphysical cause or agency (see below—the Lambertini criteria), faith must accompany science and reason in such a declaration. At the very least, the authority who declares a miracle must have a minimal belief in a providential supernatural agency (generally called "God") who is concerned with the good of humanity. Without this minimal faith, scientific authorities can declare only that a phenomenon is beyond all known naturalistic explanation and has not occurred in any context that was not itself miraculous. As we shall see, Lourdes has been the center of thousands of such phenomena since 1858, dividing the medical and scientific community into two groups—those who declare a phenomenon to be "miraculous" and those who declare it to be "scientifically inexplicable and of unknown origin".

According to the 2014 survey published in the *Journal of Religion and Health*, 65 percent of physicians are believers in God; 51 percent consider themselves religious and 24.8 percent spiritual, while

12.4 percent are agnostic and 11.6 percent are atheist.[1] Furthermore, 74 percent of physicians believe that miracles have occurred in the past, and 73 percent believe they occur in the present.[2] In view of this we might expect to find that about 75 percent of physicians will be comfortable with associating instantaneous, lifelong, scientifically inexplicable cures at Lourdes with miracles, while about 25 percent will not. There is also a group of physicians who were agnostics or atheists before going to Lourdes, but after being presented with an instantaneous cure of a severe organic malady without any previous precedent (outside the context of a miraculous phenomenon), they became believers in God and religious practitioners. Such is the case with the Nobel prize winner Dr. Alexis Carrel (see below, Section II.B) as well as Dr. van Hoestenberghe, the specialist who examined the leg of Pierre de Rudder (see below, Section II.A).

Now let us turn specifically to the miracles of Lourdes. The appearance of the Blessed Virgin Mary to Bernadette Soubirous at the Grotto of Lourdes in 1858 is probably the most well-known Marian apparition in history—not only because of the apparition itself, but also because of the thousands of scientifically unexplained cures of severe organic maladies that have taken place through the water of Lourdes (approximately seven thousand recorded in the Lourdes Office of Medical Observations[3]

---

[1] Kristin A. Robinson et al., "Religious and Spiritual Beliefs of Physicians", *Journal of Religion and Health* 56, no. 1 (February 2017), https://pubmed.ncbi.nlm.nih.gov/27071796/.

[2] Shoba Sreenlvasan and Linda E. Welnberger, "Do You Believe in Miracles? Turning to Divine Intervention When Facing Serious Medical Illness", *Psychology Today*, December 15, 2017, https://www.psychologytoday.com/us/blog/emotional-nourishment/201712/do-you-believe-in-miracles#:~:text=Even%20physicians%20believe%20in%20miracles.%20A%20national%20poll,occur%20today%20%28Poll%3A%20Doctors%20Believe%20in%20Miracles%2C%202004%29, citing a survey conducted by the Louis Finkelstein Institute for Religious and Social Studies. See also Peter P. Moschovis, "'Lord, I Need a Healing': The Uneasy Relationship between Faith and Medicine", *American Medical Association Journal of Ethics* 7, no. 5 (May 2005), https://journalofethics.ama-assn.org/article/lord-i-need-healing-uneasy-relationship-between-faith-and-medicine/2005-05#:~:text=A%202004%20national%20survey%20of,and%20necessary%20guide%20to%20life.

[3] The seven thousand recorded cures have varying degrees of tests and records to establish scientific inexplicability. According to Dr. Michael Lasalle of the Lourdes Medical Bureau, twenty-five hundred of these cures have sufficient tests and records to be classified as "medically inexplicable". This does not mean that the other forty-five hundred cures are scientifically explicable, but only that the nature, testing, and records leave the possibility open for some unknown scientific explanation. See personal communication of Dr. Lasalle to Paul Glynn in June 1997 in Paul Glynn, *The Healing Fire of Christ: Reflections on Modern Miracles—Knock, Lourdes, Fatima* (San Francisco: Ignatius Press, 2003), 73.

between 1858 and 2018[4]). Though these scientifically inexplicable cures have been well documented by multiple physicians and scientists, a strict interpretation of the Lambertini criteria for a miracle is very difficult to meet:[5]

1. The disease must be severe and incurable, or at least very difficult to treat.
2. The disease cannot be at a stage where it is likely to disappear shortly.
3. No curative treatment had been given, or if treatment has been given, it must be proved that it is not related to the cure.
4. The cure must be instantaneous—or near instantaneous.
5. The cure must be complete.
6. The cure must be permanent—no relapse.
7. The implicit underlying criterion is that all possible natural explanations—both physical and psychological—be definitively ruled out.

A strict interpretation of these criteria is exceedingly difficult to meet. For example, it is difficult to prove that a past treatment did not affect any part of the cure in *any* way (Lambertini criterion no. 3). Again, a strict interpretation of "complete" (Lambertini criterion no. 5) would mean that there could not be *any* remnant of the disease—no matter how benign—such as an impression or discoloration in the place where bone, muscle, and/or organ tissue had spontaneously regenerated. Further, a strict interpretation of "permanent" (Lambertini criterion no. 6) requires examining the person until the day of death. Finally, the implicit underlying criterion is that all possible natural explanations—both physical and psychological—be ruled out. This means considering an incredibly long list of possibilities for which there may not be definitive data to rule out even the most

---

[4] This number was disclosed by Dr. Alessandro de Franciscis, current president of the Lourdes Office of Medical Observations, in an interview given to Father Seán Connolly, "70th Miracle of Lourdes Affirmed by the Church", *Catholic World Report*, February 13, 2018, https://www.catholicworldreport.com/2018/02/13/70th-miracle-of-lourdes-affirmed-by-the-church/.

[5] Cardinal Prospero de Lambertini, *De Servorum Dei Beatificatione et Beatorum Canonizatione* (Bologna: Formis Longhi Excusoris Archiepiscopalis, 1734–1738), Liber Quartus et Ultimus, Pars Prima, Caput Octavum, 2. These are explained in Bernard François, Esther M. Sternberg, and Elizabeth Fee, "The Lourdes Medical Cures Revisited", *Journal of the History of Medicine and Allied Sciences* 69, no. 1 (January 2014).

unlikely possibility. In view of this, it should not be surprising that only 1 percent (seventy out of seven thousand)[6] of unexplained cures of severe diseases at Lourdes are *officially* declared a miracle by the competent ecclesiastical authority (after the strictest interpretation of these criteria has been fulfilled and certified by competent medical and scientific authorities). Notwithstanding this, the number of remarkable cures at Lourdes is about five to ten times higher than the seven thousand registered and recorded in the Lourdes Office of Medical Observations (thirty-five thousand to seventy thousand cures).[7] The primary reason for the significant understatement of officially recorded cures is that many of those cured do not want to go through the extensive process of being physically examined and submitting records before and after the cures, as well as multiple interviews and follow-up examinations.[8] They appear to be happy with their cure, thankful to the Lord and the Blessed Virgin, and ready to go home.

There have been steady improvements in the evaluation of inexplicable cures at Lourdes as well as advances in diagnostic technology and medical treatment of various diseases; in the midst of this, there are still approximately fifteen scientifically inexplicable cures per year presented and recorded in the Lourdes Office of Medical Observations.[9] As noted above, the true number of inexplicable cures is probably five to ten times greater than the ones presented to the Lourdes Office of Medical Observations[10]—approximately 75 to 150 inexplicable cures per year. In Section II, I took a selection of cures from different eras that were clearly rapid, physical-organic, enduring, and scientifically inexplicable. They are a mere microcosm of the

---

[6] There are several detailed medical accounts of many of these miracles, many of which were not declared "miraculous" but were nevertheless rapid, scientifically inexplicable, and permanent. See, for example, Glynn, *Healing Fire*; Ruth Cranston, *The Miracle of Lourdes* (Toronto: Galilee Trade, 1988); Édouard Le Bec, *Medical Proof of the Miraculous: A Clinic Study* (New York: P.J. Kenedy & Sons, 1923); and François Leuret, *Modern Miraculous Cures—A Documented Account of Miracles and Medicine in the 20th Century* (London: Peter Davies, 1957).

[7] François, Sternberg, and Fee, "Lourdes Medical Cures", p. 159.

[8] Ibid. For a description of the many steps in recording a cure with the Lourdes' Medical Bureau, and a history of the development of the medical bureau, see John Dowling, "Lourdes Cures and Their Medical Assessment", *Journal of the Royal Society of Medicine* 77 (August 1984), https://journals.sagepub.com/doi/pdf/10.1177/014107688407700803.

[9] Roger Pilon, *Lourdes Magazine*, July 1996, cited by Glynn, *Healing Fire*, p. 55.

[10] François, Sternberg, and Fee, "Lourdes Medical Cures".

seven thousand unexplained cures recorded by the Lourdes Office of Medical Observations (pointing to thirty-five thousand to seventy thousand actual cures not recorded by that office).

The above statistics should provoke curiosity, if not wonder, because the large number of these scientifically inexplicable cures occurring at a *single* locale (which do not occur in other nonmiraculous locales) is exceedingly improbable unless there is some *common cause* of the cures associated with the locale. Inasmuch as this common cause is consistently shown to be *nonnatural*, it is not unreasonable to infer that it is *supernatural*—as suggested by the apparition of the Blessed Virgin that gave rise to the Lourdes spring (see below, Section I). Though this does not provide *direct* scientific validation of supernatural agency at Lourdes, it does provide strongly probative inferential validation of a nonnatural/possibly supernatural common cause, if we are not to believe that this large number of inexplicable cures occurred at this one locale by sheer coincidence. As noted above, this judgment requires faith, but it is certainly not blind faith. We might say that it is faith supported by thousands of strongly probative inexplicable cases with a seemingly common nonnatural cause associated with the Blessed Virgin Mary.

As the cases discussed in Section II indicate, faith is an inextricable part of the healings that occur at Lourdes. Recall from Chapter 1 (Section IV) that faith has three stages: (1) attending to the call of the sacred divine reality, (2) openness to the call of the sacred divine reality, (3) following the call of the sacred divine reality, particularly in the revelation of Jesus. Most people who go to Lourdes are involved in the third stage, but some are engaged only in the first or second stage. It is unlikely that someone who has intentionally ignored the beginning stage of faith would go to Lourdes, except as a curiosity or some other secular motive (such as that of Dr. Alexis Carrel—see below, Section II.B).

With this understanding of "faith", it is apparent that most Lourdes miracles enhance the faith of most who come even if they are not cured—even if they are simply accompanying people who did not get cured. As noted above, many doctors were moved from unbelief to belief, and most pilgrims seem to walk away with an increased peace coming from increased trust in God and the Blessed Mother. As one studies the phenomenon of Lourdes, with its four to six million pilgrims per year,[11] the big "miracle" goes far beyond scientifically inexplicable cures, and moves outward to peace, conversion, trust, and joy for those millions who witness its transformative power emanating from the baths, the Eucharistic processions, and the deep devotion and faith of the pilgrims. If one comes to Lourdes with faith, it is very likely to deepen and grow, bringing with it the humble love, devotion, and repentance the Blessed Virgin requested. The most remarkable aspect of Lourdes is how this faith, prayer, devotion, humility, and love combine to form a healing ethos that touches the body, mind, heart, and above all, the spirit.

I. The Apparition of Our Lady of Lourdes

Let us now turn to the apparition itself. On February 11, 1858, just outside of Lourdes, France, Bernadette Soubirous (a fourteen-year-old girl without much formal education), her sister Marie Toinette, and her friend Jeanne Abadie were searching for kindling and bones in a cave. Just as she had taken off her shoes and stockings, a Lady, small in stature, dressed in white with a blue sash around her waist and holding a gold rosary, appeared to her. Bernadette tried to make the sign of the cross but was so scared she could not, at which point the Lady asked her to pray the Rosary with her, restoring her calm. Bernadette was the only one to see and hear the apparition.

When Marie Toinette returned home, she told their mother, inciting both parents to punish them for telling such a "story". Nevertheless, Bernadette was drawn back to the cave of Massabielle, and the Lady appeared to her again. Bernadette brought holy water with her and sprinkled it on the apparition to see if she would shrink from it, but the Lady only smiled, at which point Bernadette told her that if she was not of God, she would have to go away.[12] The Lady smiled and bowed, and Bernadette went into an ecstasy—sensing her holiness

---

[11] Jennifer Klimiuk and Kieran J. Moriarty, "The Lourdes Pilgrimage and the Impact on Pilgrim Quality of Life", *Journal of Religion and Health* 60, no. 6 (September 2021), https://www.ncbi.nlm.nih.gov/pmc/articles/PMC8428497/.

[12] Service Communication, "The Bernadette Year Has Begun", January 17, 2009, on the official Lourdes website, https://www.lourdes-france.org/en/the-bernadette-year-has-begun/.

and love. Her companions witnessed this ecstasy which lasted long after the apparition.

Bernadette returned a third time to the Grotto, and the Lady gave her instructions to return several times throughout the upcoming two weeks. On February 20, the Lady taught her a prayer and asked for penance for the conversion of sinners.

Bernadette returned to the Grotto on several other occasions accompanied by hundreds of people. The official Lourdes website lists the major points of the apparitions as follows:

Sunday, February 21. Early in the morning, the Lady appeared to Bernadette, who was accompanied by about a hundred people. After the apparition, she was questioned by the police commissioner, Jacomet.

Tuesday, February 23. Bernadette arrived at the Grotto, surrounded by 150 people. The Lady revealed to her a secret "only for her alone".

Wednesday, February 24. The Lady gave the following message: "Penance! Penance! Penance! Pray to God for sinners. Kiss the ground as an act of penance for sinners!"

Thursday, February 25. Three hundred people were present. Bernadette recalls: "She told me to go, drink of the spring.... I only found a little muddy water [after considerable digging with her hands]. At the fourth attempt I was able to drink. She also made me eat the bitter herbs that were found near the spring, and then the vision left and went away." This "embarrassing" incident of Bernadette trying to drink this moist mud suggested that "the Lady" was either fictitious or playing with her. However, water began to trickle from the rock near the place Bernadette was digging. Later, stone masons opened the aperture in the rock from which the water was trickling, which created a steady stream of hitherto unknown spring water from the cave of Massabielle. A few days later, this water would be found to have seemingly miraculous healing power.

Saturday, February 27. Eight hundred people were present. The Lady was silent. Bernadette drank the spring water then carried out her usual acts of penance.

Sunday, February 28. Over one thousand people were present when the ecstasy occurred. After praying, Bernadette kissed the

ground and moved on her knees as a sign of penance. She was
then brought to the home of Judge Ribes, who threatened to
put her in prison.

Monday, March 1. Over fifteen hundred people assembled, including
a priest for the first time. In the night, a friend from Lourdes who
had a dislocated arm, Catherine Latapie, went to the Grotto. She
plunged her dislocated arm into the water of the spring; her arm
and her hand regained their movement.[13]

The rest of this brief treatment of the apparition comes from Franz
Werfel's book *The Song of Bernadette*.[14] Werfel was a world-class poet
and historical writer, a believing Jew who elevated his view of God
to "the Redeemer"[15] after his experience at Lourdes. He painstak-
ingly researched the apparition and wrote the following in the pref-
ace to his profound literary-historical work:

> All the memorable happenings that constitute the substance of this
> book took place in the world of reality. Since their beginning dates
> back no longer than eighty years [now 163 years], there beats upon
> them the bright light of modern history and their truth has been con-
> firmed by friend and foe and by cool observers through faithful testi-
> monies. My story makes no changes in this body of truth.[16]

The apparitions and inexplicable cures that continued to happen
caused a great deal of controversy within both the Church and the
town. A decision was made in March to barricade the Grotto, which
had the effect of bringing it to the attention of the national press and
national government. Bernadette was not to be deterred, and so visited the barricaded Grotto at night on several occasions.

Later, the Grotto was unblocked by order of Emperor Louis
Napoleon himself, who was probably the only person who could
have done so given the public controversy throughout France and

[13] "The Apparitions", Lourdes Sanctuaire (website), accessed April 2, 2024, https://www
.lourdes-france.com/en/the-apparitions/.

[14] Franz Werfel, *The Song of Bernadette* (San Francisco: Ignatius Press, 2006).

[15] Glynn, *Healing Fire of Christ*, p. 142. "The Redeemer" may well be a subtle reference to
Jesus Christ, who not only can transform evil into good in this world, but also definitively in
the life to come.

[16] Werfel, *Song of Bernadette*, p. xiv.

secularist resistance at the time. Amazingly (or perhaps best providentially), the emperor's little son took ill, and his wife, Empress
Eugenie, had heard about the miraculous water of Lourdes; she asked
her close friend to go to Lourdes and procure some for her son.
When the son ("LouLou") quickly recovered from his illness, the
empress told Louis Napoleon that he would be ignoble if he did
not publicly acknowledge the cure and order the unblocking of the
Grotto. Her persistence eventually won out, for this is precisely what
he did.[17] The Grotto was opened, not to be closed again—except for
the Covid closure between 2020 and 2022.

On March 25, 1858, the Lady declared that she was "the Immaculate Conception"— identifying herself with her freedom from original sin at conception (the doctrine that had been declared only four
years before in 1854). The ecclesiastical authorities became quite
suspicious, believing that Bernadette must have heard this from a
teacher, her parents, or perhaps a priest. However, after multiple
interrogations, it became clear that Bernadette had not even heard
the phrase before that day, as well as did not understand it and had
never before identified "the Lady" with the Blessed Mother.[18] She
maintained the truth of this statement throughout subsequent interrogations. She joined a very rigorous religious order (the Sisters of
Charity at Nevers) until her death. As she lay on her deathbed, she
was again interrogated by Church officials under pain of sin to determine whether she was certain that the declaration came from "the
Lady". She, of course, held to her original point, and died peace-
fully.[19] Thirty-nine years after her death, her body was exhumed and
found to be almost entirely without corruption.[20]

Since the emergence of the spring, inexplicable cures began to
occur through its water. Dr. Pierre-Romain Dozous, though skeptical at first, became curious about Bernadette's claims when a patient
of his, Pierre Bouriette, claimed to have been cured in his right eye,
which was seemingly permanently damaged from an explosion in a
rock quarry where he was working.[21] Dozous decided to be a medical

[17] Ibid., pp. 412–14.
[18] Ibid., p. 298.
[19] Ibid.
[20] Ibid., p. 574.
[21] Leuret, *Modern Miraculous Cures*, p. 135.

observer of Bernadette during her apparitions, attesting that she was not emotionally excited during her ecstatic trance-like states (indicating that they were not the product of an overwrought or hysterical state of mind). He later discovered, while Bernadette was holding a candle during her ecstasy, that the flame, which was licking her fingers for ten minutes, caused no pain or damage to her skin or hand.[22] Unable to explain this medically, scientifically, or psychologically, he began to question his skepticism. Soon after, he investigated the remarkable cure of Louis Bouhohorts, an eighteen-month-old boy who was on the verge of death. The young boy was subsequently, instantaneously, and completely cured of his disease, osteomalacia (fragility of the bones), which made him incapable of standing or movement. After reviewing the medical records and events several times, Dozous could no longer deny that something truly extraordinary was happening at Lourdes through "the Lady's" communication with Bernadette.[23] As a result, he became a believer and forerunner of the Lourdes Bureau of Medical Records.[24]

The Catholic Church was concerned about the immense popularity of the Grotto and the potential for people to be misled, and so the bishop assembled an ecclesiastical committee in November 1858 to assess the veracity of Bernadette's apparitions.[25] On January 18, 1860, the bishop, following the advice of the committee, declared the apparition to be authentic. Bernadette was canonized as a saint in 1933. As mentioned above in the Introduction, four to six million pilgrims per year visit the shrine, and hundreds receive extraordinary and miraculous cures.[26]

The events surrounding the apparition of the Lady to Bernadette between 1858 and 1859 are corroborated by the thousands of scientifically unexplained cures through the spring identified by the Lady. Yet, the apparition to Bernadette itself stands on its own as something filled with grace and providence:

---

[22] Werfel, *Song of Bernadette*, p. 324.

[23] Leuret, *Modern Miraculous Cures*, p. 141.

[24] Ibid., p. 134.

[25] The Church has definite criteria for judging the authenticity of a Marian apparition. See Chapter 5, Introduction, and Sacred Congregation for the Doctrine of the Faith, *Norms regarding the Manner of Proceeding in the Discernment of Presumed Apparitions or Revelations* (February 25, 1978), I, https://www.vatican.va/roman_curia/congregations/cfaith/documents/rc_con_cfaith_doc_19780225_norme-apparizioni_en.html.

[26] Klimiuk and Moriarty, "Lourdes Pilgrimage and Impact".

- Bernadette's identification of the spring of water, which was completely hidden
- The almost immediate manifestation of scientifically inexplicable cures through the water
- The flames that licked the hand of Bernadette for ten minutes without pain or damage
- The unblocking of the Grotto by Emperor Louis Napoleon himself
- The revelation of Mary as the "Immaculate Conception"
- The virtually uncorrupted state of Bernadette's body after thirty-nine years

If Bernadette was not really seeing and hearing the Blessed Virgin Mary in her ecstasies, then how can all of this be explained, most especially the thousands of scientifically inexplicable cures that occurred through the waters of the spring that the Lady revealed to her?

II. Seven Scientifically Inexplicable Cures of Lourdes: A Review of the Medical and Scientific Investigation

Recall from above that as of 2018, the Lourdes Medical Bureau investigated and maintained records on seven thousand unexplained cures[27] of severe maladies,[28] of which only seventy (1 percent) are declared to be miracles by the competent authorities. As explained above (Introduction), this small percentage is due in great part to the Lambertini criteria.

The following seven scientifically inexplicable cures are not all classified as "miracles" according to the Lambertini criteria discussed above[29] because they fall short of one of the criteria in some *minor* way. Nevertheless, they are considered to be highly probative examples of scientifically inexplicable phenomena that are excellent candidates for the miraculous. They also manifest transformational effects

---

[27] Recall from above, that of this number of unexplained cures, Dr. Lasalle believes that about twenty-five hundred are demonstrably *scientifically* inexplicable. See Glynn, *Healing Fire of Christ*, p. 73.

[28] See Connolly, "70th Miracle of Lourdes".

[29] See de Lambertini, *De Servorum Dei*. These are explained in François, Sternberg, and Fee, "Lourdes Medical Cures".

not only on the recipients and their families, but also on the doctors, clergy, and communities with which they are associated—effects of faith, hope, compassion toward others, humility, reverence, and service. I am indebted and thankful to Father Paul Glynn (*Healing Fire of Christ*) for many of the medical details and records related below, particularly in Sections II.C through II.E.

A. Pierre de Rudder (1875)—A Cure in Belgium

The case of Pierre de Rudder, though nearly 150 years old, is very significant because it was not only an instantaneous regeneration of over one inch of bone (along with atrophied muscles), but it rules out a physical-organic explanation (explained below). Furthermore, it did not occur at the Grotto of Lourdes, but rather at the Shrine of Oostakker, a shrine of Our Lady of Lourdes in Belgium. This is significant because it means that the miraculous cures of Lourdes cannot be solely attributed to healing agents in the water of the Lourdes' Grotto, but rather to the intercession of Our Lady of Lourdes.

In brief, Pierre de Rudder was working for the viscount du Bus de Gisignies, who had an estate between Bruges and Ostend in West Flanders, Belgium. On February 16, 1867, while helping two woodcutters move the trunk of a large tree, the trunk fell on Pierre's left leg, crushing both the tibia (frontal calf bone) and fibula (thin stabilizing bone slightly to the side of the tibia) below the knee. It caused a clean fracture with a one-inch separation in the tibia.[30] The first attempt to reconnect the two parts of the tibia by Dr. Affenear failed, and when the bandages were removed, the bones had not begun to knit together and both ends were disconnected and separated.[31] In December 1874 (nearly eight years later), a specialist, Dr. van Hoestenberghe, examined Pierre's leg and noted that the bones were still separated with both ends blackened and necrosed (dead). As such, the lower part of the tibia was swinging freely and could be twisted 180 degrees without difficulty.[32]

---

[30] Glynn, *Healing Fire of Christ*, pp. 159–60.
[31] Ibid., p. 160.
[32] Ibid., p. 161.

Six different physicians recommended amputation, but Pierre refused and asked that he be allowed to go to the shrine of Our Lady of Lourdes in Oostakker. The vicount's son gave him the funds for the train and bus to the shrine.

As he was making his way to Oostakker, several witnesses testified that Pierre's leg was "freely swinging", and the bus driver ridiculed him for it.[33] When he and his wife arrived at the shrine in Belgium, he went right away to the shrine. He then began praying to our Lady, "asking pardon for my sins and begging Our Lady of Lourdes for the grace to be able to earn a livelihood for my wife and children".[34] He began to feel a strange sensation and agitation. He immediately got up, left his crutches behind, moved through the crowd, and knelt in front of the statue of Mary, all of which would have been impossible only moments before.[35] His wife was so shocked that she fainted. The cure was instantaneous, and the regeneration of bone and tissue was complete and permanent, lasting to the time of his death.[36]

As noted above, Pierre had been examined by six physicians prior to leaving on pilgrimage to the shrine. Because of this, there were several documented tests and X-rays validating the irreparable nature of Pierre's condition. This allowed Dr. Édouard Le Bec (the former vice president of the Society of Surgeons of Paris and later the president of the Lourdes Bureau of Medical Observations) to document de Rudder's scientifically inexplicable cure in his book *Medical Proof of the Miraculous*.[37] The photographs of de Rudder's leg clearly manifest a clean break,[38] and the multiple testimonies to this from six physicians and other witnesses make it undeniable. Most importantly, Le Bec's analysis shows that a naturalistic/physical explanation is virtually impossible.

Dr. Le Bec's own words best describe the events leading up to de Rudder's instantaneous generation of bone, muscle, and tissue, and his complete recovery of movement:

[33] Ibid., p. 162.
[34] Ibid.
[35] Ibid., pp. 162–63.
[36] Ibid.
[37] Le Bec, *Medical Proof of the Miraculous*.
[38] Ibid., p. 122.

Arrived at the Grotto, De Rudder rested on a seat very much fatigued from his journey.... Suddenly he realized that something was happening; he raised himself quickly and began to walk; then he knelt down and arose unaided. He next proceeded to examine his leg: the leg and foot, which some seconds before had been swollen, had resumed their normal size; the two wounds were cicatrized and the bones were solidly united. De Rudder then walked without assistance, and with his crutches, to the omnibus which went to Ghent. The following day, April 8, Dr. Affenaer came to visit him. He examined the limb, verified the cure of the wounds, and stated that the internal aspect of the tibia was quite smooth at the site of the fracture, which was consolidated. De Rudder was able to walk without the slightest lameness.[39]

After summarizing the investigations of Drs. Deschamps and Verriest of de Rudder's leg, Le Bec presented the astonishing result as follows:

Even admitting that the fragments could have come in contact and united, there would have been the recognized shortening which it would have been impossible to avoid after such a loss of bone substance. Yet nothing like this happened. The tibia shows no shortening, the osseous cavity is filled up, and the necrosed fragment is reconstituted in its totality.[40]

The investigations show a series of physical actions that have no known natural explanation. Necrosed (dead) bone does not come alive and reconstitute itself instantaneously; approximately one inch of missing bone does not regenerate itself instantaneously; an atrophied leg that had been freely swinging for eight years does not instantaneously become perfectly functional—not only to support the body, but also to perform walking, kneeling, and sustained activity immediately after the cure.

Yet, the mystery grows even deeper, because the regenerated bone could not have come from de Rudder's body. As Le Bec shows, the phosphate required to produce that much bone does not exist in any human body. This means that the regenerated bone must have come from some other source beyond de Rudder's body—a transphysical or "supernatural" source. Le Bec describes it as follows:

[39] Ibid., pp. 121–22.
[40] Ibid., p. 125.

Let us inquire now how this callus was formed, and what was the quantity of phosphate of lime necessary for the consolidation of this fracture and the regeneration of the lost fragment.... From the measurements that I have taken of De Rudder's bones, the callus was about 5 centimetres square superficially, and as the result of the separation of the fragments its thickness was considerable. It can be said without exaggeration that the weight of the phosphate of lime necessary to replace the bone lost and to form the callus that filled the gap would be about 5 grammes.... The blood, according to Schmidt, Becquerel, and Rodier—and this is admitted by all physiologists—contains on an average 1.60 grammes of phosphate of lime. This is less than a third of the weight of the callus in the case of De Rudder. From where came the other two-thirds? They did not exist in the blood of De Rudder, and there could be no deposit of these salts anywhere in his body. Were they supplied by some unknown natural force? No, for such a force could not find them in the human body.[41]

The de Rudder case comes as close as science can to showing not only the scientific inexplicability of the cure, but also the absence of any natural source of phosphate that would be needed to effect that cure. This strongly implies a supernatural creative power to regenerate the five grams of bone in de Rudder's leg. This is precisely what Le Bec concludes:

If we admit that this force has formed the phosphate suddenly, this is to admit creative power [because there is no natural source of the phosphate needed to regenerate the missing bone in de Rudder's body]. But unbelievers will never admit such a power, for this necessarily implies recognizing the existence of God [i.e., a supernatural cause].[42]

As Le Bec implies, admitting that the substance of the missing bone appeared without a natural source is tantamount to admitting the existence of a supernatural source. This implication did not go unrecognized by the unbelieving freethinker Dr. van Hoestenberghe, who was the specialist brought in to validate the irreparable nature of de Rudder's condition—and later his remarkable cure. When van Hoestenberghe came to de Rudder's house to examine him after the cure, de Rudder jumped up and down in front of van

[41] Ibid., pp. 125–26.
[42] Ibid., p. 126.

Hoestenberghe, who was, needless to say, shocked. After a time of prayer and reflection, he renounced his unbelief, became a Christian, and declared, "Now, I can affirm on my honor that I believe absolutely and that with belief I have found happiness, and an inner peace, which I had never known before."[43]

B. Marie Bailly and Alexis Carrel (1902)

Beyond the Lourdes Medical Bureau,[44] the main source for the medical assessment of Marie Bailly's cure is the Nobel Laurate physician Alexis Carrel, who witnessed the miracle and wrote about it in his book *The Voyage to Lourdes*.[45]

Alexis Carrel was twenty-nine years old in 1901 and was an unbeliever. Nevertheless, he was curious about the claims of instantaneous cures at Lourdes and hoped one day to confirm his hypothesis that all such cures could be explained psychologically. A physician colleague asked Carrel to accompany him to Lourdes, and he accepted. Unfortunately, his friend had to cancel at the last moment, leaving Carrel on his own. On the train ride to Lourdes, Carrel was summoned by a nurse for emergency help to a dying pilgrim—Marie Bailly, who was comatose and in the process of dying during the long train trip. Carrel examined her and gave his initial diagnosis as follows:

> The swelling [in Marie's abdomen] was apparently caused by solid masses, and there was a pocket of fluid under her navel. It was classic tubercular peritonitis.... The temperature above normal, legs swollen, heartbeat and breathing accelerated.[46]

While Carrel was helping to move Marie off the train in Lourdes, he spotted one of his medical colleagues to whom he refers as "A.B." They engaged in a discussion about the possibility of a miracle, with

[43] Glynn, *Healing Fire of Christ*, p. 165.

[44] *The Case of Marie Bailly*, Dossier 54 of the Archives of the Lourdes Medical Bureau. Dossier 54 is given in full in Stanley Jaki's new introduction to Alexis Carrel's *The Voyage to Lourdes*, 2nd ed., trans. Virgilia Peterson (New Hope, Ky.: Real-View-Books, 1994).

[45] See Carrel, *Voyage to Lourdes*.

[46] Ibid., p. 9.

A.B. suggesting that there have been documented cases of instantaneous organic cures like that of Pierre de Rudder (see above, Section II.A). Carrel, surprised by his colleague's "naivety", retorted (and recorded in his diary):

> Pious propaganda, lacking objective, scientific investigation!... If the de Rudder case was scientifically authenticated it would be an archetypal miracle, God's signature, supernatural.... It is my duty to approach such cases with complete skepticism.... Charcot has demonstrated how paralysis and arthritis, formerly judged to be incurable conditions, may really be nervous diseases which can be cured instantaneously.... Such remission will occur here, but not cures of truly organic sicknesses ... like that of the patient Marie in my care.... She has tubercular sores, lesions of the lungs and for the last eight months peritonitis diagnosed by a well-known surgeon from Bordeaux.... She may die at any moment.... If she were cured, I'd never doubt again; I'd become a monk![47]

Carrel did not want to bring Marie to the baths, because she was at the brink of death (which another physician validated). However, a sister indicated that it would be unconscionable not to give Marie her dying wish. They decided that her condition was far too fragile for her to be immersed in the bath, and so poured water on her distended abdomen, which resulted in her cure over the course of about one hour. At 2:20 P.M. the nurses emerged from the bath cubicles where they had poured the water on Marie's abdomen, and Carrel began to note the significant changes in Marie's condition as follows:

> [At 2:40 P.M.] The face of Marie slowly continued to change. Her eyes, so dim before, were now wide with ecstasy as she turned them toward the Grotto. The change was undeniable.[48]
> [At 2:55 P.M.] The blanket which covered Marie's distended abdomen was gradually flattening out.[49]
> [At 3:00 P.M.] A few minutes later, there was no longer any sign of distension in Marie's abdomen.[50]

[47] Ibid., p. 18.
[48] Ibid., p. 32.
[49] Ibid.
[50] Ibid.

In order to exclude the possibility of a temporary psychosomatic explanation of the cure, Carrel decided to visit Marie in her hospital room at 7:30 that evening. He not only found her pulse to be normal (eighty beats per minute), but also found the complete disappearance of the solid masses and fluid-based distension of peritonitis. He arrived at Marie's bed and saw her sitting up:

> Though her face was still gray and emaciated, it was light with life; her eyes shone; a faint color tinged her cheeks. The lines at the corners of her mouth, etched there by years of suffering, still showed. But such an indescribable serenity emanated from her person that it seemed to illuminate the whole sad ward with joy.[51]

In order to be certain of the cure, he threw her blanket back and discovered

> above the narrow hips the small, flat, slightly concave abdomen of a young, undernourished girl.... He was able to palpate the abdomen, the sides, and the pelvis, looking for traces of the distension and the hard masses he had found before. They had vanished like a bad dream. The whole region of the abdomen felt completely normal.[52]

Though Carrel did not decide to come back to the Catholic Church that evening, he was shaken out of his unbelief, and for the first time in many years, he prayed directly to God, Jesus, and especially to Mary. He concluded his diary that evening with the following:

> Gentle Virgin, Who bringeth help to the unfortunate who humbly implore Thee, keep me with Thee. I believe in Thee. Thou didst answer my prayers by a blazing miracle. I am still blind to it, I still doubt. But the greatest desire of my life, my highest aspiration, is to believe, to believe passionately, implicitly, and never more to analyze and doubt.[53]

Evidently, Dr. Carrel abandoned his agnosticism and materialism and broke through to both God and the Blessed Virgin. As he told her in his prayer, "Thou didst answer my prayers by a blazing miracle."[54]

---

[51] Ibid., p. 35.
[52] Ibid., p. 36.
[53] Ibid., p. 46.
[54] Ibid.

Marie's extraordinary cure had all the elements needed for classi-
fication as a miracle:

1. There is strong documentation of Marie's condition by several
   doctors (including Dr. Carrel) prior to having Lourdes water
   poured on her abdomen on May 27, 1901—vomiting, severe
   tubercular peritonitis with substantial abdominal distension and
   an extensive solid mass, severe abdominal pain, rapid and irregular heartbeat, and very near the point of exhaustion and death.
2. The cure occurred over the course of one hour (almost instantaneously). According to at least four physicians—Dr. Boissaire,
   Dr. Paul Geoffray, Dr. Carrel, and a doctor colleague identified by Dr. Carrel as "A.B."—there was no record of a case of
   severe tubercular peritonitis being cured so rapidly and completely as that of Marie Bailly's except at Lourdes.[55]
3. There is extensive documentation of Marie's cure at the
   Lourdes Medical Bureau. Dr. Carrel personally carried out
   rigorous physical and psychological tests even after Marie
   joined the Sisters of St. Vincent de Paul (a very hard life healing the sick).

In sum, the above physicians concluded that the case was organic
because of the presence of distension, fluid, extensive solid masses,
and other physical aspects of tubercular peritonitis. Furthermore,
it was near instantaneous as shown by the disappearance of all the
organic manifestations of tubercular peritonitis over the course of one
hour, restoring Marie to sound health. Finally, Marie displayed no
signs of emotional distress or imbalance. On November 27, 1902, Dr.
Carrel interviewed Marie in Lyons and submitted a detailed report on
the four-year genesis of her illness as well as her current physical and
psychological state (all of which he submitted to the Lourdes Medical
Bureau). Marie was in excellent physical health and ready to embrace
the rigors of the Sisters of St. Vincent de Paul. She left the next day

[55] Just prior to Marie's cure in 1901, a case of severe pulmonary tuberculosis was healed
instantaneously and completely on June 26, 1900 (Father Salvator, a Capuchin priest). Father
Salvator was on the point of death; he was instantaneously cured after being immersed in the
bath, showing no signs of tuberculosis since that time. The records of Father Salvator before
and after the cure can be found in the Lourdes Medical Bureau, Number 26, of the canonical
cures. See Glynn, *Healing Fire of Christ*, p. 106.

for Paris, joined the sisters, and worked in the laundry and other strenuous activities for thirty-five years, dying of natural causes in 1937. Her cure was documented as being instantaneous, permanent, and inexplicable by the laws of nature (as determined by competent, scientific authorities, and analysis).

The priest and physicist Stanley Jaki fills in the rest of Alexis Carrel's story in his article "Two Lourdes Miracles and a Nobel Laureate".[56] Marie Bailly's cure became well known and came to the attention of the faculty of the Lyons Medical School. Carrel was hoping that no one would associate him with the miracle, but a newspaper published an article "implying that Carrel refused to believe in the miracle".[57] This forced Carrel to respond to the article, indicating that he did not refuse to believe in the miracle, but only that he was not willing to make the declaration of it. This led several professors in the medical school to advocate his departure from the school for implying the possibility of a miracle. One of the professors noted, "With such notions you seem to be entertaining, Dr. Carrel, I believe it is my duty to inform you that we have no place for you here."[58] From there, the brilliant young physician went to Paris, then to Montreal, then the University of Chicago, and then to Johns Hopkins University, and finally to the Rockefeller Institute in New York.[59] Eight years later (1912), he won the Nobel Prize for Medicine for his discovery of the technique to suture blood vessels (vascular anastomosis).[60] This gives the University of Lyons Medical School the rare distinction of being the only university to dismiss a future Nobel Prize winner solely for acknowledging the *possibility* of a miracle.

Though Carrel had believed in the miraculous causation of Marie's healing while at Lourdes, he was not yet persuaded enough to go back to the Catholic Church. In 1910, Carrel had the privilege of witnessing a second miraculous cure at Lourdes—the instantaneous cure of an eighteen-month-old boy who was born blind.[61] Though

[56] Stanley Jaki, "Two Lourdes Miracles and a Nobel Laureate", *Linacre Quarterly* 66, no. 1 (February 1999), https://www.catholicculture.org/culture/library/view.cfm?id=2866.

[57] Ibid.

[58] Glynn, *Healing Fire of Christ*, p. 114.

[59] Jaki, "Two Lourdes Miracles".

[60] Ibid.

[61] Ibid.

this affected him, he still did not return to the practice of faith. He delayed his conversion until 1939, when he met a Trappist monk, Father Alexis Presse. He developed a relationship with him, which opened him to deeper spiritual reflection. When he was about to die in 1944, he sent for Father Presse, who traveled from Brittany on a freight train to go to his bedside. Carrel received the last sacraments and died in the Catholic Church.[62]

C. Gabriel Gargam (1901)

The case of Gabriel Gargam occurred in 1901.[63] He was born to practicing Catholic parents, but lost his faith at the age of fifteen and no longer practiced it. He became a postal sorter, and during the course of his work in 1899, the train on which he had been sorting was in a collision with another train traveling at fifty miles per hour. He was thrown fifty-two feet from the train and was badly injured. After eight months, he was at the point of death—a mere seventy-eight pounds with gangrenous feet, unable to take solid food. He could be fed only once every twenty-four hours by a tube and required two nurses to take care of him. His condition was well-attested not only by his physicians, but also by those involved in the lawsuit he filed against the railroad—the court records and physicians' testimonies still exist today.[64] His attending physician, Dr. Decrassaac, testified as follows:

> Gabriel Gargam is a cripple for life and a physical wreck, unable to do anything, requiring constant nursing care—hardly susceptible of improvement; more likely to terminate fatally.[65]

Gargam spent two years in bed—unable to be moved from his room. Though his aunt (a religious sister) and his mother begged him to go to Lourdes, he refused to do so, preferring to suffer his fate in

[62] Ibid.

[63] Elaine Jordan, "The Lourdes' Miracle of Gabriel Gargam", Tradition in Action (website), April 13, 2013, http://www.traditioninaction.org/religious/h106_Lourdes.htm. See also Glynn, Healing Fire of Christ.

[64] Jordan, "Gabriel Gargam".

[65] Glynn, Healing Fire of Christ, p. 168.

his room. Finally, he relented and consented to the trip, but being moved on a stretcher and riding on the train almost killed him.

He was then brought to the waters in the Grotto. The strain was so great that he lost consciousness and his attendants believed him to be dead, so they put him on a carriage, put a cloth on his face, and began to wheel him back to the hotel. On the way there, a Eucharistic procession was passing by. The bishop leading the procession saw the sorrowful crowd around Gargam, and then he blessed them with the Holy Eucharist, at which point Gargam gripped the sides of the stretcher and called out, "Help me, I can walk, I feel I can walk."[66] He had not been able to speak out loud for twenty months. He then sat upright by his own power (which he had not been able to do for two years), and then proceeded to get off the stretcher and walk around. The astonished crowd accompanied him back to the hospital, where he sat down to eat a hearty meal (though he had not taken solid food for two years).[67] The next morning (August 21, 1901), Gargam was examined by sixty-three physicians for two hours, all of whom pronounced him completely cured.

Among the physicians present was Dr. Gustave Boissarie, who became the director of the Lourdes Medical Bureau and wrote a history of it up to that time.[68] He testified that Gargam looked like a specter, and his leg muscles were completely atrophied after the injury and twenty months without use. The instantaneous restoration of the muscles is scientifically inexplicable.[69] He used Gargam's case (with its substantial physician verifications, X-rays, and medical tests) to challenge the allegations by Émile Zola and Jean-Martin Charcot, who tried to discredit Lourdes miracles as mere auto-suggestion. Boissarie showed that Gargam's leg muscles had completely atrophied, and within seconds, they grew back so firmly that he could stand erect and walk without difficulty—quite beyond the "powers" of auto-suggestion.[70] Gargam's case still remains scientifically inexplicable today.

Gargam also underwent a spiritual metamorphosis, consecrating himself to the Blessed Virgin Mary and the service of the sick at

[66] Ibid.

[67] Jordan, "Gabriel Gargam".

[68] Gustave Boissarie, *Lourdes: Histoire Medicale 1858–1891* (Paris: Librairie Victor Lecoffre, 1891).

[69] Glynn, *Healing Fire of Christ*, p. 171.

[70] Ibid.

Lourdes. He lived a normal healthy life until his death at eighty-three years of age (fifty-one years after his cure).[71] He worked at Lourdes as a stretcher bearer during his vacations every year throughout those fifty-one years, giving witness to the intercessory power of the Blessed Virgin at Lourdes.

D. John Traynor (1923)

John Traynor's cure occurred in 1923. Traynor was raised a Catholic and was a bonafide World War I hero who was severely injured on three occasions during the war—in Bruges, Belgium, on October 8, 1914; early in 1915, hit with a bullet through the right knee in a battle at Suez; and April 25, 1915, wounded in the arm at Gallipoli, Turkey.[72]

With respect to the first injury, Traynor was carrying his commanding officer to a medical station near Bruges, Belgium, when a piece of shrapnel hit him in the head and lodged there. He lost consciousness for five weeks, during which time the shrapnel was surgically removed. He regained consciousness in a hospital in Deal, England. At first, Traynor appeared to be healed, but in August 1915, he experienced his first epileptic attack. These attacks grew in frequency and severity throughout 1915 to 1920. In April 1920, Major Montserrat (surgeon) used a trephine (a small circular saw) to cut a one-inch diameter hole in Traynor's skull to relieve pressure. However, the procedure did not relieve the epileptic attacks, which actually grew in severity—sometimes up to three attacks per day.[73] Since the brain's pulsations were visible, a metal plate was inserted. In addition to the epileptic attacks, Traynor developed partial paralysis of the legs and incontinence of urine and feces, all of which led to progressive loss of strength.[74]

With respect to the third wound, Traynor went to Gallipoli, Turkey, on April 25, 1915, as a machine gun captain. On May 8, 1915, during a bayonet charge, the Turks sprayed him with machine gun fire. A bullet lodged under his collarbone, which severed the brachial plexus and the large nerves in the axilla.[75] As noted by Dr. Francis

[71] Ibid.
[72] Ibid., p. 60.
[73] Ibid., pp. 61–62.
[74] Ibid.
[75] Ibid., p. 60.

Izard, the nerves in his arm were severed, and as a consequence, he suffered loss of sensation and paralysis, and eventually, the muscles atrophied (muscle tissue loss and death).[76] He subsequently underwent four surgeries to suture the nerves, but all of them failed. After the fourth unsuccessful surgery, the naval surgeon recommended amputation, but Traynor refused.[77]

Traynor returned to Liverpool, where he was cared for by his wife, while living on a full pension. He was in a continuously deteriorating condition not only with atrophied muscles in his right arm, but also with paralysis of the legs, incontinence, and four epileptic attacks per day. He was not able to do anything and had to be moved from his bed to his wheelchair.[78] In 1923, Traynor's diocese of Liverpool organized a pilgrimage to Lourdes. Traynor, who had a sincere devotion to the Blessed Virgin, wanted to go, though his physicians, wife, the government ministry of pensions, and even the priest organizing the pilgrimage begged him to stay home. They thought the trip would be suicide, and they were almost correct. Traynor was wheeled to the train in Liverpool and suffered tremendously on the trip to Lourdes. When he arrived, he was almost dead, and one woman wrote to his wife indicating that he would be buried at Lourdes.

The physicians who initially examined him during and after the train ride to Lourdes noted the following in a signed statement, the original of which can still be accessed at the Lourdes Bureau of Medical Observation:

John Traynor of 121 Grafton St., Liverpool, is suffering from the following conditions:

(1) Epileptic (We ourselves saw several attacks during his journey to Lourdes).
(2) Paralysis involving the median, musculo-spinal and ulnar nerves of the right arm, the hand is "en griffe" (clawlike), deformity and wasting of all the muscles of the right upper limb, wasting of the right pectoral and axillary muscles, wrist drop and muscular atrophy.

[76] Ibid.
[77] Ibid., p. 61.
[78] "Miracles of Lourdes", Our Lady of the Rosary Library (website), accessed March 18, 2024, https://olrl.org/stories/lourdes.shtml.

(3) The trephine opening in the right parietal region is about 2.5 c.m. The pulsations of the brain are visible. A metal plate covers the trephine orifice.

(4) There is absence of voluntary movement of the legs, with loss of sensibility.

(5) Incontinence of urine and feces.

[Signed by] Drs. Finn, Marley and Azurdia.[79]

During his stay at Lourdes, Traynor was taken to the baths nine times, and on the occasion of his tenth time (July 25, 1923), his legs felt agitated in the bath. After the bath, he was placed in the wheelchair to receive a Eucharistic blessing from the bishop of Rheims, who was passing by in a Eucharistic procession. After being blessed by the Host, his arm (which had been paralyzed for eight years) grew so strong that he was able to burst through his bandages. He then regained the use of his legs (which had been partially paralyzed for eight years, preventing him from standing and walking). He got out of his chair and walked several steps, but his attendants put him to bed for the evening because they were afraid he might hurt himself. During the night, he leapt out of his bed, knelt down to finish a Rosary, and ran out his door to go to the Grotto—to the utter amazement of everyone watching. He knelt down in the Grotto to finish his prayers, but seemed to suffer a temporary lapse of memory about his condition prior to being cured. The healing not only cured his paralysis and epilepsy; it seemed to mask the memory of his former misery.

On July 27, the above three doctors signed a second statement certifying Traynor's instantaneous cure, which is still preserved at the Lourdes Medical Bureau. It read as follows:

(1) He can walk perfectly.

(2) He has recovered the use and function of his right arm.

(3) He has recovered sensation in his legs.

(4) The opening in his skull has diminished considerably.[80]

Two days later, while riding on the train back to Liverpool, Archbishop Keating of Liverpool came into his compartment and reminded

[79] Glynn, *Healing Fire of Christ*, p. 68.
[80] Ibid., pp. 68–69.

him of his former condition—only then was his memory revived, and both he and the archbishop broke down in tears.[81]

On July 7, 1926, three years after Traynor's cure, the president of the Lourdes Bureau of Medical Observation, Dr. Vallet, and the three original physicians—Drs. Finn, Marley, and Azurdia—attested to the cure's scientific and natural inexplicability. It was witnessed by Drs. Harrington of Preston, Lancashire, and Moorkens of Antwerp, Belgium. They noted that there were some minor remnants of Traynor's maladies still present (such as a slight impression where once there was a hole in the skull). However,

> no atrophy of the pectoral or shoulder muscles now exists. The trephine opening has been obliterated; with the finger a slight depression only is found in the bone. There have been no epileptic seizures since the cure in 1923.[82]

This cure is quite remarkable, because it constitutes an instantaneous regeneration of the muscles in the arm and the pectoral area as well as the severed nerves and the bone in the skull. Additionally, it brought the epileptic seizures (about four times per day) to a complete end and restored the muscles in the legs and the capacity to regulate urine and feces. As a result, on October 2, 1926, the Lourdes Medical Bureau stated that "this extraordinary cure is beyond and above natural causes".[83]

Traynor's cure was so complete that he went into the coal and hauling business (lifting two-hundred-pound sacks of coal), pledged himself to service at the Grotto of Lourdes every summer, and died on the eve of the Solemnity of the Immaculate Conception in 1943 (twenty years after his cure). A large number of conversions occurred in Liverpool as a result of the obvious miracle.[84]

E. Two Children: François Pascal (1937) and Guy Leydet (1946)

The records certifying the instantaneous, scientifically inexplicable cure of François Pascal are considerable and are recounted not only at

[81] Ibid., p. 69.
[82] Ibid., p. 73.
[83] Ibid.
[84] Ibid.

the Lourdes Office of Medical Observation, but also in Paul Glynn's
Healing Fire of Christ.[85] In 1937–1938, François Pascal, a four-year-
old boy from Beaucaire, France, was diagnosed with spinal meningitis by local Dr. Darde, which caused permanent paralysis in all four
limbs and complete blindness. The organic cause of meningitis was
confirmed by a lumbar puncture done by Dr. Lesbros of Avignon in
December 1937.[86] The total blindness was confirmed by the presence of lesions on the eyes by oculist Dr. Polger of Arles in June
1938.[87] As a result, Dr. Darde discontinued any further treatment
as pointless.[88]

François' mother signed up for the annual Lourdes pilgrimage and
boarded the train with the other pilgrims in August 1938. Dr. Roman
certified that François traveled on the train "lying down and quite
blind".[89] When they arrived in Lourdes, François' mother carried
him to the Grotto, where he was immersed in the bath for the first
time, but without effect. Undeterred, his mother brought him back
to the baths a second time, almost immediately after which François
began to move his limbs, pointing to various objects and identifying
them accurately.[90] He continued to do this on the way back to the
hospital, where the remediation of his paralysis and blindness was
confirmed by Dr. Roman.

The family returned to Beaucaire four days later, and his mother
took François to see Dr. Darde, who was stunned to see the child
walking, moving his arms, and evidently seeing well. He continuously observed him until November 9, 1938, when he wrote his
signed statement validating the remarkable cure. It ran as follows:

I, the undersigned, declare I had under my care from December 18,
1937, to June 14, 1938, Francois Pascal, age four. The child, seen in
consultation with Drs. Julian of Tarascon, Barre of Avignon, Dufoix
Fils of Nines and Polge of Arles, had been suffering from lymphocytic
meningitis (analysis of Dr. Lesbros of Avignon). At the end of July the
patient was paralyzed in all four members. Visual acuity was nil. He

[85] Ibid., pp. 144–46.
[86] Ibid., p. 143.
[87] Ibid., pp. 143–44.
[88] Ibid., p. 144.
[89] Ibid.
[90] Ibid.

did not perceive even light and could not distinguish day from night. Before the child was taken to Lourdes I was asked to examine him (for a medical certificate). The condition I recorded previously in June was exactly the same: paralysis of four limbs, vision nil. Returning from Lourdes (August 28) Madame Pascal brought the child to me, leading him by the hand. He was walking! I recorded the disappearance of paralysis and the return of vision. From that date the improvement has been maintained ... medically speaking one cannot explain such a result (Signed, Dr. Darde, November 9, 1938).[91]

World War II broke out in France before François and his mother could return to Lourdes with the medical dossiers to validate the organic cause of his spinal meningitis, paralysis, and blindness. They waited until the end of the war, and on October 2, 1946, they returned to Lourdes—eight years later. They presented the dossiers to the twelve physicians at the Lourdes Medical Bureau. François was examined by all twelve doctors, who noted that he was a strong-muscled boy who was attending a normal school, and that his disease was organic and not merely psychological (as confirmed by the lumbar puncture and the lesions on the eyes). The conclusion of the report signed by all twelve doctors reads as follows:

> When the development of the illness left no hope of improvement, there was an abrupt cessation of symptoms. The cure is confirmed by complete recovery of vision, of the ability to walk and of all other functions.... The cure is now of ten years' duration. There is no medical explanation for the instantaneous disappearance of the malady and its symptoms.[92]

As with the above Lourdes' cures, François Pascal's cure influenced the doctors who examined him (including Dr. Auguste Vallet, who later became president of the Lourdes Office of Medical Observation[93]), his local community, and the popular press in France. François himself became deeply religious, returning to Lourdes every year to be a stretcher bearer and a spokesman for Lourdes.

[91] Ibid., pp. 144–45.

[92] Lourdes Medical Bureau, Dossier for Canonical Miracle #45: François Pascal, cited in Glynn, Healing Fire of Christ, pp. 145–46.

[93] Glynn, Healing Fire of Christ, p. 146.

The case of Guy Leydet is similar to that of François Pascal's. It is significant because of the virtual impossibility of explaining his cure scientifically. He had documented brain damage (destruction) from encephalitis so severe that he could not recognize his mother and was reduced to animalic guttural sounds—an IQ between 0 and 25. This was accompanied by frequent epileptic fits. Additionally, the encephalitis caused paralysis in all four limbs, and he was incontinent.[94] After fruitlessly pursuing a medical cure for Guy's condition and being told that "medical science had exhausted all possibilities",[95] they went to Lourdes.

When they arrived in Lourdes, his mother brought him to the baths and immersed him in the cold water, hoping it would not bring on an epileptic fit. It certainly did not. Instead, he called for his mother and began to speak intelligently, and his paralyzed limbs began to move, and he regained continence.

Guy's speech began to improve quite rapidly, making up for the lost two years after his bout of encephalitis. From all accounts, Guy's brain damage had been instantaneously and completely reversed (along with the epileptic fits, paralysis, and incontinence). Guy caught up with his peers in speech and reasoning before the year was out.

The cure was so unprecedented that it provoked a vibrant discussion among the forty physicians examining Guy at the Lourdes Medical Office of Observation. Paul Glynn describes the conversation that proceeded along the following lines. One of the physicians asked, "What brain does the child think with? A new brain, or the one partially destroyed by meningoencephalitis?"[96] This led to a discussion about whether the instantaneous restoration of a partially destroyed brain could be naturalistically explained, which in turn led to the observation that even though such a cure had never occurred previously, there might be some unknown natural force that produced it. According to Glynn, Professor Lelong of Paris replied, "If anyone has come across a case of a postencephalitic idiot [0 to 25 IQ] regaining normality as this child has, I vow never to sign another Lourdes medical report."[97]

94 Ibid., p. 127.
95 Ibid.
96 Ibid., p. 128.
97 Ibid.

The discussion among the forty doctors examining Guy Leydet may be broadened to most of the seven thousand unexplained cures at Lourdes. It can be set out in six steps and provides the basis for reasonably and responsibility concluding to the presence and action of the Blessed Virgin Mary in the natural world today. This argument is presented below in Section III (Conclusion).

F. Brother Leo Schwager (1952)

My reason for including this case is focused on the efficacy of the Holy Eucharist as an integral part of the healing process (along with the water from the Grotto) in many of the scientifically inexplicable cures at Lourdes. In addition to this case, we have also seen this phenomenon in the cases of Gabriel Gargam (Section II.C) and John Traynor (Section II.D). As we shall see, being in the presence of the Eucharist after being immersed in the baths caused a most remarkable and immediate healing of Brother Leo. This efficacious role of the Holy Eucharist in healing will be explained in the Conclusion (Section III).

Brother Leo Schwager, a Benedictine monk who belonged to the Swiss monastery of Fribourg, was diagnosed with advanced multiple sclerosis (MS) at the age of twenty-seven in 1951. The diagnosis was first made by Brother Leo's physician, Chief Doctor Ott, and confirmed by multiple tests by Dr. Kreienbuhl at the Clinic of Neurosurgery in Zurich.[98] Brother Leo's condition worsened very rapidly, and all of his organs were in a state of decline. At the beginning of 1952, he weighed only 103 pounds (five foot nine) and was unable to speak and function on his own, requiring constant nursing care. His doctor estimated that he would not live till summer at his current rate of decline.[99]

Brother Leo's superior, Dom Notker Mannhart, strongly encouraged him to go to Lourdes. When he arrived on April 28, 1952, he was semiconscious and barely able to swallow some liquid nourishment spooned into his mouth.[100] On April 30 he was taken to the

[98] Ibid., p. 77.
[99] Ibid., pp. 77–78.
[100] Ibid., p. 78.

baths, but did not experience any relief from the MS, organ decline, inability to speak, and his weakness and pain. At 4:00 P.M. he was taken again to the Grotto and reimmersed into the bath, but experienced only pain without improvement. He was then wheeled to the Rosary Esplanade to receive the blessing of the bishop with the Eucharistic Host.[101]

As Brother Leo was wheeled up to receive the Eucharistic blessing from the bishop, something quite remarkable occurred. In Brother Leo's own words:

> I was in such a miserable state that I could not even join in the invocations. I was simply there. Next the bishop gave the Blessing of the Sick right in front of me. It was as if an electric shock went over my whole body, from head to foot! I thought, "Good, now I can die in peace." I think I lost consciousness, and everything became blurred. Then all of a sudden I found myself on my knees in front of the bishop carrying the Blessed Sacrament! I felt wholly well, as if reborn, and all my pain had vanished. Dr. Jeger from Chur, a member of the Swiss pilgrimage, immediately rushed up, took me by the shoulder and asked, "Brother Leo, what has happened?" I promptly answered, "I'm well, I'm healthy!" He knelt down beside me. I myself prayed interiorly: "I love you, though very unworthy, O hidden Lord" and then recited Mary's prayer when Elizabeth praised her, the Magnificat.[102]

It seems that when Brother Leo experienced what he took to be an electrical shock and lost consciousness, he was catapulted out of his wheelchair into a kneeling position in front of the bishop holding the Holy Eucharist. When he came back to consciousness, he knew he was cured. This is quite similar to the cure of Gabriel Gargam, who was on a stretcher and thought to be on the brink of death. Recall that when he received the Eucharistic blessing, he spoke out loud for the first time in twenty months, promptly sat up, and got off his stretcher and began to walk around for the first time in two years (see above, Section II.C).

Brother Leo's case also resembles that of John Traynor's. Recall from Section II.D that John Traynor remained effectively paralyzed

---

[101] Ibid., pp. 78–79.
[102] Testimony of Brother Leo Swagger recorded in Glynn, *Healing Fire of Christ*, p. 79.

even after his tenth immersion in the bath, but when he received the Eucharistic blessing from the bishop of Rheims, his right arm that had atrophied over the course of eight years regained its strength. He was then able to stand and walk (for the first time in eight years). In the evening he ran back to the Grotto and knelt before it, giving thanks to the Blessed Virgin.

It should not be surprising that the Blessed Mother and her Son are present together in the healing of these three men and many other cases. After all, she reflects His glory in every imaginable way. Furthermore, we should not be surprised that the Holy Eucharist has miraculous healing power, for if it is the real Body of Christ as Jesus, the apostles, and the early Church insisted—and as the Eucharistic miracles discussed in Chapter 4 imply—then why shouldn't the Real Presence of Christ have the same miraculous healing effect it did throughout His ministry? The Holy Eucharist heals us not only spiritually, but as the above healings suggest, physically as well.[103]

Patients who have faith and pray for themselves have significantly higher rates of health and healing than those who do not. Dr. Harold Koenig, the co-director of the Center for Spirituality, Theology, and Health at Duke University Medical Center, indicates that religious people experience significantly better health than those who do not.[104]

III. Conclusion

The unique character of Lourdes—namely, persistent, instantaneous, scientifically inexplicable healings—presents us with the possibility of formulating an inductive inferential argument in favor of a transnatural or supernatural force, persistently breaking into the natural world at this special place and through the intercession of Our Lady of Lourdes. The argument may be set out as follows:

---

[103] Harold G. Koenig et al., "Does Religious Attendance Prolong Survival? A Six-Year Follow-Up Study of 3,968 Older Adults", *Journal of Gerontology, Medical Sciences* 54A, no. 7 (1999). Though this has not been studied, we might suppose that the Real Presence of Christ in the Holy Eucharist will have remarkable healing power even beyond regular prayer (assessed by Koenig).

[104] Ibid.

1. Inasmuch as a malady is proven to be organic, such as brain destruction in the case of Guy Leydet, spinal encephalitis and eye lesions in François Pascal, hard masses and fluid deposits of tubercular peritonitis in Marie Bailly, an atrophied right arm and a hole in the skull in the case of John Traynor, or a dangling limb in the case of Pierre De Rudder, and

2. Inasmuch as an instantaneous cure occurs through the regeneration of tissue and bone or the complete removal of foreign agents (e.g., cancer or hard masses), and

3. Inasmuch as these instantaneous scientifically inexplicable healings happen persistently at Lourdes, but not in other nonreligious healing environments (e.g., hospitals),

4. We may then infer that this proliferation of exceedingly unusual events that occur in one locale (or through association with that locale)—Lourdes—has a common cause.

5. Inasmuch as the healings are instantaneous and cannot be naturalistically explained, we might further infer that this common cause is transnatural or supernatural (beyond nature).

6. This is confirmed by the multiple apparitions of the Lady who showed Bernadette Soubirous the site of the water that became the instrument of this supernatural power—the Blessed Virgin Mary.

Evidently, each of these conclusions (nos. 4–6) can be challenged on the grounds that the inference is inductive (inferred from special observational data that does not occur at other healing locales, like hospitals) rather than deductive (derived from self-evident first principles). This should not be too disturbing to scientists because their method also proceeds from induction—inferences made on the basis of quantitative and qualitative characteristics of observational data— which is not deduced from first principles. The only real difference between the above chain of inferences and scientific inferences is that the conclusion results in a nonnatural or supernatural causative agency rather than a natural one. This supernatural conclusion cannot be validated scientifically, precisely because it is neither natural nor observable (as scientific conclusions are). Yet, this does not render supernatural conclusions false, but only "incapable of being observationally validated". Notwithstanding this, supernatural conclusions are falsifiable through observational data because if a natural cause is found, the supernatural explanation is falsified.

So where does this leave us? It enables believers to have considerable probative evidence to ground their belief reasonably and responsibly that the Blessed Virgin Mary is present and active in the natural world, and that her words to Bernadette Soubirous about remaining faithful to her Son, repenting for sins, and praying should be taken seriously.

As noted above, many of the scientifically inexplicable cures at Lourdes (including the cures of Gabriel Gargam, John Traynor, and Brother Leo) occurred through a Eucharistic blessing, which shows not only the efficacy of the Holy Eucharist for spiritual and physical healing, but also the close relationship that the Blessed Virgin and her divine Son have in the economy of salvation.

Chapter Seven

The Catholic Church and Natural Science

Introduction

Though popular culture often caricatures the Catholic Church as the enemy of science, nothing could be further from the truth. Hundreds of priests and Catholic clerics contributed to the development of all branches of science, including the discoverer of the Big Bang theory—Monsignor Georges Lemaître. Furthermore, some academics have implied that the Church's treatment of Galileo shows her bias against science and scientists. As will be explained, the controversy with Galileo was centered not on *confirmed* science, but on the *lack* of confirmation for heliocentrism that Galileo, contrary to his promise, published as a fact. The Church is also challenged by many of the faithful who believe that she adheres to biblical fundamentalism and is against scientific evolution. As we shall see, the Church has stated precisely the opposite in papal encyclicals going back to 1943. The following chapter is devoted to correcting these misimpressions and to setting the record straight. To do this, we will consider four major ways in which the Church has interacted with natural science throughout the centuries:

1. The Catholic Church's Contribution to Natural Science (Section I)
2. The Galileo Controversy (Section II)
3. Scientific Discovery versus Biblical Revelation (Section III)
4. Scientific Evolution versus the Biblical Account of Creation (Section IV)

I. The Catholic Church's Contribution to Natural Science

The Catholic Church was integral to the development of science throughout the centuries. As noted above, there are hundreds of priests and clerics instrumental in the development of most branches of physics, chemistry, biology, and applied mathematics. Following are a few of the most well-known priest/cleric scientists:[1]

- Roger Bacon, a Franciscan friar, is acknowledged to be the father of contemporary scientific method.
- Nicolaus Copernicus, a Catholic cleric, developed the first mathematical heliocentric model of the solar system (where the earth and other planets revolve around the sun).
- Gregor Mendel, an Augustinian monk and abbot, is acknowledged to be the founder of modern quantitative genetics.
- Nicolas Steno, a Danish Catholic bishop, is acknowledged to be one of the founders of modern stratigraphy and geology.
- Georges Lemaître, a Belgian diocesan priest with a Ph.D. in physics, is acknowledged to be the founder of contemporary cosmology after discovering the Big Bang theory in 1927—a revolutionary, rigorously established comprehensive theory of universal origin. Though Einstein initially resisted Lemaître's expanding universe theory, he later acknowledged it to be true after examining Hubble's survey of the Heavens. When Einstein and Lemaître co-presented at a conference at Mt. Wilson in 1933, Einstein said, "This is the most beautiful

---

[1] For additional information about well-known priests and clergy who were scientists, see Stephen M. Barr and Andrew Kassebaum, "Important Catholic Scientists of the Past", Society of Catholic Scientists, 2024, https://catholicscientists.org/scientists-of-the-past/?view -style=list; Matthew E. Bunson, "Fathers of Science", *Catholic Answers* (magazine), September 1, 2008, https://www.catholic.com/magazine/print-edition/fathers-of-science; Stephen Beale, "These 5 Catholic Scientists Shaped Our Understanding of the World", *Aleteia*, May 25, 2018, https://aleteia.org/2018/05/25/these-5-catholic-scientists-shaped-our-under standing-of-the-world/; and Filip Mazurczak, "Ten Catholic Scientists and Inventors Everyone Should Know", *Catholic World Report*, July 21, 2022, https://www.catholicworldreport .com/2022/07/21/ten-catholic-scientists-and-inventors/.

and satisfactory explanation of creation to which I have ever listened."[2]

- Pierre Gassendi, a French diocesan priest, discovered the orbits of Mercury and Venus, the diameter of the moon, and accurate measurement of speed of sound.
- Roger Boscovich, a Croatian Jesuit priest, produced a precursor to modern mathematical atomic theory and made several astronomical discoveries.
- Marin Mersenne, a French diocesan priest and mathematician, discovered Mersenne's prime numbers and Mersenne's laws of vibrating strings.
- Bernard Bolzano, an Italian/German diocesan priest and mathematician, discovered the idea of "limit" in mathematics (used in calculus) and gave the first analytic proof of the fundamental theorem of algebra.
- Francesco Grimaldi and Giovanni Riccioli, Italian Jesuit priests, confirmed gravitational law (distance of fall is proportional to square of time) and discovered gravity on earth's surface.
- Athanasius Kircher (Jesuit Catholic priest) made multiple geological contributions (volcanoes and fossils) and biological contributions (one of the first to observe microbes through a microscope and attribute the plague to a microbial infection); he invented the first magnetic clock and the first megaphone.

Additionally, the Jesuits were integral to the development of seismology throughout the world. They made a substantial number of organizational, experimental, and theoretical contributions to the discipline, and started thirty-eight seismographic stations, some of which were the only stations in vast regions (e.g., South America, Africa, and Asia).[3] They maintain their work of education in five major institutions—St. Louis University, the Weston Observatory,

[2]David Topper, *How Einstein Created Relativity out of Physics and Astronomy* (New York: Springer, 2013), p. 175, and Simon Singh, "Even Einstein Had His Off Days", *New York Times*, January 2, 2005, http://www.nytimes.com/2005/01/02/opinion/02singh.html.

[3]Augustn Udías and William Stauder, "The Jesuit Contribution to Seismology", *Seismological Research Letters* 67, no. 3 (May/June 1996), https://www.seismosoc.org/inside/eastern-section/jesuit-contribution-seismology/.

Instituto Geofisico (Bogota), Observatorio S. Calixto (La Paz), and Manila Observatory. They continue to make research contributions in the theory of plate tectonics, earthquake mechanisms, and earthquake detection.[4]

The Catholic Church is the only church to have an academy of sciences with Nobel Prize winners from every area of science. Its lineage goes back to 1601 and is very active today. The Pontifical Academy of Sciences currently has eighty members (twenty of whom are Nobel laureates),[5] and throughout its history has had ninety Nobel laureates.[6] In addition to sharing scientific expertise, scientists study the interrelationship between faith, philosophy, and natural science.

The Catholic Church also supports departments of natural science in 1,861 universities.[7] These departments produce outstanding research and publications by esteemed professors in every scientific field. The Church also supports contemporary science departments in approximately fifty thousand Catholic secondary schools.[8] Additionally, the Church supports scientific research institutes (independent of and attached to universities), such as the Pontifical observatories and Jesuit observatories in Rome, Tucson (Arizona), and Santiago (Chile). It also supports some of the largest contemporary and historical scientific libraries. Recently, Dr. Stephen Barr and other colleagues started the Society of Catholic Scientists in the United States, and in two years it had eight hundred members, all of whom have Ph.D.'s in all areas of natural science.

In view of the above, it can hardly be said that the Church is antiscience. There has never been an antiscience movement in the Catholic Church throughout the seven-hundred-year history of scientific

[4] Ibid.

[5] Colleen Dulle, "Pope Francis Appointed Three Women to the Pontifical Academy of Sciences This Summer. What's Their Role at the Vatican?", *America: The Jesuit Review*, August 13, 2021, https://www.americamagazine.org/faith/2021/08/13/pope-francis-pontifcal-academy-science-women-241214.

[6] "Nobel Laureates", Pontifical Academy of Sciences (website), accessed March 18, 2024, https://www.pas.va/en/academicians/nobel.html.

[7] United States Council of Catholic Bishops, "Catholic Education", https://www.usccb.org/offices/public-affairs/catholic-education.

[8] Quentin Wodon, *Global Catholic Education Report 2023: Transforming Education and Making Education Transformative* (Washington, D.C.: Global Catholic Education, 2022), p. 98, https://www.globalcatholiceducation.org/_files/ugd/b9597a_b54239f33dec48ddb4f2d735d10cba7c.pdf.

development. Quite the contrary—wherever science has developed, there also Catholic scientists (including priests and religious brothers and sisters) have abounded and contributed.

If there has not been an antiscientific movement in the Church, how can the Galileo controversy be explained?

II. The Galileo Controversy

It should be noted at the outset that the Galileo controversy was not about the veracity of scientific method, or established scientific conclusions, but rather about disregard for a Vatican warning, a broken promise, and an insult. Yes, there was a controversy concerning what we today accept as an established scientific conclusion—for example, heliocentrism (the theory that the earth and other planets revolve around the sun rather than the sun and planets around the earth—geocentrism). [9] Nevertheless, during Galileo's time, heliocentrism was not an established fact, but only one of two possible models of the solar system (along with geocentrism—the earth at the center of the solar system). The Catholic Church was open to both models, and indeed, the heliocentric model was originally justified by a Catholic cleric—Nicolaus Copernicus (see above). So, what was the problem?

The problem for both the Church and the scientific community was that neither heliocentrism nor geocentrism could be *proven* as a fact. The only technique to do this at the time was stellar parallax—examining the sun's position relative to other stars' positions as the earth rotates on its own axis. [10] The sun's position relative to the rotating

[9] The heliocentric model stood in opposition to the geocentric model justified by Ptolemy around A.D. 150. During Galileo's time the geocentric model was preferred not only because it was long held, but also because it gave the earth (and mankind) a primacy over all other heavenly bodies. The opposition between heliocentrism and geocentrism lay at the heart of the Galileo affair.

[10] Professor Thomas Woods gives an excellent analogy to explain stellar parallax—a person on a merry-go-round looking at the angles of two different objects as the merry-go-round rotates. If one of the objects is the sun and the other a distant star (or stars), one can infer whether the earth is orbiting around the sun, or whether the sun is orbiting around the earth. For more information about the Galileo affair, see Thomas E. Woods, "The Church and Science", in *How the Catholic Church Built Western Civilization* (Washington, D.C.: Regnery Publishing, 2005), pp. 67, 69–74.

earth was not a problem, but it had to be compared to another object or objects—distant stars. These stars are *very* distant from the earth compared to the sun, which requires very powerful and highly sensitive astronomical instruments (e.g., a very large telescope) to make an accurate determination of whether the earth is orbiting around the sun or the sun around the earth. At the time, astronomical instrumentation was not nearly accurate enough to make a conclusive finding. Indeed, the poor instrumentation in the early seventeenth century seemed to indicate geocentrism as a more likely model. Importantly, the Church did not rush to assert geocentrism as fact on the basis of this evidence. It reserved its judgment until stellar parallax could confirm it conclusively. Friedrich Bessel was the first to make this conclusive confirmation with much more accurate astronomical instrumentation in 1839—over two hundred years later.

The view of Saint Robert Bellarmine (the Jesuit cardinal who oversaw much of the dialogue with Galileo, but unfortunately died in 1621, twelve years before Galileo's trial) represents well the Church's (and the pope's) view of heliocentrism. In a letter to Paolo Foscarini (a friend of Galileo), Bellarmine cautions that the heliocentric model should not be published as fact until it is conclusively proven, because it breaks with many theologians' previous interpretations of the Scriptures. Nevertheless, he says if science should conclusively prove the truth of heliocentrism, this scientific truth should be accepted as conclusive, and that previous scriptural interpretation should be revised to reflect it:

> I say that if there were a true [scientific] demonstration that the sun is at the center of the world and the earth in the third heaven, and that the sun does not circle the earth but the earth circles the sun, then one would have to proceed with great care in explaining the Scriptures that appear contrary; and say rather that we do not understand them than that what is [scientifically] demonstrated is false. But I will not believe that there is such a demonstration, until it is shown me.[11]

---

[11] Robert Bellarmine, quoted in Luca Arcangeli, "At the Roots of the 1616 Decree: Robert Bellarmine's Letter to Paolo Foscarini", Interdisciplinary Encyclopedia of Religion and Science (website), accessed March 18, 2024, https://inters.org/copernicanism-bellarmine-foscarini, citing Galileo Galilei, *Edizione Nazionale delle Opere*, ed. A. Favaro (Florence: Giunti-Barbera, 1968), 12:172.

Bellarmine and others within the Church were concerned to protect the proper interpretation of Scripture while maintaining the integrity of scientifically demonstrated fact. If Cardinal Bellarmine represented the Church's thinking on this issue, why did the controversy start and escalate? The problem was not concerned with the appropriateness of scientific method, with the acceptance of scientific facts (even those which *seem* to conflict with Scripture), or with heliocentrism itself (the first person to justify heliocentrism mathematically was Nicolaus Copernicus—a Catholic cleric who was not suspected of anything inappropriate). Rather, the problem was *proof*. The Catholic Church (and specifically the pope) was being criticized by Protestants and Catholic constituencies for interpreting Scripture too loosely to accommodate science, and so Pope Urban VIII wanted Galileo to have positive proof that heliocentrism was scientifically valid before he allowed a metaphorical interpretation of Genesis. The pope and Bellarmine knew well that Galileo did not have scientific proof of heliocentrism. Though Galileo promised the pope he would not publish heliocentrism as fact before there was scientific proof, he nevertheless published his theory in 1632 as fact in his new tractate *Dialogue concerning the Two Chief World Systems*. Though Galileo presented some evidence suggestive of heliocentrism (e.g., the orbits of moons around Jupiter), he was unable to provide definitive proof. The trial centered on Galileo's broken promise, which was the main issue. He publicly insulted the pope for not subscribing to his position,[12] which probably did not help his position. In any event, the court ruled that he should not be allowed to publish without ecclesial check, and exiled him to his villa outside of Florence. Galileo expected a less severe sentence than he received because he had a previous friendship with the pope as well as his friendship with several clerics. Nevertheless, the court believed that Galileo had recklessly undermined the veracity of Scripture without the appropriate proof, and so rendered its decision. In retrospect, Galileo's sentence was too harsh, because the Church did not have proof of the Ptolemaic earth-centered view.

---

[12] In his *Dialogue concerning the Two Chief World Systems*, Galileo portrayed the pope's position through a character named "Simplicio", loosely translated "Simpleton". Pope Urban VIII was anything but a simpleton, having a doctorate in law as well as proficiency in Latin Scriptures.

These circumstances motivated Saint John Paul II in 1992 to issue a public apology to the Pontifical Academy of Sciences (directed to the whole scientific community) for the Church's actions during the Galileo sentencing. He admitted that the theologians of the Church got their interpretation of Scripture's meaning wrong, and that this errant interpretation of Scripture caused them to bring a matter of scientific inquiry into the realm of sacred doctrine:

> This [treatment of scriptural interpretation] led [the Church's theologians] unduly to transpose into the realm of the doctrine of the faith a question which in fact pertained to scientific investigation.[13]

In her vigilance to protect the traditional meaning of Scripture, the Church overreacted to Galileo's breach of promise and punished Galileo as if he had really violated Church doctrine. In fact, Galileo had proclaimed his unproven theory only to be true, but had not violated Church doctrine. Though this caused tremendous pressure to be put on the pope to discipline Galileo for undermining the traditional interpretation of Scripture, this traditional interpretation had not yet been proven to be consistent with scientific fact, and so the penalty was excessive. Pope John Paul II wanted to apologize for this excessive penalty, but in so doing, he was not saying that Galileo was correct in proclaiming the truth of an unproven fact or that the Church had been against heliocentrism or science.

In retrospect, the whole matter could have been avoided if Saint Robert Bellarmine had lived twelve more years (to Galileo's trial in 1633), as well as if Galileo had not violated his promise and the Church had not been overly zealous to protect a traditional interpretation of Scripture that had not yet been shown to be consistent with proven scientific fact. Was the Church antiscience? She was not; rather, she was overly zealous to protect an unproven traditional interpretation of Scripture—as a reaction to the pressures of the time.

---

[13] Pope John Paul II, Address to the Plenary Session on "The Emergence of Complexity in Mathematics, Physics, Chemistry and Biology" (plenary session of the Pontifical Academy of Sciences, Vatican City, October 31, 1992), https://www.pas.va/en/magisterium/saint-john-paul-ii/1992-31-october.html.

III. Scientific Discovery versus Biblical Revelation

The biblical and scientific accounts of creation are quite different, but does that mean they are in conflict? As we shall see, it does not, because theology and science have different objectives, methods, and legitimate conclusions. As Saint Thomas Aquinas discerned, faith and science cannot be in conflict, for they come from the *same source*—the mind of God:

> Although the truth of the Christian faith which we have discussed surpasses the capacity of the reason, nevertheless that truth that the human reason is naturally endowed to knowcannot be opposed to the truth of the Christian faith. For that withwhich the human reason is naturally endowed is clearly most true; so much so, that it is impossible for us to think of such truths as false. Nor is it permissible to believe as false that which we hold by faith, since this is confirmed in a way that is so clearly divine.[14]

There *seems* to be discrepancies between biblical claims and scientific discoveries. The Bible tells us that God created the universe, life, and even human beings in six days, while science holds that from the time of the Big Bang to earthly microbial life was about 9.9 billion years. Again, the Bible implies that the universe is about five thousand years old, while science again contends that it is about 13.8 billion years old. Can these two viewpoints be reconciled?

In his 1943 encyclical[15] *Divino Afflante Spiritu*, Pope Pius XII stated that the purpose of the Bible (in which God speaks through inspired authors) is to present sacred truths needed for salvation. In contrast to this, the purpose of science is to give empirical-mathematical descriptions and explanations of the physical universe. Thus, when the Bible talks about creation, its purpose is not to give a correct observationally based explanation of the physical universe. Instead, it is meant to impart truths that will be beneficial for salvation—such as

---

[14]Thomas Aquinas, *Summa Contra Gentiles* 1, 7, 1, trans. Anton C. Pegis (1975; repr., Notre Dame, Ind.: University of Notre Dame Press, 2014).

[15]Papal encyclicals are public letters written by the pope to the Church. One purpose of an encyclical is to clarify and communicate the Church's teaching on important matters.

there is one God, not many; that everything else in reality besides the one god is the creation of God; that God is just and makes things good. However, science is not primarily concerned with God, His relationship with creation, and His benevolence toward it, but rather, with discovering the universe's physical laws, expressing them mathematically, and using observational evidence to determine how and when the physical universe originated.

In view of this, the Bible can be correct about the six days of creation insofar as it gives us truths needed for our salvation according to God's revelation—such as the one true God created human beings in His own image and likeness. However, truths that have no salvific value and fall within the scientific purview (such as "the universe was created in six days") are outside the realm of biblical interpretive method and purpose.

In contrast to this, science legitimately operates within the domain of the physical universe and empirical mathematical explanation. Hence, it can legitimately assert that the universe is 13.8 billion years old, that it went through several eras of creation—from a quantum cosmological beginning (where all four universal forces were unified), to a space-time era (described by the general theory of relativity), to an inflationary era (where the universe experienced hyperacceleration for a fraction of a second), to the separation of the other three forces (strong nuclear force, weak force, and electromagnetic force), to a plasma era (in which positive and negative charges were unified in a plasma field), to the Lepton-Hadron era (where the breakdown of the plasma led to separate positive and negative particles to which the Higgs field imparted rest mass), to the era of stellar nucleosynthesis (in which stars condensed and exploded). Though this is a very important description for scientists, it has little value for aiding us in our salvation (in other words, we do not have to know about stellar nucleosynthesis to be saved).

Just as biblical authors did not have expertise in scientific method (and therefore could not competently give scientific judgments), so also most scientists do not have an expertise in God's self-revelation, His relationship with creation, and the path to eternal salvation (and cannot give competent judgments about these areas). Each discipline has its proper purpose, method to accommodate that purpose, and criteria for correctness in its conclusions. If we let science be science and let biblical interpretation be biblical interpretation, both disciplines can be correct without contradiction. However, if we try to make biblical interpretation give scientific judgments, or we try to make science give judgments appropriate to biblical interpretation, we will soon get into trouble—the very trouble that Pope Pius XII was trying to avoid, the trouble that leads to needless conflict between science and revelation.

There are some remarkable coincidences where the biblical narrative (written between 500 and 600 B.C.) uncannily agrees with the scientific judgment (discovered about twenty-four hundred years later) that the first moment of creation was the creation of light (energy)— "Let there be light" (Gen 1:3). However, these coincidences are not the primary purpose of biblical interpretation or scientific discovery, and we should not try to force the respective conclusions to agree with each other beyond what they do according to their proper methods.

Pope Pius XII also made another important clarification. He indicated that God's inspiration of the various human biblical authors was intertwined with the thoughts and capacities of those authors—God and the biblical authors were "partners", so to speak, in the writing of the biblical text:

> For having begun by expounding minutely the principle that the inspired writer, in composing the sacred book, is the living and reasonable instrument of the Holy Spirit, [Catholic authorities and exegetes] rightly observe that, impelled by the divine motion, [the biblical author] so uses *his* faculties and powers, that from the book composed by him all may easily infer "the special character of each one and, as it were, *his personal traits*." Let the interpreter then, with all care and without neglecting any light derived from recent research, endeavor to determine the peculiar character and circumstances of the sacred writer, the age in which he lived, the sources written or oral to which he had recourse and the forms of expression he employed.[16]

God is clearly the source of inspiration, and so the primary source of revelation; however, He uses the biblical author to produce a

---

[16]Pope Pius XII, Encyclical on Promoting Biblical Studies *Divino Afflante Spiritu* (September 30, 1943), no. 33 (emphasis added), http://www.vatican.va/content/pius-xii/en/encyclicals/documents/hf_p-xii_enc_30091943_divino-afflante-spiritu.html.

work that can be understood by the biblical author's audience situated within a particular culture and time.

Notice that this "partnership" theory of inspiration is quite different from the "dictation" theory. The dictation theory holds that God simply spoke to the mind of the biblical author, who in turn wrote down what he "heard"—verbatim. In this view, the biblical author plays only a transcriber's role, while God does everything else, so that every word in the Bible is the literal truth of God—which must in turn be taken literally. Though some Christian denominations hold this view, Catholicism does not. This was clarified by Pope Pius XII's encyclical.

The "partnership" theory of divine inspiration holds that the biblical author plays a role in the production of the revealed text. He brings His thinking patterns, His culture, His sense of history, and His categories to the writing process. Why would God allow this? Because He wants to communicate with the people in the biblical author's audience. The author and audience of Genesis 1 (in about 500 B.C.) could not possibly have understood a scientific explanation of creation (as we understand it today). They did not understand the method and mathematics of science—nor did they have the instrumentation necessary to discover scientific data. According to Pope Pius XII, God was really not concerned with giving a proper scientific account of creation when he inspired the biblical author. He was concerned only to give—through the author's and audience's own categories and culture—sacred truths necessary for salvation. As Galileo Galilei put it, "The intention of the Holy Ghost [the author of the Bible] is to teach us how one goes to heaven, not how heaven goes."[17] The sacred truths necessary for salvation all concern the nature of God's relationship with mankind and how all of creation, especially human beings, bears the imprint of their Maker.

This partnership theory of divine inspiration is a long-held belief within the Catholic Church and was summed up by Saint Thomas Aquinas in the thirteenth century, when he said, "Whatever is received into something is received according to the condition of

[17]Galileo Galilei, "Letter to the Grand Duchess Christina of Tuscany, 1615", Internet History Sourcebooks Project (website), 1997, https://sourcebooks.fordham.edu/mod/galileo -tuscany.asp.

the recipient."[18] Thus, if God wants to communicate His truth to a sixth-century B.C. Israelite audience, He will have to use the categories and mindset for that same time period and audience—and what better way to do it than to "work with" a sixth-century B.C. Israelite author? By doing this, He communicates effectively with those audiences and does not impede communication with future audiences, for those audiences would be able to understand the categories and mindsets of a less sophisticated, nonscientific time and culture through historical research.

Today we can understand the salvific truths in the Genesis narrative as easily as the biblical author's audience in 500 B.C. If we do not confuse the salvific intention and content of God's revelation with the method and content of the natural sciences, there will be no contradiction between the biblical and scientific accounts of creation. Each account has its own purpose with its own method and its own content. They are talking about two different things. Conflating them is a misunderstanding of God's intention in revealing Himself to us through the Bible. It is also vital to note that the human authors of the Bible, especially the Old Testament, often made their intended points in nonliteral genres, such as stories, allegories, poems, and songs, which make use of figurative tools like metaphor and symbol. This of course might pose a problem if the biblical authors were trying to write science—but they were not! Making use of symbolic, nonliteral language is often a great way to make a point about the nature of God and the meaning of His relationship with mankind. And it is precisely these sorts of truths that the Bible communicates to us for our salvation.

So what are the "sacred truths necessary for salvation" conveyed in the "days of creation" narrative? In order to answer this question, we will first want to understand the problem faced by the inspired biblical author in the sixth century B.C. At that time, several surrounding countries—Babylon, Assyria, and Egypt, among others—wrote myths of creation that presented themselves as rival explanations to the revelation of Yahweh to Israel. The most well known of these myths was the Babylonian epic Enuma Elish and the Sumerian-Akkadian

---

[18] *The Summa Theologica of St. Thomas Aquinas*, trans. Fathers of the English Dominican Province, vol. 1 (New York: Benziger Brothers, 1947), I, q. 75, art. 5, p. 367.

epic Gilgamesh. These myths, and others like them, diverged from God's revelation to Israel in several major ways:

1. The rival myths presented many gods (polytheism), while the biblical author asserts that there is only one God.
2. Rival myths spoke of nature gods (such as a moon god and an ocean god) who influenced how the world functioned, while the biblical author asserts that the one God created and sustained everything in being. Hence, all natural objects were viewed by Israel as creations of God—not as gods.
3. Rival myths presented human beings as mere "playthings" and "slaves" for the gods, while the biblical author asserts God created human beings in His own image and likeness—giving human beings an ultimate and transcendent dignity.
4. Rival myths believed that matter was fundamentally evil and disordered, while the biblical author held that matter—since it was a deliberate creation of the all-intelligent (omniscient) God—was fundamentally good and ordered.

So, are these four revelations necessary for salvation today? Absolutely—as necessary today as in the sixth century B.C. All of them are very fundamental—without them we would still be immersed in polytheism and nature worship, and we would have no sense of our true ultimate dignity, our life-giving relationship with God, or even the blessedness of creation. Can we understand these truths as well as a sixth-century B.C. Israelite audience? Of course—and now we see them in an even greater context of world history and culture. The truths of salvation in the Bible still stand firm, guiding our minds and hearts to the God of Jesus Christ.

None of the four core-revealed truths communicated by the "days of creation" story contradict the Big Bang theory or any other truth of science. Anyone can hold both the "days of creation" story and the Big Bang theory to be true without contradiction. The Bible tells us that God created an ordered world according to His intelligent plan, with mankind having a special place in that plan, while science tells us about the physical explanation of how that creation happened. Far from there being any contradiction, we can see that what science tells us about the universe in terms of the Big Bang and its fine-tuning is best explained in light of the fact that God caused the Big Bang and fine-tuned the universe! Only if all the details of the "days of creation" story are interpreted literally could there be a contradiction between it and science. It is important to bear in mind that it would not have occurred to the human author(s) of Genesis that what they were doing was writing a literal scientific account of precisely how the world began. The biblical authors were well aware that they were using the term "day" in a figurative, symbolic manner; after all, there cannot be a day without the sun, and yet in the "days of creation" story, the sun is created only on the fourth "day" (Gen 1:14–19). Rather, the biblical authors were self-consciously writing a theology about the relationship between God, mankind, and the world in order to correct the false theologies they encountered in the epics of other cultures at the time.

The Church Fathers recognized that the literary genre of the first chapters of Genesis is neither science nor history. For example, Origen (184–253) and Saint Augustine (354–430) both wrote of how the Book of Genesis uses imagery (symbols, metaphors, etc.) to make theological points about the ultimate nature of God and His relationship with creation and humanity. In *The Literal Interpretation of Genesis* (written in A.D. 408), Saint Augustine set out a hermeneutical principle that governed much of biblical interpretation throughout the centuries, affecting Saint Robert Bellarmine in his relationship with Galileo and Saint Pius XII in the encyclical cited above. He stated plainly that there should not be a literal interpretation of Scripture that runs contrary to the rational perception of nature (i.e., to well-grounded science). If Scripture seems to contradict such things, we should seek another interpretation (e.g., symbolic) that shows how this passage helps our salvation. Saint Augustine put it succinctly:

With the Scriptures it is a matter of treating about the faith. For that reason, as I have noted repeatedly, if anyone, not understanding the mode of divine eloquence, should find something about these matters [about the physical universe] in our books, or hear of the same from those books, of such a kind that it seems to be at variance with the perceptions of his own rational faculties, let him believe that these other things are in no way necessary to the admonitions or accounts or predictions of the Scriptures. In short, it must be said that our authors knew the truth about the nature of the skies; but it was not the intention of the Spirit of God, who spoke through them, to teach men anything that would not be of use to them for their salvation.[19]

In an earlier passage from the same work, Saint Augustine speaks of how foolish Christian writers might sound, if they preach literal interpretations of Scripture that contradict well-established facts about nature:

> It not infrequently happens that something about the earth, about the sky, about other elements of this world, about the motion and rotation or even the magnitude and distances of the stars, about definite eclipses of the sun and moon, about the passage of years and seasons, about the nature of animals, of fruits, of stones, and of other such things, may be known with the greatest certainty by reasoning or by experience, even by one who is not a Christian. It is too disgraceful and ruinous, however, and greatly to be avoided, that he [the non-Christian] should hear a Christian speaking so idiotically on these matters, and as if in accord with Christian writings, that he might say that he could scarcely keep from laughing when he saw how totally in error they are.[20]

As can be seen, it has long been the practice of the Catholic Church to respect well-established facts and laws of nature, and to interpret Scripture in a manner that does not contradict them. The *seeming* exception to this practice is the trial of Galileo, but this was explained above. The trial was not about the validity of science or heliocentrism, but rather Galileo's breach of a promise that increased pressure on the pope by both Protestants and Catholic constituencies. The Church overreacted to these pressures and unfortunately penalized Galileo excessively for his breach of promise. As unfortunate as this was, it was *not* antiscience.

In view of the above, it can scarcely be said that the Catholic Church was against science, scientific method, heliocentrism, or any other scientifically demonstrated fact. As can be seen, by the work of clerical

---

[19]Saint Augustine, *The Literal Interpretation of Genesis* 2, 9, 20, trans. W. A. Jurgens, in vol. 3 of *The Faith of the Early Fathers*, ed. W. A. Jurgens (Collegeville, Minn.: Liturgical Press, 1979).

[20]Ibid.,1, 19, 39.

scientists, the papal encyclical *Divino Afflante Spiritu*, and the centuries of teaching proper scientific method in secondary schools and universities, the Catholic Church has in the main adhered to the principle of Robert Bellarmine: If a fact that seems contrary to Scripture is scientifically demonstrated, we should reinterpret the Scriptures rather than contradict a scientifically demonstrated fact.

IV. Scientific Evolution versus the Biblical Account of Creation

Evolution is a scientific theory that explains the development of species using evidence from genetic similarities among species, fossil delineation, and geographic distribution. Very basically, the theory of evolution states that biological species physically evolved to their present state slowly and over a long period of time. Evolution of species is driven by how small changes (mutations) in offspring make them more or less suited to survival than offspring that do not possess these changes; changes that are better suited to survival are more likely to be passed from one species' generation to the next. And so a particular species evolves along the path of changes better suited to survival. Evolution is considered by the vast majority of scientists to explain biological (physical-organic) development within and among biological species. There are three points at which neo-Darwinian materialistic evolution is challenged to explain advances in physical organic activities and processes:

1. The leap from nonliving to living beings
2. The leap from nonconscious (e.g., vegetative) beings to beings with sensate consciousness (e.g., self-moving animals)
3. The leap from sensate consciousness to rational self-conscious beings

Scientists such as Michael Polanyi believe that each categorical leap requires a whole new system of organization that cannot be reduced to the lower one. In his seminal article "Life's Irreducible Structure", Polanyi, a senior researcher in physical chemistry at the universities of Manchester and Oxford, argued that living systems are "dual control", constituted by lower-level physical-chemical laws and activities as well as higher-level boundary conditions that harness these activities toward objectives that *cannot be reduced* to those lower-level laws and activities.[21] This applies just as much to the DNA molecules that contain the instructions for the development of such systems as the systems that arise out of those instructions.[22] As Polanyi puts it:

> A boundary condition is always extraneous to the process which it delimits.... Their structure cannot be defined-in terms of the laws which they harness.... Therefore, if the structure of living things is a set of boundary conditions, this structure is extraneous to the laws of physics and chemistry which the organism is harnessing. Thus the morphology of living things transcends the laws of physics and chemistry.[23]

This irreducible hierarchy applies to all three "leaps" in the evolutionary development of human beings:

1. The higher-level boundary conditions of living things cannot be reduced to the lower-level physical and chemical activities being harnessed by them.
2. The higher-level boundary conditions of sensate consciousness cannot be reduced to the lower-level physical, chemical, and biological activities being harnessed by them.[24]
3. The higher-level boundary conditions of mind (rational self-consciousness) cannot be reduced to the lower-level physical, chemical, biological, and sensate activities harnessed by them.[25]

Polanyi explains that the higher-level activities manifest "design" that cannot arise out of random assemblages of lower-level activities, which seems to put the three leaps in human development beyond merely materialistic evolution. He does not indicate what the

---

[21] Michael Polanyi, "Life's Irreducible Structure", *Science* 160, no. 3834 (June 21, 1968): 1308–9. Also, see Michael Polanyi, "Transcendence and Self-transcendence", *Soundings* 53, no. 1 (Spring 1970): 88–94.
[22] Polanyi, "Life's Irreducible Structure", 1308–9.
[23] Ibid., p. 1309.
[24] Ibid., pp. 1311–12.
[25] Ibid., p. 1312.

explanation is for the occurrence of these seemingly designed higher organizational boundary conditions, but strongly asserts that it is *not* attributable to the lower-level activities being harnessed by them.[26] In so doing, he implies that the rational self-consciousness of human beings cannot be fully explained by physical, chemical, biological, and sensate activities. This corresponds to the philosophical argument for the existence of a transphysical soul from the activities of conceptual intelligence and self-consciousness that cannot be explained by physical structures, systems, and processes alone.[27]

There are other difficulties with materialistic evolutionary explanation of the development of sensate consciousness and rational self-consciousness beyond the above three ontological leaps—namely, the exceedingly short time period of 4.6 billion years (the age of the earth) to explain literally thousands of significant evolutionary developments. One such case is made by professed atheist Thomas Nagel, who candidly states:

> The more we learn about the intricacy of the genetic code and its control of the chemical processes of life, the harder [the problems of merely natural selection and mutation explanations] seem. With regard to evolution, the process of natural selection cannot account for the actual history without an adequate supply of viable mutations, and I believe it remains an open question whether this could have been provided in geological time merely as a result of chemical accident, without the operation of some other factors determining and restricting the forms of genetic variation. It is no longer legitimate simply to imagine a sequence of gradually evolving phenotypes, as if their appearance through mutations in the DNA were unproblematic—as Richard Dawkins does for the evolution of the eye.[28]

The above problems apply only to materialistic evolutionary explanations, but if one allows for an evolutionary process interacting with or involving a transphysical cause (e.g., involving

[26] Ibid., pp. 1308–9.

[27] See Robert Spitzer, *Science at the Doorstep to God: Science and Reason in Support of God, the Soul, and Life after Death* (San Francisco: Ignatius Press, 2003), Chapter 5, Sections I.A and II.B.

[28] Thomas Nagel, *Mind and Cosmos: Why the Materialist Neo-Darwinian Conception of Nature Is Almost Certainly False* (New York: Oxford University Press, 2012), p. 9.

a higher-level designing cause) or a transphysical agency (e.g., a soul), then evolution in this broader sense may provide an explanation for the development of continuously advancing speciation as well as the leaps from nonliving to living beings, nonconscious to conscious beings, and nonrationally self-conscious to rationally self-conscious beings. This is discussed below with respect to three conceptions of theistic evolution—nomogenesis, orthogenesis, and nomo-orthogenesis.

Before examining theistic evolution, we must first take account of the Catholic Church's position on it, which may be different from the views of other Christian denominations. In his well-known 1950 encyclical, *Humani Generis*, Pope Pius XII asserted that Catholics are not forbidden from adhering to an evolutionary explanation of human embodiment so long as it does not preclude the existence of a unique transphysical soul in every human being created by God:

> For these reasons the Teaching Authority of the Church does not forbid that, in conformity with the present state of human sciences and sacred theology, research and discussions, on the part of men experienced in both fields, take place with regard to the doctrine of evolution, in as far as it inquiries into the origin of the human body as coming from pre-existent and living matter—for the Catholic faith obliges us to hold that souls are immediately created by God.[29]

Pope Saint John Paul reinforced Pope Pius XII's teaching in a letter to the Papal Academy of Sciences in 1996, declaring that the scientific validation of evolution moves it from the status of hypothesis to a well-grounded theory:

> Today, more than a half-century after the appearance of [*Humani Generis*], some new findings lead us toward the recognition of evolution as more than a hypothesis. In fact it is remarkable that this theory has had progressively greater influence on the spirit of researchers,

---

[29] Pope Pius XII, Encyclical concerning Some False Opinions Threatening to Undermine the Foundations of Catholic Doctrine *Humani Generis* (August 12, 1950), no. 36, https://www.vatican.va/content/pius-xii/en/encyclicals/documents/hf_p-xii_enc_12081950_humani-generis.html.

following a series of discoveries in different scholarly disciplines. The convergence in the results of these independent studies—which was neither planned nor sought—constitutes in itself a significant argument in favor of the theory.[30]

The above declaration clearly indicates that Catholics can believe in the theory of evolution so long as they do not deny the creation of a unique transphysical soul by God in every human being. This means that Catholics cannot believe in a purely *materialistic* view of evolution, because evolution is a physical-biological process that cannot explain a transphysical soul that is beyond such processes.[31] Thus, Catholics can believe in any view of evolution that will accommodate a transphysical soul, which would have to be created by a transphysical cause—like God. The explanation of how the soul and body can interact through a Bohmian quantum reduction process is given in a previous work.[32]

Can Catholics believe that the physical-biological dimension of human beings evolved from other species? Yes. Can they believe that even the cerebral cortex came from an evolutionary process—from great apes to *Homo habilis* to *Homo erectus* to *Homo sapiens*? Yes. In fact, it is most reasonable to believe on the basis of all the evidence that the human frontal and cerebral cortices evolved in a remarkable way to meet all the conditions required to accommodate a transphysical soul capable of thinking activities completely beyond physical processes—for example, conceptual ideas, syntactically significant language, abstract mathematics, self-consciousness, and transcendental and religious thought.[33] As noted in a previous work, the transphysical soul outside of the physical body (during near-death experiences) manifests sensation, imagination, and memory, as well as its unique activities of conceptual ideas, syntactically significant language, mathematics,

---

[30]Pope John Paul II, Message to the Pontifical Academy of Science on Evolution (October 22, 1996), no. 4, https://www.ewtn.com/catholicism/library/message-to-the-pontifical-academy-of-science-on-evolution-8825. EWTN's website has a note on the mistranslation of "more than a half-century", which should read "almost a half-century": "the original French text correctly notes that 1996 is 'almost' a half-century after 1950 when the encyclical was published."

[31]See Spitzer, *Science at the Doorstep to God*, Chapter 6, Section V.

[32]See ibid.

[33]See ibid., Chapters 5–6.

self-consciousness, and transcendental and religious consciousness.[34] Interestingly, the physical brain also has the capacity for sensation, imagination, and memory. Why the overlap between the soul's and the physical brain's capabilities? The Nobel Prize–winning physiologist Sir John Eccles believes the overlap constitutes a point of contact that allows for quantum reduction mediation between soul (mind) and body.[35] In light of this, Eccles proposes a model for how the cerebral and frontal cortices might have developed to accommodate a soul/mind in the final stages of the evolutionary process.[36]

As shown in previous work, our rational self-conscious transphysical soul is precisely what distinguishes us from all other animals.[37] Our souls, which make us capable of reason, morality, love, art, and free will, demonstrate that we are not just any kind of animal. We are animals capable of knowing and loving God (only human beings are capable of religion). The fact that all of us equally possess spiritual souls is what gives us our equal human dignity— that is, our special value as human beings (a special value that is at the basis of human rights).

As noted above, acceptance of evolutionary theory does not require a materialist, reductionist interpretation of evolutionary processes. Indeed, the strictly materialistic view of evolution is rejected by an increasing number of outstanding theorists—even those who profess agnosticism and atheism (e.g., Thomas Nagel, as noted above). So how do nonreductionistic philosophers and scientists view evolution? The most prevalent view is theistic evolution, which holds that there is a God; that God is the creator of the material universe, and all life within it; and that biological evolution is a natural process within that creation for which God is ultimately responsible.[38] As previously

---

[34] See ibid., Chapter 4, and Spitzer, *The Soul's Upward Yearning* (San Francisco: Ignatius Press, 2015), Chapter 5.

[35] Sir John Eccles and Karl Popper, *Evolution of the Brain: Creation of the Self* (New York: Routledge, 1989), pp. 195–99.

[36] Ibid., pp. 121–78.

[37] See Spitzer, *Science at the Doorstep to God*, Chapter 5, Sections I.B and II.C.

[38] See the following report prepared by YouGov and the University of Birmingham: Kate Gosschalk et al., *Science and Religion Exploring the Spectrum: A Multi-Country Study on Public Perceptions of Evolution, Religion and Science* (London: YouGov, 2023), https://scienceand beliefinsociety.org/wp-content/uploads/2023/12/UoB-YouGov.-Science-and-Religion -Survey-Report.-8-Dec-2023-.pdf.

mentioned, the vast majority of physicians (65 percent) are theists, believers in God or a higher supernatural power (Chapter 6, Introduction),[39] and as shown in the previous volume, 51 percent of all scientists (and 66 percent of young scientists) also profess belief in God or a higher spiritual reality.[40] The vast majority of scientists also believe in evolution, so we may infer that the majority of scientists and physicians subscribe to theistic evolution, which is the view of the Catholic Church as well as mainline Protestant churches.[41]

There are several different theories of theistic evolution, but three major schools represent the majority:

1. *Nomogenesis.* In this view, evolution occurs according to finely tuned fixed laws and constants infused in the universe by God at creation (God acting as efficient cause). God "frontloads" His creation with fine-tuning of universal constants and initial conditions as well as chemical, biological, and sensitive psychological laws and conditions (which come into play sometime after creation) so that the evolutionary process can proceed naturally without ongoing direction from Him. A large number of religious scientists and philosophers subscribe to this view, such as Francis Collins (the celebrated director of the Genome project in the United States), who asserts that God "frontloads" all that's required for an autonomous ongoing evolutionary process that can proceed naturally without intermittent influence from Him.[42]

---

[39] Kristin A. Robinson et al., "Religious and Spiritual Beliefs of Physicians", *Journal of Religion and Health* 56, no. 1 (February 2017), https://pubmed.ncbi.nlm.nih.gov/27071796/.

[40] Pew Research Center, "Religion and Science in the United States: Scientists and Belief", November 5, 2009, https://www.pewresearch.org/religion/2022/12/21/key-findings-from-the-global-religious-futures-project/attachment/7/, cited in Spitzer, *Science at the Doorstep to God*, Introduction, Section I.

[41] See Eugenie Scott, "Antievolution and Creationism in the United States", *Annual Review of Anthropology* 26 (October 1997): 263–89. For a list of different religious groups' statements on this topic, see Pew Research Center, "Religious Groups' Views on Evolution", February 13, 2014, https://www.pewresearch.org/religion/2009/02/04/religious-groups-views-on-evolution/.

[42] Francis Collins, *The Language of God* (New York: Free Press, 2007), p. 200. It should be noted that Collins does believe that God interacts with human beings both interiorly and providentially, particularly through conscience and the moral law as well as through spiritual yearning and satisfaction. See ibid.

2. *Orthogenesis.* In this view, evolutionary processes are directed by God acting as a goal (final cause). The most famous proponent of this view was the celebrated Jesuit priest and paleontologist Pierre Teilhard de Chardin, who viewed evolution as directed toward an ultimate goal—"an omega-point"—within the mind of God.[43]

3. *Nomo-Orthogenesis* (God acting as both efficient cause and final cause). In this view, God "frontloads" creation with the laws, constants, and initial conditions required for an ascending evolutionary process, but also remains present to it (e.g., as a transcendent mind). The physical chemist Michael Polanyi (cited above) combines both positions, advocating for a robust "fine-tuning" of initial constants and conditions at the creation of the universe with a transcendent mind that pervades and influences the ongoing physical evolutionary process.[44] This enables him to combine natural processes of evolution with his nonreductionistic multilevel view of nature. As noted above, Polanyi held that higher-level boundary conditions are necessary to explain the three leaps in the evolutionary development of human beings—the leap from nonliving to living beings, the leap from nonconscious to conscious living beings, and the leap from nonrational to rational self-conscious beings.[45] Bernard Lonergan (a Jesuit priest, metaphysician, philosopher of science and theologian) implicitly argues a similar position to explain human knowing as well as a nonreductionistic multilevel ontology.[46]

There are a variety of nuances within these major schools. For example, Sir John Eccles holds a *physical* evolutionary development

[43] Pierre Teilhard de Chardin, *The Phenomenon of Man* (New York: Harper Perennial Modern Classics, 2008). See also Pierre Teilhard de Chardin, *The Divine Milieu* (New York: Harper Perennial Modern Classics, 2001).

[44] Vincent Smiles, "Transcendent Mind, Emergent Universe, in the Mind of Michael Polanyi", *Open Theology* 1 (November 6, 2015), https://www.degruyter.com/document/doi/10.1515/opth-2015-0030/html?lang=en.

[45] See Polanyi, "Life's Irreducible Structure".

[46] Bernard Lonergan, *Collected Works of Bernard Lonergan*, vol. 3, *Insight: A Study of Human Understanding*, ed. Frederick E. Crowe and Robert M. Doran (Toronto: University of Toronto Press, 1992), pp. 154–57, 288–93, and Chapters 8 and 19.

of species (including the development of the brain), as well as a special creation of each individual soul ("self") by God (a transmaterial cause). He explains the interaction between his dualistic principles (the material brain and the transmaterial soul) by appealing to quantum processes affecting the release of neurotransmitters.[47] As noted in previous work, the famous physicist David Bohm also sets up a nuanced quantum reduction theory of body-soul interaction.[48]

We might sum up by noting that the majority of scientists, physicians, Catholics, and mainline Protestants subscribe to some form of theistic evolution allowing for a divine intelligence to influence physical processes toward higher and higher levels of activity and being. Though religious belief influences many theistic evolutionists, they are also influenced by the inadequacy of purely materialistic neo-Darwinian theories, which cannot explain the origin of transphysical activities in human intelligence (implying a soul) or the astronomically high improbability of the ascendency of species by physical reductionistic random processes over 4.6 billion years. In view of this, Catholics are justified in holding this position, which is consistent not only with their religious beliefs, but also with sound evidence and scientific reasoning.

On this issue—as well as all other issues of science and faith— Catholics should always seek the truth, for there can be no contradiction between reason and faith. After all, faith and reason come from the same source—the all-knowing God. We conclude by again referring to Pope John Paul II's message to the Papal Academy of Sciences:

> In celebrating the 60th anniversary of the re-foundation of the Academy, it gives me pleasure to recall the intentions of my predecessor, Pius XII, who wished to bring together around him a chosen group of scholars who could, working with complete freedom, inform the Holy See about the developments in scientific research and thus provide aid for reflections.

[47] Eccles and Popper, *Evolution of the Brain*.

[48] This is explained in Spitzer, *Science at the Doorstep to God*, Chapter 6, Section IV. See also Paavo Pylkkänen, "Henry Stapp vs. David Bohm on Mind, Matter, and Quantum Mechanics", *Activitas Nervosa Superior* 61 (April 7, 2019): 49, https://link.springer.com/content/pdf/10.1007/s41470-019-00035-2.pdf.

To those whom he enjoyed calling the Scientific Senate of the Church, he asked simply this: that they serve the truth. That is the same invitation, which I renew today, with the certainty that we can all draw profit from "the fruitfulness of frank dialogue."[49]

V. Conclusion

Does science conflict with Catholic teaching? As can be seen from the above, it does not—and it will not because faith and science come from the same source: the unrestricted intelligence and love of God. Recall the principle of Saint Robert Bellarmine: If a fact that seems contrary to Scripture is scientifically demonstrated, we should reinterpret the Scriptures rather than contradict a scientifically demonstrated fact.[50]

Recall further that the purpose of Scripture is not to give an empirical-mathematical description and explanation of the physical universe, but rather to give sacred truths necessary for salvation, and the purpose of science is not to give sacred truths necessary for salvation, but rather to discover an empirically based hypothetico-deductive mathematical explanation for the physical universe. If we allow each domain to articulate its truth by means of its own proper method, and acknowledge the limits of those methods, there would be virtually no conflicts between faith and science.

[49]John Paul II, "Message on Evolution", no. 1, quoting Discourse to the Academy of Sciences (October 28, 1986), no. 1.

[50]In Arcangeli, "Letter to Paolo Foscarini", citing Galileo, Edizione Nazionale delle Opere.

CONCLUSION

This study of the scientific investigation of Christ sheds light on the veracity of Jesus' Passion and Resurrection; His Real Presence in the Holy Eucharist; and the presence of the Blessed Virgin in the economy of salvation and the Catholic Church herself. We have indicated multiple times that our faith does not rest on such scientific evidence, but as may now be clear, it certainly enhances, corroborates, and strengthens an already solid scriptural, historical, and ecclesial foundation. In the Introduction to this book, we mentioned Saint John Henry Newman's informal inference, which assembles multiple individually antecedently probable sets of evidence that point to the same conclusion. In this book, the common conclusion to which independent sets of evidence point is that Jesus Christ was historically real, crucified, and risen in glory, and is present to us through his Holy Spirit, the Holy Eucharist, and His Blessed Mother. There are seven major sets of data that support this conclusion, all of which have multiple complementary and corroborating features. A brief summary will suffice to show the probative strength of our general conclusion:

1. Scriptural and historical (nonscientific) evidence (Chapter 2): Historical, exegetical, and archaeological scholars have developed several methods for confirming the likelihood of Jesus' historicity, miracles, Passion, Resurrection, and gift of the Spirit:
   a. The historicity of the five major independent sources— Mark, John, Q, Luke special, and Matthew special
      1) The likelihood of eyewitnesses embedded in specific Gospel narratives (Richard Bauckham)
      2) Criteria of historicity applied to the five independent sources and the four Gospels (Joachim Jeremias, Raymond Brown, John P. Meier, and N. T. Wright)
   b. Hostile contemporaneous testimonies to the historical Jesus (Tacitus, Josephus, and Talmud)

   c. The Pauline list of witnesses (and his argument that they
      had everything to lose and nothing to gain testifying to the
      Resurrection)

   d. N. T. Wright's first argument for the historicity of Jesus'
      Resurrection—the inexplicable exponential growth of the
      Christian messianic movement after the public humiliation
      and execution of its messiah

   e. N. T. Wright's second argument for the historicity of Jesus'
      Resurrection, Christian mutations of Second Temple Juda-
      ism's doctrine of the Resurrection—the Christian Church
      did not want to separate herself from the synagogue and
      her doctrine, but did so almost exclusively with respect to
      the Resurrection.

   f. The historicity of Jesus' miracles by His own authority from
      the testimony of His adversaries (e.g., the testimony of Jose-
      phus, and scripturally, "He casts out demons by the prince
      of demons"—Mt 9:34; cf. 12:24; Mk 3:22; Lk 11:15)

   g. The apostles' capacity to perform miracles in the name of
      Jesus—a power that is very evident in healing Masses and
      ceremonies today. Why would God work through the
      name of Jesus if the apostles were lying about Jesus and His
      Resurrection?

2. Scientific evidence of Jesus' historicity, Passion, and Resurrection from the Shroud of Turin (Chapter 3): There
   are several features on the Shroud that indicate its authenticity
   and two features that are beyond naturalistic explanation and
   human production:

   a. The strong likelihood of a date of origin in the mid-first
      century—wide-angle X-ray scattering dating, which shows
      A.D. 55–74 (peer-reviewed 2022); Fourier transform infra-
      red spectroscopy; Raman spectroscopy; vanillin levels in
      the linen; 120 points of congruence with bloodstains on the
      Shroud and the Sudarium of Oviedo, which has a prove-
      nance going back to A.D. 616; and large numbers of pollen
      grains, some of which are unique to Jerusalem and northern
      Judea; as well as the debunking of the 1988 carbon dating by
      Raymond Rogers' thermochemical analysis of the sample
      and Tristan Casabianca's statistical analysis of C-14 raw data.

b. Definitively established human blood with real hemoglobin, bilirubin, AB+ blood type, plasma–serum differentiation, human albumin, human whole blood serum, human immunoglobins, and synthesis of ferritin and creatinine (Aland Adler, John Heller, P. L. Bollone, Elvio Carlino, and A. van der Hoeven)

c. The presence of bloodstains before the image on the Shroud was formed, which is almost impossible for a forger to produce (Kitty Little)

d. The perfect matching of the bloodstains with the anatomy of the body in the image (Pierre Barbet, Frederick Zugibe, Robert Bucklin, and Mateo Bevilacqua et al.)

e. The portrayal of bloodstains with great anatomical accuracy a crucifixion very similar to the unique Crucifixion of Jesus in the Gospels: *Roman* flagrum (three-thronged whip), *Roman* legionary spear, crown of thorns (unique to Jesus' Crucifixion), spear wound in the side with blood and water marks near aperture (unique to Jesus' Crucifixion), contusions and lowering of right shoulder consistent with carrying the Cross, and high levels of ferritin and creatinine, indicating a polytrauma

f. An image that cannot have been produced by liquids, dyes, vapors, or scorching—which has thirty-two enigmas that can be explained only by a collimated source of radiation coming from the body; there is no known natural cause for this, apparently requiring supernatural causation:

   1) Vacuum ultraviolet radiation several billion watts for 1/40 billionth of a second (John P. Jackson, Paolo Di Lazzaro, et al.)

   2) Particle radiation—every stable atomic nucleus in the body simultaneously undergoes nuclear disintegration in a low-temperature nuclear reaction, creating flows of trillions of heavy charged particles (protons, alpha particles, and deuterons), uncharged heavy particles (e.g., neutrons), and non-heavy charged particles (electrons and gamma rays)—Kitty Little, John Baptiste Rinaudo, Arthur C. Lind, and Mark Antonacci

g. Requirement for the body to become mechanically transparent (spiritual) for the linen cloth to penetrate so that three-dimensional proportional imaging between the inside and outside of the body could occur on the frontal and dorsal parts of the Shroud—also naturalistically inexplicable, apparently requiring supernatural causation (John P. Jackson, Giulio Fanti, et al.)

3. Transmutated Eucharistic hosts (Chapter 4): There are naturalistically inexplicable common elements in the three transformed Eucharistic hosts—from Tixtla, Sokółka, and Buenos Aires (Ricardo Castañón Gómez, Frederick T. Zugibe, Maria Elżbieta Sobaniec-Łotowska, Stanislaw Sulkowski, et al.).
   a. The absence of notable macroscopic decomposition in the tissue (though there is autolysis in the cells and fine tissue areas)
   b. The presence of living white blood cells in human blood
   c. The presence of *living* cardiac tissue growing out of (or having grown out of) the substance of a consecrated Host, which is molecularly distinct from it
   d. Indications of thrombi (Buenos Aires), fragmentation, segmentation (Sokółka), and macrophages phagocytizing lipids (Tixtla), all of which indicate distress to the heart—perhaps from trouble breathing, a blow to the heart, or a heart attack
   e. The presence of genetic material without a DNA profile on amplifiable polymer chains

4. The tilma and image of Our Lady of Guadalupe (Chapter 5, Section I): There are several features of the tilma and image that are beyond naturalistic explanation and human production:
   a. The longevity of the agave tilma, which should have decomposed over four hundred years ago (Philip Serna Callahan et al.)
   b. The absence of degradation in the pigment and brightness of color on the original figure of the Madonna, which should have suffered degradation and color change over four hundred years ago (Philip Serna Callahan et al.)
   c. No known natural or synthetic pigment (Richard Kuhn)
   d. The absence of sizing, brushstrokes, and undersketching in the production of the original image of the Madonna, which many artists consider beyond human capability (Daniel Castellano, Philip Serna Callahan, et al.)

e. The use of the coarse fabric and weave to present depth (e.g., in the eyes of the Madonna) or add artistic enhancement (e.g., on the lips of the Madonna) appears to be beyond human capability—at least certainly in the sixteenth century (Daniel Castellano, Philip Serna Callahan, et al.)

f. The appearance of the Purkinje-Sanson triple reflection on the corneas of the Madonna, which is beyond human capability (José Aste Tonsmann, Javier Torroella-Bueno, Rafael Torrija Lavoignet)

g. The appearance of depth in the eyes of the Madonna (detected through ophthalmoscope), though they are painted on a flat surface (Rafael Torrija Lavoignet)

h. The likely appearance of multiple figures in a scene of Juan Diego showing his tilma to Bishop Zumárraga and others, which is coordinated between the right and left eyes through expected transition functions and coefficients calculated by Aste Tonsmann, which is beyond human capability or natural causation (José Aste Tonsmann)

5. The scientifically inexplicable phenomena of the sun at Fatima (Chapter 5, Section II): There are several factors validating the witnesses of this phenomenon, which has strong indications of supernatural agency:

a. Factors validating the witnesses:

1) Hostile and skeptical witnesses who were scientists, journalists, attorneys, politicians, police working for the government, educators, artists, historians, poets, and authors all admit that the Miracle of the Sun occurred. No one who was at the Cova da Iria denied the occurrence of the miracle. Furthermore, hostile and skeptical witnesses agreed that there were at least thirty thousand to fifty thousand people at the Cova who witnessed the events. Nonhostile professionals estimated there were between fifty thousand and one hundred thousand (John de Marchi, John Haffert, G. Almeida Garrett, Pio Scatizzi, and Stanley Jaki).

2) Avelino de Almeida was the editor of the secular masonic newspaper O Seculo, who was against miracles

and published and predicted that the miracle would not occur in his morning article on October 17. He risked his secular journalistic career to assert and then reassert (in two articles) that the "dancing of the sun" really did occur in front of thirty thousand people.

3) There are many photographs of people falling, crying out, and staring at the sun. It is quite evident that they were reacting to some commonly experienced phenomena (John de Marchi, Gonçalo Almeida Garrett, and Stanley Jaki).

4) Multiple witnesses documented by John Haffert attest to the rapid drying of the ground and clothes (soaked because of rain the day before and on the morning of October 17) during the time the sun fell toward the earth and then resumed its place in the sky.

b. Supernatural character of the event (Gonçalo Xavier de Almeida Garrett, José Maria Pereira Gens, Pio Scatizzi, Stanley Jaki, and Ignacio Ferrín):

1) The phenomenon of the sun was not a solar event, but rather an atmospheric or meteorological event that occurred through a gigantic air lens with ice crystals (Stanley Jaki) or a small comet (Ignacio Ferrín).

2) Six virtually unique factors of the phenomenon:
   i. The reappearance of the sun with a pearl/silver rim
   ii. The gazing at the sun by fifty thousand people for about ten minutes without any reports of pain or retinal damage
   iii. The "dancing of the sun"
   iv. The segmentation and different colors of the rays coming from the sun
   v. The descending sun
   vi. The drying of the ground and clothing

c. The phenomenon of the sun very probably requires supernatural causation:

1) All simple natural explanations fail to explain most of the phenomenon.

2) The two complex naturalistic explanations explain the phenomenon, but the odds against them happening at

the date, time, and place predicted by Lucia are thousands of trillions to one *against*.

3) Therefore, the phenomenon is very probably supernaturally caused (Stanley Jaki and Ignacio Ferrín).

6. **Seven thousand scientifically inexplicable cures at Lourdes (Chapter 6):** There are seven thousand unexplained cures documented by the Lourdes Office of Medical Observation, formerly the Lourdes International Medical Bureau:

a. There are many more cures at Lourdes than are documented by the Lourdes Office of Medical Observation because most cured individuals leave the Grotto without staying additional time to document the cures and provide all medical tests before and after, as well as multiple medical follow-ups over the long term.

b. There are several cures where atrophied bone and tissue were instantaneously regenerated without the chemical constituents needed for such generation in the blood stream:

1) Pierre de Rudder—left tibia completely fractured (completely swinging). Fracture completely conjoined, and the quantity of phosphate of lime needed to fuse conjunction was not present in the body (van Hoestenberghe, Édouard Le Bec, et al.).

2) Gabriel Gargam—due to a train accident, Gargam's leg muscles were completely atrophied; he had not eaten solid food for two years (using feeding tube) and had not been able to speak out loud for twenty months. After Lourdes water *and* Eucharistic blessing, Gargam was instantaneously cured. His leg muscles were completely regenerated within a matter of seconds; he was able to speak and then eat a hearty meal (Gustave Boissarie and Dr. Decrassaac).

3) John Traynor—several bullet wounds from World War I caused severe damage to his right arm and pectoral muscles and ancillary nerves (virtually completely atrophied). He had epileptic seizures about four times per day, so a hole was drilled in his skull to relieve pressure and epileptic seizures, which actually grew in severity.

He developed partial paralysis of his leg muscles and irremediable incontinence. After Lourdes water *and* a Eucharistic blessing, his arm and pectoral muscles were instantaneously regenerated along with the nerves, the hole in his skull diminished considerably, his legs became functional, and his incontinence was instantaneously cured (Drs. Francis Izard, Finn, Marley, and Azurdia).

c. There are several cures during which all diseased elements disappeared within a matter of a few hours, leading to complete recovery:

1) Marie Bailly—very advanced tubercular peritonitis with solid masses significantly distending abdomen, on the point of death. Future Noble laureate Alexis Carrel testified that after three pourings of Lourdes water, all solid masses disappeared without any fluid discharge; tubercular peritonitis was cured, and physical health was completely restored—Marie took the train back on her own; she joined a rigorous religious order and died thirty-six years later.

2) François Pascal (four years old)—spinal meningitis, complete irreparable paralysis of all four limbs, complete irreparable blindness from lesions. After two immersions in the Lourdes bath, there was an instantaneous cure of both the irreparable paralysis and blindness (Drs. Polger, Darde, and Roman).

3) Guy Leydet (five years old)—severe encephalitis resulting in brain damage, reducing IQ to between 0 and 25, and frequent epileptic fits; this was accompanied by complete paralysis in all four limbs. After a single immersion in the Lourdes baths, he experienced an instantaneous regeneration of his brain, enabling him to speak intelligently (moments after immersion) as well as experience complete healing of paralysis and epileptic fits (Lelong).

4) Brother Leo Schwager (twenty-seven years old)—Severe multiple sclerosis, rendering him unable to speak and causing his organs to shut down gradually. He went to the baths twice, but nothing happened. So he went to the rosary grotto for a Eucharistic blessing. The moment he

received the blessing, he felt an electric shock throughout his body, which catapulted him and placed him kneeling in front of the bishop with the Host. He was instantaneously cured of the multiple sclerosis, organ damage, and speech loss, the very moment he found himself kneeling before the bishop and the Host (Drs. Jeger, Kreienbuhl, and Otto).

Conclusion to the Informal Inference

As can be seen, the above six sources of historical and scientific evidence constitute a complementary and mutually corroborative inference that Jesus Christ is risen from the dead and His supernatural power is manifest today along with His Real Presence in the Holy Eucharist and the supernatural power of His mother. Each part of the six parts of this inference is also grounded in a subordinate informal inference having multiple manifestations of scientifically inexplicable phenomena (indicating supernatural agency). The combination of historical-scriptural evidence, the scientific evidence of the Shroud, transformed Eucharistic hosts, the tilma and image of Guadalupe, the Miracle of the Sun at Fatima, and the inexplicable cures of Lourdes all corroborate the same conclusion—that the Risen Christ is really present throughout the world, in the Holy Eucharist, and along with His mother.

If Jesus is risen from the dead and His supernatural power and presence continues in the world till this day, then God has been giving overwhelming approbation to the universal preaching of the apostles throughout the world that Jesus is the Lord and His only begotten Son. If Jesus is not divine (the Lord) and the Son of God, then why would God work miracles through the name of Jesus? Why would He supernaturally manifest His Son's Crucifixion and Resurrection on what appears to be His burial Shroud? Why would He supernaturally manifest His Son's Real Presence through transformed Eucharistic hosts on multiple occasions? Why would He manifest His supernatural power through the agency of the Blessed Virgin at Guadalupe? Why would He supernaturally manifest His cosmic power through the agency of the Blessed Virgin and the Miracle of the Sun at Fatima? And why would He manifest His supernatural power through the agency of the Blessed Virgin Mary and thousands of scientifically inexplicable cures at Lourdes? If Jesus is not truly the Son of God, then all the scientifically investigated and validated evidence spoken of in this book should never have occurred; but it has occurred, and it has been open to science throughout the twentieth and twenty-first centuries.

This interrelated corroboration of historical and scientific evidence points to one further conclusion: Jesus Christ founded the Catholic Church and remains present to her as the Gospels report. If Jesus did not found and remain in the Catholic Church, then why do these scientifically tested miracles concern distinctively Catholic doctrines, such as the Real Presence of Christ in the Eucharist, the intersession of the Blessed Virgin in the economy of salvation, and the Immaculate Conception? These apparent miracles not only point to the divinity of Jesus, but also to the authenticity of the Catholic Church's teaching—the teaching certified by Saint Peter and his successors. If this is the case, then why would we not seriously consider the Catholic Church as the definitive representative and interpreter of our Lord Jesus Christ recorded in the Gospel of Matthew?

> And I tell you, you are Peter, and on this rock I will build my Church, and the gates of Hades shall not prevail against it. I will give you the keys of the kingdom of heaven, and whatever you bind on earth shall be bound in heaven, and whatever you loose on earth shall be loosed in heaven. (Mt 16:18–19)

Readers interested in examining the historical arguments for the foundation of the Catholic Church by Jesus Himself may want to study the appendix that follows.

APPENDIX

The Foundation of the Catholic Church—
Jesus Christ

The purpose of this appendix is to give a summary of the evidence for Jesus as the founder of the Catholic Church. Chapters 4 through 6 of this book were devoted to topics connected to the Catholic Church that showed the truth of two of her most controversial teachings—the Real Presence of Christ in the Eucharist and the Blessed Virgin Mary as integral to the economy of salvation. If, as the miracles in Chapters 4 through 6 show, Christ is really present in the Holy Eucharist and Mary is integral to the economy of salvation, then why would we not want to become Catholics? Why would we want to practice our Christian faith without these two powerful sources of grace, inspiration, and help? Indeed, if the Catholic Church is correct about these two doctrines, then why would we believe that she is wrong about other less central teachings? Why would we believe that Christ is not still working through the Catholic Church as He promised?

If some non-Catholic readers are asking themselves these questions, I want to provide a short appendix summarizing the evidence for what are likely to be their next questions:

1. What is the evidence that Jesus is founder of the Catholic Church? (Section I)
2. What is the evidence that Jesus continues to work in that Church? (Section II)
3. What are the benefits of the Catholic Church not found in other Christian churches? (Section III)

If this evidence is sufficiently probative to render a positive judgment of the Catholic Church, then why not seek to become an active participant in that Church as soon as possible?

313

I. Was Jesus Really the Founder of the Catholic Church?

There can be little doubt that Jesus intended to start a hierarchical church that would succeed Him in His ministry after His Resurrection. He initiated three levels of ministerial authority:

1. Peter—the highest and supreme authority with the power of "the keys" (Mt 16:17–19; explained below)
2. The twelve apostles—whom He gave the authority equivalent to today's bishops (see Mt 10:1–5; 18:15–18)
3. Another seventy disciples with the power to preach the Gospel, heal the sick, and exorcise demons (see Lk 10:1–12)

Jesus was aware of the imminence of His Passion and Resurrection, which He predicted three times (Mk 8:31; 9:30–31; 10:33–34). He did not believe that the end time would follow immediately upon His Resurrection as indicated by the two stages of the eschatological discourse—the fall of the Temple, and then sometime later, the end of the world (Mk 13; Mt 24–25; Lk 21). Furthermore, He said that He did not know when the end time would be: "But of that day or that hour no one knows, not even the angels in heaven, nor the Son, but only the Father" (Mk 13:32; cf. Mt 24:36). Furthermore, His Eucharistic words, "Do this in remembrance of me" (Lk 22:19; 1 Cor 11:24), suggest that there will be some length of time after His death to relive and receive His Body and Blood. In view of this, Jesus' intention to start an authoritative, and even a hierarchical church to shepherd His flock seems evident.

Did Jesus intend to give supreme authority only to Peter—or also to His successors? The answer to this question lies in Jesus' rationale for giving supreme authority to Peter. We can then ask if that same reason would have to extend also to His successors. So why did Jesus give the power of the "keys" (explained below) to Peter? The evident answer is that He anticipated disagreements, disputes, and factions about His words and the direction of the Church after His death and Resurrection. Jesus was raised in a Jewish community, where factions and parties were everywhere present—for example, Pharisees, Sadducees, and Essenes—and divisions within those parties, such as followers of Hillel and Shammai. It was not difficult to surmise this would happen to His Church, too. Unwilling to allow His Church to be fragmented or weakened by internal dissension, He intentionally gave supreme authority to Peter so that all disputes could be definitively resolved. Without a supreme authority, disputes would inevitably undermine the new Church. Would this apply to the successors of Peter? Of course—why would disputes and internal dissension stop with the death of Peter? It would be nonsensical to affirm anything else. The inevitability of confusion, in-fighting, power politics, and the factions coming from them are ingredient to the whole of human history, and the only way to prevent these things from happening is legitimate definitive authority. Jesus not only could foresee this in the religious milieu around Him but could hear it daily in the bickering among His disciples. Hence, He gave a commission to Peter, and by implication to Peter's successors, imparting legitimate, definitive authority:

> Blessed are you, Simon Bar-Jona! For flesh and blood has not revealed this to you, but my Father who is in heaven. And I tell you, you are Peter, and on this rock I will build my Church, and the gates of Hades shall not prevail against it. I will give you the keys of the kingdom of heaven, and whatever you bind on earth shall be bound in heaven, and whatever you loose on earth shall be loosed in heaven. (Mt 16:17–19)

This passage has several indications validating its origin in Jesus Himself:

- Five Semitisms, indicating an early Palestinian origin, uttered in the spoken language of Jesus (discussed below).
- Four parallels between Matthew 16:17–19 and the commissioning narrative in Galatians 1 and 2, indicating Saint Paul's awareness of a narrative similar to Matthew's narrative of the commissioning of Peter. This shows the likelihood of another source of the commissioning of Peter earlier than Matthew's narrative (discussed below). This means that Matthew did not author his commissioning passage. It was a tradition known to both him and Saint Paul.

- There are four unique expressions of Jesus, indicating authorship of the commissioning passage by Him, in Matthew 16:17–19:
    - The use of "Abba" ("my Father") as an address to God.[1] This address is typical of Jesus, but there are very few instances of this use of "Abba" in the whole of Jewish literature, meaning that the expression in verse 17 very likely originated with Jesus.
    - The use of the emphatic *ego* in 16:18: "I say to you"—"*ego* lego" (ESV).[2] This method of solemnizing a pronouncement is typical of Jesus and very rare in other Jewish biblical literature. Hence, the expression in verse 18 likely originated with Jesus.
    - Peter's change in name (from Simon to Cephas/Peter). Changing a name requires a greater authority than that possessed by parents. Jesus is the only one in Peter's company that could have formally proclaimed this in verse 18.[3]
    - The conveyance of divine authority to Peter in verse 19 ("Whatever you bind on earth shall be bound in heaven"). The only one in Peter's company who would or could convey divine authority is Jesus.

Why is this passage, apparently authored by Jesus, so important? The phrase "keys of the kingdom of heaven" signifies the power and authority particular to the *office* of prime minister—the highest office with the power to act on behalf of the king himself. This is shown in the parallel passage of Isaiah 22:20–23. In this passage, the prophet Isaiah takes the keys of the office of prime minister from Shebna and gives them to Eliakim with words of commissioning similar in form and content to Jesus' commission of Peter: "He shall open, and none shall shut; and he shall shut, and none shall open" (Is 22:22, paralleling

---

[1] See Joachim Jeremias, *New Testament Theology*, vol.1, *The Proclamation of Jesus* (New York: Charles Scribner's Sons, 1971), 65.

[2] Adding the pronoun "I" ("*ego*") before a verb such as "*lego*". In Semitic languages and Greek, the pronoun is unnecessary because it is contained in the conjugation of the verb, but as Jeremias indicates, when Jesus is making a declaration tantamount to a law or a doctrine, He adds the pronoun: "ego lego—*I* say." See ibid., pp. 252–54.

[3] See W. D. Davies and Dale C. Allison, *International Critical Commentary*, vol. 2, *Matthew 8–18* (New York: T&T Clark, 1991), pp. 626–29.

Mt 16:19—"Whatever you bind on earth shall be bound in heaven, and whatever you loose on earth shall be loosed in heaven"). In view of this, it is likely that Jesus created an *office* of prime minister on earth to bind on behalf of Him (the King) in Heaven.

Inasmuch as Jesus did initiate such a supreme office of teaching and juridical authority applicable to Peter and his successors, we may properly infer that Jesus intended the Catholic Church (acting under the authority of the holder of that supreme office) to be the definitive interpreter of Jesus' doctrine and moral teaching as well as the definitive authority to apply that moral teaching to new ethical context and situations. This was the view of first- and second-century popes and bishops.

In addition to the scriptural evidence, there is considerable evidence that the early popes and Church Fathers believed that the successors to Peter (occupying his Chair in Rome) had supreme teaching and juridical authority over the whole Church. Pope Clement (fourth holder of the supreme office—A.D. 88–99) believed that he had the power to order rebellious members of the Corinthian church to restore their bishop and clergy to their offices and to be "obedient to the things which we have written through the Holy Spirit" *under pain of sin.*[4] If he did not have supreme juridical power, how could he have made such a declaration expecting to be obeyed? Furthermore, Saint Ignatius of Antioch, bishop of Antioch, in about A.D. 102, wrote a letter to the Church of Rome acknowledging that it was superior to—and presided over—all other Christian churches.[5] Saint Irenaeus (writing around A.D. 189) declared that the Church of Rome (whose presiding bishop is the pope) is owed obedience in matters of teaching by all other Christian churches.[6] Saint Cyprian of Carthage, one of the greatest Latin apostolic Fathers and bishop of Carthage, wrote an important treatise on the unity of the Catholic Church in A.D. 251, where he declared that

[4]Clement of Rome, *Letter to the Corinthians* 1, 58–59, 63, trans. Kirsopp Lake, in *The Apostolic Fathers*, ed. T. E. Page and W. H. D. Rouse, Loeb Classical Library (London: William Heinmann; New York: MacMillan, 1912).

[5]Ignatius of Antioch, *Epistle to the Romans* 2, 6, in *The Epistles of St. Ignatius*, Early Church Classics, ed. J. H. Srawley (London: Society for Promoting Christian Knowledge, 1910).

[6]Saint Irenaeus, *Against Heresies* 3, 3, 2, in *Five Books of S. Irenaeus, Bishop of Lyons, against Heresies*, trans. John Keble (London: James Parker, 1872).

The rest of the apostles were also the same as was Peter, endowed with a like partnership both of honour and power; but the beginning proceeds from unity.... Does he who does not hold this unity of the Church think that he holds the faith?[7]

These testimonies indicate that Saint Peter and the other apostles communicated clearly to the second generation of disciples that the successors to the office of Peter (located at his See in the heart of the Roman Empire) shared his supreme teaching and juridical authority. In light of all the above, it is highly likely that Jesus intended to give the supreme power of the office of the keys of the Kingdom of Heaven to Peter's successors.

II. Jesus Continues to Protect the Church

Jesus has been true to His promise to protect the Catholic Church from the "powers of death" throughout history. The great historian of culture and civilization Arnold Toynbee testified to the unique nature of the Catholic Church to endure beyond any other institution throughout history—a conviction that brought him from agnosticism to the light of Christ:

The Church in its traditional form thus stands forth armed with the spear of the Mass, the shield of the Hierarchy, and the helmet of the Papacy ... [and] the divine intention ... of this heavy panoply of institutions in which the Church has clad herself is the very practical one of outlasting the toughest of the secular institutions of this world, including all the civilizations. If we survey all the institutions of which we have knowledge in the present and in the past, I think that the institutions created, or adopted and adapted, by Christianity are the toughest and the most enduring of any that we know and are therefore the most likely to last—and outlast all the rest.[8]

---

[7] Cyprian of Carthage, On the Unity of the Church 4, trans. Robert Ernest Wallis, in Ante-Nicene Fathers, vol. 5, ed. Alexander Roberts, James Donaldson, and A. Cleveland Coxe (Buffalo, N.Y.: Christian Literature Publishing, 1886), revised and edited for New Advent by Kevin Knight, https://www.newadvent.org/fathers/050701.htm.

[8] Arnold J. Toynbee, Civilization on Trial (New York: Oxford University Press, 1948), pp. 242–43.

The Church has certainly had her ups and downs. She has suffered from persecution, heretical sects, scandals, wars, and other calamities since the time of Judas' betrayal and Jesus' Crucifixion. Yet she seems to defy what would have destroyed any other institution. The more actively she was persecuted, the more quickly her ranks swelled, and the more she was threatened by dissension, the stronger her unity and teaching authority became. Every time she encountered a crisis, exceedingly holy and talented men and women would come out of nowhere not only to turn the tide of crisis but to advance the cause of evangelization. The history of the Church has not been neat and tidy, but it certainly has demonstrated her incredible capacity not just to recover, but to grow and prosper through crisis. This remarkable resiliency, which relentlessly defies entropy over the course of time, is not attributable solely to "good luck" and human ingenuity—for the Church is all too human. There must be something else that counters historical entropy, human weakness, and bad fortune—something beyond natural tendency itself, which Toynbee calls "divine intention" and Catholics call "divine providence", "the Holy Spirit", and the "communion of saints". When we retrospectively view Jesus' commission to Peter, we may be assured that He kept His promise to him, when He said to Peter, "I tell you, you are Peter, and on this rock I will build my Church, and the gates of Hades shall not prevail against it" (Mt 16:18). He remains present in the Church, present in the Holy Eucharist, present in His Word, present in the Church's living tradition, present in her teaching, present in her unity, and present in Peter's successor.

III. The Benefits of the Catholic Church

The Catholic Church shares many benefits in common with other Christian churches—the Word of God (the Bible), the gift of the Holy Spirit through Baptism, a Church community, and some basic doctrines (such as the Incarnation and the Trinity). However, since the time of the Reformation, four of Jesus' intended benefits (graces) were either abandoned or substantially weakened—benefits that virtually every Catholic would call vital to spiritual life and salvation:

1. The Holy Eucharist, which Jesus intended to be His real crucified and risen Body and Blood to forgive sins and lead us to eternal life (Mt 26:26–28; Mk 14:22–24; Lk 22:19–20; Jn 6:35–69)[9]
2. The Sacrament of Reconciliation, whose power was given by
   Jesus to the apostles and their successors (Jn 20:23)
3. The other five sacraments of the Church—four of which (with
   the exception of Baptism) are unique to the Catholic (and
   Orthodox) Church—Confirmation, Anointing of the Sick,
   Matrimony, and Holy Orders[10]
4. The definitive doctrinal and moral teaching authority vested in
   the office of Saint Peter and his successors

Let us consider the first benefit—the *Holy Eucharist*. The following
five points summarize more extensive previous work:[11]

1. Contemporary scholarship on John 6:35–69 shows Jesus' intention to give us His *real* Body and Blood, particularly the central passage—"I am the living bread which came down from
   heaven; if any one eats of this bread, he will live for ever; and
   the bread which I shall give for the life of the world is my flesh"
   (6:51).[12] When this is combined with the Jews' question, "How
   can this man give us his flesh to eat?" (6:52), and the fact that
   His words produced so much scandal that everyone left Him
   (6:66–69), it is difficult to believe that this reaction was provoked by Jesus referring to a mere symbol of His Body.
2. The Aramaic background of Jesus' Eucharistic words indicates
   an *identification* of the bread in His hand with His Body, and the
   wine in the cup with His Blood.[13]

[9] These graces of the Holy Eucharist are explained in detail in Robert Spitzer, *Escape from Evil's Darkness: The Light of Christ in the Church, Spiritual Conversion, and Moral Conversion* (San Francisco: Ignatius Press, 2021), Chapter 2, Section I.

[10] A detailed explanation of these sacraments is given in ibid., Chapter 2, Sections 2–6.

[11] See ibid., Chapter 2.

[12] Despite some attempts to call this verse a later Christian interpolation, contemporary scholars hold that the identification of the Eucharistic bread with Jesus Himself in John 6:51 follows from the preceding passage, and that the continued repetition with respect to blood shows that this is the Johannine author's intention throughout the passage. See Bruce Vawter "The Gospel According to John", in *The Jerome Biblical Commentary*, ed. R. Brown, J. Fitzmyer, and R. Murphy (Englewood Cliffs, N.J.: Prentice Hall, 1968), 2:437–38.

[13] Joachim Jeremias, *The Eucharistic Words of Jesus* (London: SCM Press, 1966), pp. 223–24.

3. The phrase "Do this in remembrance of me" (Lk 22:19; 1 Cor 11:24) does not mean "call this event to mind". It is an exhortation to relive the event in a prophetic way that collapses the past Eucharistic species into the present bread and wine in the hands of the priest.[14]

4. The unanimous belief of the early Church Fathers[15]—particularly first-century witnesses such as the *Didache* (9, 10, and 4) and Saint Ignatius of Antioch (*Letter to the Smyrnaeans* 7)—about the *Real* Presence of Jesus in the Eucharist indicates that the apostles must have unanimously taught this to their disciples.

5. The Eucharistic miracles of Tixtla, Sokółka, and Buenos Aires imply the Real Presence of Christ in the Eucharist (discussed in Chapter 4).

There are five major graces of the Holy Eucharist that facilitate the believer's entrance into the Kingdom of Heaven (enabling him to help others to do so):

1. Forgiveness of venial sins and healing from the effects of evil
2. Transformation in the Heart of Jesus
3. Companionship within the Mystical Body (the Church)
4. Spiritual peace
5. Everlasting life

These graces are explained in detail in previous work.[16] The grace arising out of Jesus' crucified risen Body and Blood in the Eucharist is so powerful that it became the central activity for the spiritual life of the Catholic Church from the first century to the present. It is so spiritually fruitful that Jesus virtually guarantees its worthy reception will lead to eternal life—"If any one eats of this bread, he will live for ever; and the bread which I shall give for the life of the world is my

---

[14]See the double collapse of time in Spitzer, *Escape from Evil's Darkness*, Chapter 2, Section I.A. Note the sources for this interpretation, particularly Johannes Betz, "Eucharist", in *Sacramentum Mundi: An Encyclopedia of Theology*, ed. Karl Rahner, vol. 2, *Contrition to Grace and Freedom* (London: Burns & Oates, 1970), p. 257.

[15]See the list of references to virtually all early Church Fathers in *The Catholic Encyclopedia* (New York: Robert Appleton, 1909), s.v. "Eucharist", by J. Pohle, New Advent (website), https://www.newadvent.org/cathen/05572c.htm.

[16]See Spitzer, *Escape from Evil's Darkness*, Chapter 2, Section I.C.

flesh.... He who eats my flesh and drinks my blood has eternal life, and I will raise him up at the last day" (Jn 6:51, 54). With such a gift and assurance, why would anyone willingly ignore or refuse it?

The second major benefit, the Sacrament of Reconciliation, was given by Jesus to the apostles and their successors with the proclamation "If you forgive the sins of any, they are forgiven; if you retain the sins of any, they are retained" (20:23). This sacrament provides definitive absolution for even the most serious of sins, breaks the grip of the evil one, and reunites us with the Lord and His Mystical Body (the Church). The following is a summary of its five major graces:[17]

1. Definitive absolution for mortal and venial sins
2. Spiritual solidification of a turning point in life
3. Healing of the damage of sin and release from the grip of the evil spirit
4. Graced resolve for continued conversion
5. The peace of Christ

If we can receive definitive absolution for even the most serious sins (and be certain of this absolution and restoration to grace), and this sacrament can break the grip of the evil spirit, help break sinful habits, purify our intentions, and lead us to salvation, why would we willingly ignore or refuse it?

The third benefit of the Catholic Church—the other four Catholic sacraments (Confirmation, Anointing of the Sick, Matrimony, and Holy Orders)—are explained extensively in previous work.[18]

The fourth benefit—the supreme teaching and juridical authority vested in Saint Peter and his successors—is partially explained above. In sum, the supreme authority of Peter's successors has underpinned and strengthened the unity of the Church for two millennia. It has also led to consistency and clarity of doctrine (the interpretation of Scripture and tradition), which has given trustworthy guidance to the faithful throughout the centuries.[19] When one considers that in five hundred years, at least two hundred major denominations of Protestants have been initiated, and that within those major denominations there are

---

[17] For a complete explanation of these graces, see ibid., Chapter 7.
[18] See ibid., Chapter 2, Sections II–VI.
[19] See ibid., Chapter 1.

several thousand distinct doctrinal groups,[20] it is striking that the Catholic Church should be unified and have a consistent doctrinal viewpoint (set out in the *Catechism of the Catholic Church*) throughout her two thousand years. The primary reason for this is the supreme teaching and juridical authority of the pope (Peter's successor) throughout the centuries. Evidently, Jesus really knew what He was doing when He gave Peter and His successors supreme authority to maintain the unity of the Church and consistency and constancy of her doctrine.

These sacraments and the supreme teaching authority of the Church are proclaimed by virtually all the thousands of canonized saints to be the source and summit of their lives of heroic virtue and self-sacrificial love. If these graces of Jesus given to the Catholic Church produced such remarkable benefits for both the saints' salvation and that of the millions of people they served, why wouldn't they do the same for all of us if we receive them as Jesus intended? If the objective of our life is heroic self-sacrificial love and virtue leading to our and others' salvation, why wouldn't we seek these gifts of Jesus and the Church as the most important dimensions of our lives?

There is a fifth benefit of the Catholic Church that falls outside the domain of sacraments, unity, and constancy of teaching—devotion to the Blessed Virgin Mary. Some Christians feel that Catholics have so accentuated this devotion that it is tantamount to divinizing her. The official Church teaching has never done anything of the kind. Rather, the Church has continually emphasized that Mary is the *reflection* of her Son's divine love, grace, and light. Though the Trinity is the source of God's love and power, Mary reflects these supernatural gifts in a very distinctive motherly, feminine, and gentle way. She is at once

---

[20] Stephen Beale, "Just How Many Protestant Denominations Are There?", *National Catholic Register* (blog), October 31, 2017, https://www.ncregister.com/blog/just-how-many -protestant-denominations-are-there. In a 2013 report by the Center for the Study of Global Christianity, it was projected that the total number of Protestant denominations in the world by 2020 would be around forty thousand when denominational churches in different countries are counted separately. Center for the Study of Global Christianity at Gordon-Conwell Theological Seminary, "Christianity in Its Global Context, 1970–2020: Society, Religion, and Mission" (South Hamilton, Mass.: Center for the Study of Global Christianity, June 2013), p. 18, https://www.gordonconwell.edu/wp-content/uploads/sites/13/2019/04/2Christianity initsGlobalContext.pdf. The same center reported that there were forty-five thousand Christian denominations in the world in 2019. See the response to the frequently asked question "How Do You Define a 'Denomination'?" on their website, accessed March 22, 2024, at https://www.gordonconwell.edu/center-for-global-christianity/research/quick-facts/.

our motherly and feminine protector, motherly and feminine com-
forter, motherly and feminine teacher, and the mother of the Mystical
Body and communion of saints. Her motherly and feminine presence
represents the power, love, and grace of the Trinity with a qualitatively
and aesthetically distinct sense of gentleness, care, and compassion.

The influence of Mary on culture is perhaps best brought into focus
by the *secular* historian William Lecky, who recognized the need for
Jesus' teaching and spirit to be given by Mary's gentle, motherly, feminine, humble, and Spirit-filled voice. The Christian Church's recognition of Mary's glorification and co-participation in the unfolding of
salvation past and present, blends her gentleness, simplicity, and humility with the eternal glory and power of God, which in turn inspired
the transformation of Western culture to elevate the genuinely feminine to the highest dimensions of cosmic and divine significance:

> The world is governed by its ideals, and seldom or never has there been
> one which has exercised a more profound and, on the whole, a more
> salutary influence than the medieval concept of the Virgin. For the first
> time woman was elevated to her rightful position, and the sanctity of
> weakness was recognised as well as the sanctity of sorrow. No longer the
> slave or toy of man, no longer associated only with ideas of degradation
> and of sensuality, woman rose, in the person of the Virgin Mother, into
> a new sphere, and became the object of reverential homage of which
> antiquity had no conception.... A new type of character was called into
> being: a new kind of admiration was fostered. Into a harsh and ignorant
> and benighted age this ideal type infused a conception of gentleness and
> purity unknown to the proudest civilisations of the past. In the pages
> of living tenderness, which many a monkish writer has left in honour
> of his celestial patron, in the millions who, in many lands and in many
> ages, have sought to mold their characters into her image, in those holy
> maidens who, for love of Mary, have separated themselves from all the
> glories and pleasures of the world, to seek in fastings and vigils and hum-
> ble charity to render themselves worthy of her benediction, in the new
> sense of honour, in the chivalrous respect, in the softening of manners,
> in the refinement of tastes displayed in all walks of society: in these and in
> many other ways we detect its influence. All that was best in Europe
> clustered around it, and it is the origin of many of the purest elements
> of our civilisation.[21]

[21] W. E. H. Lecky, *History of the Rise and Influence of the Spirit of Rationalism in Europe* (London: Longmans, Green, 1865), 1:234–35.

This description of the influence of Mary in Western culture intro-
duces a sixth benefit of the Catholic Church, which is not so much
a benefit to individual Catholics, but a benefit to world culture and
every person touched by it. In Chapter 1 (Section III.C), we noted
that the Catholic Church was by far the greatest agent of social and
cultural change in world history. She not only progressively undermined the slavery and social inequities of Rome, but became throughout her history the largest healthcare, welfare, and educational system
in the world. Today she is still the largest of these systems:

- With respect to nonstate education, recent data shows that the
  Church provides services in approximately fifty thousand secondary schools and one hundred thousand primary schools.[22]
- With respect to healthcare, the Church oversees approximately
  a quarter of all worldwide healthcare facilities and hospitals.[23]
- With respect to public welfare, the Church provides services in
  approximately fifteen thousand homes for the elderly, chronically ill, and disabled, as well as approximately ten thousand
  orphanages, marriage counseling centers, and nurseries—not
  including any healthcare facilities.[24]

In addition to this, the Catholic Church was at the forefront of
the development of science—not just in medieval universities but
throughout her history. In Chapter 7 (Section I), we discussed the
contributions of some of the hundreds of Catholic priests and clergy
to the development of all scientific disciplines—such as Brother
Roger Bacon (originator of scientific/empirical method), Nicolaus

[22] Quentin Wodon, *Global Catholic Education Report 2023: Transforming Education and Making Education Transformative* (Washington, D.C.: Global Catholic Education, 2022), p. 98,
https://www.globalcatholiceducation.org/_files/ugd/b9597a_b54239f33dec48ddb4f2d735
d10cba7c.pdf.

[23] Matt Moran, "The Role of the Catholic Church in Healthcare Provision Globally",
Independent Catholic News, October 11, 2023, https://www.indcatholicnews.com/news
/48212#:~:text=The%20contribution%20of%20the%20Catholic,services%20through
out%20Africa%20is%20inestimable. This same percentage was reported in 2010; see Catholic
News Agency, "Catholic Hospitals Comprise One Quarter of World's Healthcare, Council
Reports", February 10, 2010, https://www.catholicnewsagency.com/news/catholic_hospitals
_represent_26_percent_of_worlds_health_facilities_reports_pontifical_council.

[24] Agenzia Fides (Vatican news agency), "Vatican—Catholic Church Statistics 2023",
October 20, 2023, https://fides.org/en/news/74319-VATICAN_CATHOLIC_CHURCH
_STATISTICS_2023.

Copernicus (first mathematical justification of heliocentrism), Father Gregor Mendel (father of quantitative genetics), Bishop Nicolas Steno (father of stratigraphy and modern geology), Father Roger Boscovich (precursor to atomic and quantum theories), and Father Georges Lemaître (the discoverer of the Big Bang theory and the expanding universe). The Catholic Church also made important contributions to the theory of justice and human rights, particularly Saint Augustine (justice as higher than the positive law), Saint Thomas Aquinas (natural law theory), Father Francisco Suárez (natural and inalienable rights theory[25]), and Father John Courtney Murray (religious freedom and church-state separation).

The combined effect of the Catholic Church on education, healthcare, social welfare, justice and rights theory, science, and social egalitarianism has, to say the least, been utterly transformative. At the very least, we may conclude that the Catholic Church has made the world a much more humane, compassionate, educated, and hopeful place to live. Of course, the sinfulness of some of her members and the scandals of some of her leaders cannot be ignored, but the blood of her martyrs and the light of her saints (both heralded and unheralded) who created and maintained the greatest educational, healthcare, and social welfare systems throughout two thousand years, outshines those moments of darkness. Its seems not only that Jesus kept His promise to protect the Church from the powers of death; He also manifested through the Church His unfailing light, love, and life for the world. If we judge the Church by her tens of thousands of saints rather than her comparatively few sinners, she emerges not as an anachronistic institution, but as the living Body of Christ.

---

[25] Father Francisco Suárez (School of Salamanca) was the first to articulate the theory of inalienable rights. Hugo Grotius (the father of international law) read Suárez's volume, *De Legibus* (on the laws), and applied inalienable rights to international law. John Locke read Grotius and integrated natural rights into *The Second Treatise on Government*, which in turn was read by Thomas Jefferson and the Founding Fathers, who integrated it into the U.S. Declaration of Independence. The inalienable rights theory formed the heart of the United Nations' Universal Declaration of Human Rights.

BIBLIOGRAPHY

Adler, Alan. "The Nature of the Body Images on the Shroud of Turin". Shroud of Turin website, 1999. https://www.Shroud.com/pdfs/adler.pdf.

Agenzia Fides (Vatican news agency). "Vatican—Catholic Church Statistics 2023". October 20, 2023. https://fides.org/en/news/74319-VATI CAN_CATHOLIC_CHURCH_STATISTICS_2023.

Aleteia. "Eucharistic Miracle Beheld by Pope Francis?", April, 22, 2016. https://aleteia.org/2016/04/22/eucharistic-miracle-beheld-by-pope-francis/.

Anonymous. "A Map of the Constellations?" https://www.myguadalupe .com/blog/a-map-of-the-constellations.

Antonacci, Mark. "Can Contamination Be Detected on the Turin Shroud to Explain Its 1988 Dating?" Paper presented at the International Workshop on the Scientific Approach to the Archeiropoietos Images, Frascati, Italy, May 4–6, 2018.

———. "Particle Radiation from the Body Could Explain the Shroud's Images and Its Carbon Dating". *Scientific Research and Essays* 7, no. 29 (2012): 2613–23. https://academicjournals.org/article/article1380798649 _Antonacci.pdf.

———. *Test the Shroud: At the Atomic and Molecular Levels*. Brentwood, Tenn.: Forefront Publishing, 2016.

Aquinas, Thomas. *Summa Contra Gentiles*. Translated by Anton C. Pegis. 1975. Reprint, Notre Dame, Ind.: University of Notre Dame Press, 2014.

———. *The Summa Theologica of St. Thomas Aquinas*. Vol. 1, translated by Fathers of the English Dominican Province. New York: Benziger Brothers, 1947.

Arcangeli, Luca. "At the Roots of the 1616 Decree: Robert Bellarmine's Letter to Paolo Foscarini". Interdisciplinary Encyclopedia of Religion and Science (website). Accessed March 18, 2024. https://inters.org /copernicanism-bellarmine-foscarini.

Aste Tonsmann, José. *Our Lady of Guadalupe's Eyes*. Digital Discoveries, 2012. Apple Books. https://books.apple.com/us/book/our-lady-of -guadalupes-eyes/id548965040.

Augustine. *Confessions*. Edited and translated by Henry Chadwick. New York: Oxford University Press, 1991.

————. *The Literal Interpretation of Genesis*. Translated by W. A. Jurgens. In Vol. 3 of *The Faith of the Early Fathers*, edited by W. A. Jurgens. Collegeville, Minn.: Liturgical Press, 1979.

Baima Bollone, P. L. "The Forensic Characteristics of the Blood Marks". In *The Turin Shroud: Past, Present, and Future; International Scientific Symposium, Torino, 2–5 March, 2000*. Edited by Silvano Scannerini and Piero Savarino, pp. 125–35. Torino, Italy: Effata Editrice, 2000.

Baldacchini, Giuseppe, Paolo Di Lazzaro, Daniele Murra, and Giulio Fanti. "Coloring Lines with Excimer Lasers to Simulate the Body Image of the Turin Shroud". *Applied Optics* 47, no. 9 (2008): 1278–85.

Barber, Janet. *A Handbook on Guadalupe*. New Bedford, Mass.: Franciscan Friars of the Immaculate, 2001.

Barbet, Pierre. *A Doctor at Calvary: The Passion of Our Lord Jesus Christ as Described by a Surgeon*. P. J. Kenedy & Sons, 1953.

Bargellini, Clara. "Echave Orio, Baltasar de". Encyclopedia.com (website), March 14, 2024. https://www.encyclopedia.com/humanities/encyclo pedias-almanacs-transcripts-and-maps/echave-orio-baltasar-de-c -1558-c-1623.

Barr, Stephen M., and Andrew Kassebaum. "Important Catholic Scientists of the Past". Society of Catholic Scientists, 2024. https://catholic scientists.org/scientists-of-the-past/?view-style=list.

Barthas, Canon. *Fatima: Merveille inouïe; Les Apparitions; Le Pèlerinage; Les Voyants; Des Miracles; Des Documents*. Toulouse, France: Fatima Editions, 1943.

Bauckham, Richard. *Jesus and the Eyewitnesses: The Gospels as Eyewitness Testimony*. 2nd ed. Grand Rapids, Mich.: Eerdmans, 2017.

Beale, Stephen. "Just How Many Protestant Denominations Are There?" *National Catholic Register* (blog), October 13, 2017. https://www.nc register.com/blog/just-how-many-protestant-denominations-are-there.

————. "These 5 Catholic Scientists Shaped Our Understanding of the World". *Aleteia*, May 25, 2018. https://aleteia.org/2018/05/25/these -5-catholic-scientists-shaped-our-understanding-of-the-world/.

Bec, Édouard Le. *Medical Proof of the Miraculous: A Clinic Study*. New York: P. J. Kenedy & Sons, 1923.

Betz, Johannes. "Eucharist". In *Contrition to Grace and Freedom*, edited by Karl Rahner, pp. 257–67. Vol. 2 of *Sacramentum Mundi: An Encyclopedia of Theology*. London: Burns & Oates, 1970.

Bevilacqua, Mateo, and Michele D'Arienzo. "Medical News from Scientific Analysis of the Turin Shroud". *MATEC Web of Conferences* 36 (2015). https://www.researchgate.net/publication/307773215_Medical_News _From_Scientific_Analysis_of_the_Turin_Shroud.

Bevilacqua, Matteo, Giulio Fanti, Michele D'Arienzo, and Raffaele De Caro. "Do We Really Need New Medical Information about the Turin Shroud?" *Injury* 45 (2014): 460–64. https://www.injuryjournal.com /article/S0020-1383(14)00115-6/fulltext.

Białous, Adam. *Hostia: Cud eucharystyczny w Sokółce*. Translated by Jakub Juszczyk. Częstochowa: Edycja Świętego Pawła, 2015.

Bible.org. "Archaeology and the Synoptic Gospels: Which Way Do the Rocks Roll", 2024. https://bible.org/seriespage/archaeology-and-synoptic -gospels-which-way-do-rocks-roll.

Boissarie, Gustave. *Lourdes: Histoire Medicale 1858–1891*. Paris: Librairie Victor Lecoffre, 1891.

Bonelli, Raphael, Rachel E. Dew, Harold G. Koenig, David H. Rosmarin, and Sasan Vasegh. "Religious and Spiritual Factors in Depression: Review and Integration of the Research". *Depression and Research Treatment* (2012): 1–8. https://www.hindawi.com/journals/drt/2012 /962860/.

Borrini, Matteo, and Luigi Garlaschelli. "A BPA Approach to the Shroud of Turin". *Journal of Forensic Sciences* 64, no. 1 (2018): 137–43.

Brown, Raymond. *An Introduction to New Testament Christology*. New York: Paulist Press, 1994.

Bucklin, Robert. "An Autopsy on the Man of the Shroud". Shroud of Turin website, 1997. https://www.Shroud.com/bucklin.htm.

Bunson, Matthew E. "Fathers of Science". *Catholic Answers* (magazine), September 1, 2008. https://www.catholic.com/magazine/print-edition /fathers-of-science.

Burger, John. "New Data Questions Finding That Shroud of Turin Was Medieval Hoax". *Aleteia*, July, 22, 2019. https://aleteia.org/2019/07/22 /new-data-questions-finding-that-shroud-of-turin-was-medieval-hoax/.

Callahan, Philip S. *The Tilma under Infra-Red Radiation*. CARA Studies on Popular Devotion. Vol. 2, Guadalupan Studies, no. 3. Washington, D.C.: Center for Applied Research in the Apostolate, 1981.

Campbell, Steuart. "The Miracle of the Sun at Fatima". *Journal of Meteorology* 14, no. 142 (October 1989): 334–38. http://www.ijmet.org/wp -content/uploads/2014/09/142.pdf.

Carlino, Elvio, and Giulio Fanti. "Atomic Resolution Studies Detect New Biologic Evidences on the Turin Shroud". *PLOS One* 12, no. 6 (2017). https://www.ncbi.nlm.nih.gov/pmc/articles/PMC5493404/#:~:text =Indeed%2C%20a%20high%20level%20of,wrapped%20in%20the%20 Turin%20Shroud.

Carrel, Alexis. *The Voyage to Lourdes*. 2nd ed. Translated by Virgilia Peterson. New Hope, Ky.: Real-View-Books, 1994.

Casabianca, Tristan, Emanuela Marinelli, Giuseppe Pernagallo, and Benedetto Torrisi. "Radiocarbon Dating of the Turin Shroud: New Evidence from Raw Data". *Archaeometry* 61, no. 5 (2019): 1223–31. https://onlinelibrary.wiley.com/doi/full/10.1111/arcm.12467.

Castellano, Daniel J. "The Codex Escalada". Part XI in *Historiography of the Apparition of Guadalupe*. ArcaneKnowledge (website), 2013. https://www.arcaneknowledge.org/catholic/guadalupe11.htm.

———. "The Image in the Twentieth Century". Part XIII in *Historiography of the Apparition of Guadalupe*. ArcaneKnowledge (website), revised 2018. https://www.arcaneknowledge.org/catholic/guadalupe13.htm.

Catholic News Agency. "Catholic Hospitals Comprise One Quarter of World's Healthcare, Council Reports", February 10, 2010. https://www.catholicnewsagency.com/news/catholic_hospitals_represent_26_percent_of_worlds_health_facilities_reports_pontifical_council.

Center for the Study of Global Christianity at Gordon-Conwell Theological Seminary. "Christianity in Its Global Context 1970–2020: Society, Religion, and Mission". South Hamilton, Mass.: Center for the Study of Global Christianity, June 2013. https://www.gordonconwell.edu/wp-content/uploads/sites/13/2019/04/2ChristianityinitsGlobalContext.pdf.

Charlesworth, James H., ed. *Jesus and Archaeology*. Grand Rapids, Mich.: William B. Eerdmans Publishing, 2006.

———. "Jesus Research and Archaeology: A New Perspective". In *Jesus and Archaeology*, edited by James H. Charlesworth, pp. 11–63. Grand Rapids, Mich.: William B. Eerdmans Publishing, 2006.

Chavez, Carlos Salinas, and Manuel de la Mora. *Descubrimiento de un Busto Humano en los Ojos de la Virgen de Guadalupe*. Mexico: Editorial Tradición, 1980.

Chilton, Brian. "Resurrection Defense Series: The Testimony of Women". Crossexamined.org (blog), March 29, 2021. https://crossexamined.org/resurrection-defense-series-the-testimony-of-women/.

Choi, Charles. "Shroud of Turin Is a Fake, Bloodstains Suggest". Live Science (website), July 18, 2018. https://www.livescience.com/63093-shroud-of-turin-is-fake-bloodstains.html.

Clement of Rome. *Letter to the Corinthians*. Translated by Kirsopp Lake. In *The Apostolic Fathers*, edited by T. E Page and W. H. D. Rouse. Loeb Classical Library. London: William Heinmann; New York: MacMillan, 1912.

Códice 1548 o "Escalada", Insigne y Nacional Basílica de Santa María de Guadalupe (website), 2013, https://web.archive.org/web/20150108213938/http://basilica.mxv.mx/web1/-apariciones/Documentos_Historicos/Mestizos/Codice_1548.html.

Collins, Francis. *The Language of God*. New York: Free Press, 2007.

Connolly, Seán. "70th Miracle of Lourdes Affirmed by the Church". *Catholic World Report*, February 13, 2018. https://www.catholicworldreport.com/2018/02/13/70th-miracle-of-lourdes-affirmed-by-the-church/.

Cranston, Ruth. *The Miracle of Lourdes*. Toronto: Galilee Trade, 1988.

Cyprian of Carthage. *On the Unity of the Church*. Translated by Robert Ernest Wallis. In *Ante-Nicene Fathers*, vol. 5, edited by Alexander Roberts, James Donaldson, and A. Cleveland Coxe. Buffalo, N.Y.: Christian Literature Publishing, 1886. Revised and edited for New Advent by Kevin Knight. https://www.newadvent.org/fathers/050701.htm.

da Capelinha, Maria. "Interview with de Marchi". Reproduced in *The True Story of Fatima*. St. Paul, Minn.: Catechetical Guild Educational Society, 1956.

Damon, P.E., D.J. Donahue, B.H. Gore, A.L. Hathaway, A.J.T. Jull, T.W. Linck, P.J. Sercel, et al. "Radiocarbon Dating of the Shroud of Turin". *Nature* 337 (1989): 611–15.

Darcy, Peter, "Why the Stars on Guadalupe's Mantle are Miraculous". SacredWindows.com. https://sacredwindows.com/why-the-stars-on-guadalupes-mantle-are-miraculous/.

Davies, W.D., and Dale C. Allison. *International Critical Commentary*. Vol. 2, *Matthew 8–18*. New York: T&T Clark, 1991.

Dawson, Christopher. *The Formation of Christendom*. New York: Sheed & Ward, 1965.

de Almeida, Avelino. "The Dance of the Sun". *O Seculo*, October 18, 1917.

de Caro, Liberato, Teresa Sibillano, Rocco Lassandro, Cinzia Giannini, and Giulio Fanti. "X-ray Dating of a Turin Shroud's Linen Sample". *Heritage* 5, no. 2 (2022): 860–70. https://www.mdpi.com/2571-9408/5/2/47/htm.

de Chardin, Pierre Teilhard. *The Divine Milieu*. New York: Harper Perennial Modern Classics, 2001.

———. *The Phenomenon of Man*. New York: Harper Perennial Modern Classics, 2008.

de Lambertini, Prospero. *De Servorum Dei Beatificatione et Beatorum Canonizatione*. Bologna: Formis Longhi Excusoris Archiepiscopalis, 1734–1738.

de Marchi, John. *The True Story of Fatima*. St. Paul, Minn.: Catechetical Guild Educational Society, 1956. https://www.ewtn.com/catholicism/library/true-story-of-fatima-5915.

DeMarest, Donald, and Coley Taylor. *The Dark Virgin: The Book of Our Lady of Guadalupe—A Document Anthology*. Fresno, Calif.: Academy Guild Press, 1957.

Dervic, Kanita, Maria A. Oquendo, Michael F. Grunebaum, Steve Ellis, Ainsley Burke, and J. John Mann. "Religious Affiliation and Suicide Attempt". *American Journal of Psychiatry* 161, no. 12 (2004): 2303–8. https://ajp.psychiatryonline.org/doi/full/10.1176/appi.ajp.161.12.2303.

di Lazzaro, Paolo. 2012. "Could a Burst of Radiation Create a Shroud-like Coloration? Summary of 5-Years Experiments at ENEA Frascati". Paper presented at 1st International Congress on the Holy Shroud in Spain, Centro Español de Sindonologia (CES), Valencia, Spain, April 28–30, 2012. https://www.academia.edu/4028955/Could_a_burst_of_radiation _create_a_Shroud-like_coloration_Summary_of_5-years_experiments _at_ENEA_Frascati.

———. "Shroud-like Coloration of Linen by Ultraviolet Radiation". Shroud of Turin website, May 2, 2015. https://www.Shroud.com/pdfs /duemaggioDiLazzaroENG.pdf.

Diaz, Ary Waldir Ramos. "The Future Pope Francis Was in Charge of Dealing with This Reported Eucharistic Miracle". *Aleteia*, June 13, 2020. https://aleteia.org/2020/06/13/the-future-pope-francis-was-in-charge -of-dealing-with-this-reported-eucharistic-miracle/.

Dodd, C. H. *The Apostolic Preaching and Its Development*. New York: Harper and Brothers, 1962.

Dowling, John. "Lourdes Cures and Their Medical Assessment". *Journal of the Royal Society of Medicine* 77 (August 1984): 634–38.

Dulle, Colleen. "Pope Francis Appointed Three Women to the Pontifical Academy of Sciences This Summer. What's Their Role at the Vatican?" *America: The Jesuit Review*, August 13, 2021. https://www.america magazine.org/faith/2021/08/13/pope-francis-pontifcal-academy -science-women-241214.

Eccles, Sir John, and Karl Popper. *Evolution of the Brain: Creation of the Self*. New York: Routledge, 1989.

Editorial. "New Study Shows Man of the Shroud Had 'Dislocated' Arms". *La Stampa*, May 8, 2014. https://www.lastampa.it/vatican-insider/en /2014/05/08/news/new-study-shows-man-of-the-shroud-had-dislocated -arms-1.35751980.

Editorial. *O Dia*. October 13, 1917.

Esparza, Daniel. "Shroud of Turin Coins May Finally Have Been Identified". *Aleteia*, April 26, 2017. https://aleteia.org/2017/04/26/Shroud -of-turin-coins-may-finally-have-been-identified/.

Faccini, Barbara, and Giulio Fanti. "New Image Processing of the Turin Shroud Scourge Marks". Proceedings of the International Workshop on the Scientific Approach to the Acheiropoietos Images, ENEA Frascati, Italy, May 4–6, 2010. http://www.acheiropoietos.info/proceed ings/FacciniWeb.pdf.

Fanti, Giulio. "Optical Features of Flax Fibers Coming from the Turin Shroud". *SHS Web of Conferences* 15 (2015), 2014 Workshop on Advances in the Turin Shroud Investigation. https://www.shs-conferences.org

/articles/shsconf/abs/2015/02/shsconf_atsi2014_00004/shsconf_atsi2014
_00004.html.

Fanti, Giulio, Pietro Baraldi, Roberto Basso, and Anna Tinti. "Non-destructive Dating of Ancient Flax Textiles by Means of Vibrational Spectroscopy". *Vibrational Spectroscopy* 67 (2013): 61–70. http://dx.doi .org/10.1016/j.vibspec.2013.04.001.

Fanti, Giulio, and Pierandrea Malfi. "Multi-parametric Micro-mechanical Dating of Single Fibers Coming from Ancient Flax Textiles". *Textile Research Journal* (2013): 714–27. http://dx.doi.org/10.1177/0040517513 507366.

Fanti, Giulio, Pierandrea Malfi, and Fabio Crosilla. "Mechanical and Opto-chemical Dating of the Turin Shroud". *MATEC Web of Conferences* 36 (2015), Workshop of Paduan Scientific Analysis on the Shroud. https:// www.researchgate.net/publication/287294012_Mechanical_ond_opto -chemical_dating_of_the_Turin_Shroud.

Ferrín, Ignacio. *An Astronomical Explanation of the Fatima Miracle.* Amazon Kindle Edition, 2021.

Filas, Francis. *The Dating of the Shroud of Turin from Coins of Pontius Pilate.* Youngtown, Ariz.: Cogan Productions, 1982.

Fitzmyer, Joseph, S.J. "Pauline Theology". In *The New Jerome Biblical Commentary*, pp. 1382–416. Englewood Cliffs, N.J.: Prentice Hall, 1990.

François, Bernard, E. Sternberg, and E. Fee. "The Lourdes Medical Cures Revisited". *Journal of the History of Medicine and Allied Sciences* 69, no. 1 (January 2014): 135–62.

Freeman, Bill. "Science or Miracle?; Holiday Season Survey Reveals Physicians' Views of Faith, Prayer and Miracles". WorldHealth.Net, December 22, 2004. https://www.worldhealth.net/news/science_or _miracle_holiday_season_survey/.

Frei, Max. "Il passato della Sindone alla luce della palinologia". In *La Sindone e la Scienza, Atti del II Congresso Internazionale di Sindonologia*, Turin, October 7–8, 1978. Turin: Edizioni, 1979.

———. "Nine Years of Palynological Studies on the Shroud". *Shroud Spectrum International* 3 (1982): 2–7.

———. "Note a seguito dei primi studi sui prelievi di polvere aderente al lenzuolo della S. Sindone". *Sindon* 23 (1976): 5–9.

Fu, Xiaorong, Ge Liu, Alexander Halim, Yang Ju, Qing Luo, and Guanbin Song. "Mesenchymal Stem Cell Migration and Tissue Repair". *Cells* 8, no. 8 (2019). https://www.ncbi.nlm.nih.gov/pmc/articles/PMC6721 499/.

Fuller, Reginald. *The Formation of the Resurrection Narratives.* New York: Macmillan, 1971.

Galilei, Galileo. *Edizione Nazionale delle Opere*. Vol. 12, edited by A. Favaro. Florence: Giunti-Barbera, 1968.

————. "Letter to the Grand Duchess Christina of Tuscany, 1615". Internet History Sourcebooks Project (website), 1997. https://sourcebooks .fordham.edu/mod/galileo-tuscany.asp.

Garrett, Goncalo Almeida. "Letter to Cannon Formigao concerning Observations at Fatima on October 13th". In *Documents on Fatima and Memoirs of Sister Lucia*, translated by Antonio Maria Martins, 173–75. Alexandra, S. Dak.: Fatima Family Apostolate, 1992.

Gens, José Maria Pereira. *Fátima: Como Eu A Vi e Como A Sinto (Memoria dum Médico)*. Vol. 1. Leiria, Portugal: Oficinas da "Graffica de Leiria", 1967.

Gilbert, Roger, Jr., and Marion Gilbert. "Ultraviolet-Visible Reflectance and Fluorescence Spectra of the Shroud of Turin". *Applied Optics* 19, no. 12 (1980): 1930–36.

Glynn, Paul. *The Healing Fire of Christ: Reflections on Modern Miracles—Knock, Lourdes, Fatima*. San Francisco: Ignatius Press, 2003.

Goldoni, Carlo. "The Shroud of Turin and the Bilirubin Blood Stains". Paper presented at the Shroud Science Group International Conference, "The Shroud of Turin: Perspectives on a Multifaceted Enigma", Columbus, Ohio, August 14–17, 2008.

Gómez, Ricardo Castañón. *Crónica de Un Milagro Eucarístico: Esplendor en Tixtla Chilpancingo, México*. Mexico: Grupo Internacional Para La Paz (GIPLAP), 2014. Translated by English Coaching (Fiverr) for the author.

————. Interview on 1996 Eucharistic Miracle of Buenos Aires. In *The Eucharistic Miracles of the World: Catalogue Book of the Vatican International Exhibition*. Presented by the Real Presence Eucharistic Education and Adoration Association. Bardstown, Ky.: Eternal Life, 2009. http://www .therealpresence.org/eucharst/mir/english_pdf/BuenosAires1.pdf.

Gonella, Luigi. "Scientific Investigation of the Shroud of Turin: Problems, Results, and Methodological Lessons". In *Turin Shroud—Image of Christ? Proceedings of a Symposium Held in Hong Kong, March 1986*, edited by William Meacham, pp. 29–40. Hong Kong: Turin Shroud Photographic Exhibition Organizing Committee, 1987.

Gosschalk, Kate, Laura Piggott, Honor Gray, James Riley, and Fern Elsdon-Baker. *Science and Religion Exploring the Spectrum: A Multi-Country Study on Public Perceptions of Evolution, Religion and Science*. London: YouGov, 2023. https://scienceandbeliefinsociety.org/wp-content/uploads/2023/12 /UoB-YouGov.-Science-and-Religion-Survey-Report.-8-Dec-2023-.pdf.

Guerra, Giulio Dante. "La Madonna di Guadalupe: Un Caso Di 'Inculturazione' Miracolosa". *Christianita*, nos. 205–6, June 7, 1992. https://alleanza cattolica.org/la-madonna-di-guadalupe-un-caso-di-inculturazione -miracolosa/.

Guscin, Mark. "The Sudarium of Oviedo: Its History and Relationship to the Shroud of Turin". In *Actes Du III—Symposium Scientifique International Du C.I.E.L.T., Nice 1997*, pp. 197–99. Paris: C.I.E.L.T, 1998. https://www.Shroud.com/guscin.htm#top.

Habermas, Gary R. "Mapping the Recent Trend toward the Bodily Resurrection Appearances of Jesus in Light of Other Prominent Critical Positions". In *The Resurrection of Jesus: John Dominic Crossan and N.T. Wright in Dialogue*, edited by Robert B. Stewart, pp. 78–92. Minneapolis: Fortress Press, 2006.

Haffert, John M. *Meet the Witnesses of the Miracle of the Sun*. Spring Grove, Penn.: American Society for the Defense of Tradition, Family, and Property—TFP, 1961.

Heiler, Friedrich. "The History of Religions as a Preparation for the Cooperation of Religions". In *The History of Religions*, edited by Mircea Eliade and J. Kitagawa, pp. 142–53. Chicago: Chicago University Press, 1959.

Heller, John, and Alan Adler. "Blood on the Shroud of Turin". *Applied Optics* 19, no. 16 (1980): 2742–44.

Hermosilla, Alfonso Sánchez. "Answer to the Article 'A BPA Approach to the Shroud of Turin', by Matteo Borrini and Luigi Garlaschelli". Shroud of Turin website, July 18, 2018. https://www.Shroud.com/pdfs/Hermosilla%20EN.pdf.

Ignatius of Antioch. *Epistle to the Romans*. In *Ante-Nicene Fathers*. Vol. 1, *The Apostolic Fathers with Justin Martyr and Irenaeus*, edited by Philip Schaff. Grand Rapids, Mich.: Christian Classics Ethereal Library, 1985. https://archive.org/details/ante-nicene-fathers-vol-1/page/n247/mode/1up.

———. *Epistle to the Romans*. In *The Epistles of St. Ignatius*. Early Church Classics. Vol. 2, edited by J. H. Srawley. London: Society for Promoting Christian Knowledge, 1910.

Insigne y Nacional Basílica de Santa María de Guadalupe (website). "Códice 1548 o 'Escalada'" ["Codex 1548 or 'Escalation'"]. 2013. https://web.archive.org/web/20150108213938/http://basilica.mxv.mx/web1/-apariciones/Documentos_Historicos/Mestizos/Codice_1548.html.

Irenaeus of Lyons. *Against Heresies*. In *Ante-Nicene Fathers*. Vol. 1, *The Apostolic Fathers with Justin Martyr and Irenaeus*, edited by Philip Schaff. Grand Rapids, Mich.: Christian Classics Ethereal Library, 1985. https://archive.org/details/ante-nicene-fathers-vol-1/page/n1346/mode/1up.

———. *Against Heresies*. In *Five Books of S. Irenaeus, Bishop of Lyons, against Heresies*. Translated by John Keble. London: James Parker, 1872.

Jackson, John P. "Is the Image on the Shroud Due to a Process Heretofore Unknown to Modern Science?" *Shroud Spectrum International* 34 (1990): 3–29.

————. "An Unconventional Hypothesis to Explain all Image Character-
istics Found on the Shroud Image". In *History, Science, Theology and the
Shroud*, edited by Aram Berard, pp. 325–44. St. Louis: Man in the Shroud
Committee of Amarillo, 1991.

Jackson, John P., Eric J. Jumper, and William R. Ercoline. "Correlation of
Image Intensity on the Turin Shroud with the 3-D Structure of a Human
Body Shape". *Applied Optics* 23, no. 14 (1984): 2244–70.

Jaki, Stanley L. *God and the Sun at Fatima*. Royal Oak, Mich.: Real View
Books, 1999.

————. Introduction to *The Voyage to Lourdes* by Alexis Carrel. 2nd ed.
Translated by Virgilia Peterson. New Hope, Ky.: Real-View-Books, 1994.

————. *A Mind's Matter: An Intellectual Autobiography*. Grand Rapids,
Mich.: Eerdmans, 2002.

————. "Two Lourdes Miracles and a Nobel Laureate". *Linacre Quarterly*
66, no. 1 (1999): 65–73.

Jeremias, Joachim. *The Eucharistic Words of Jesus*. London: SCM Press 1966.

————. *Heiligengräber in Jesu Umwelt*. Göttingen: Vandenhoeck & Ruprecht,
1958.

————. *New Testament Theology*. Vol. 1, *The Proclamation of Jesus*. New
York: Charles Scribner's Sons, 1971.

————. *The Parables of Jesus*. 2nd rev. ed. London: Pearson, 1972.

John Paul II. Address to the Plenary Session on "The Emergence of Complexity in Mathematics, Physics, Chemistry and Biology". Plenary session of the Pontifical Academy of Sciences, Vatican City, October 31,
1992. https://www.pas.va/en/magisterium/saint-john-paul-ii/1992-31
-october.html.

————. "Message to the Pontifical Academy of Science on Evolution".
October 22, 1996. https://www.ewtn.com/catholicism/library/message
-to-the-pontifical-academy-of-science-on-evolution-8825.

Johnson, Luke Timothy. *The Gospel according to Luke*. Vol. 3 of Sacra Pagina
Series, edited by Daniel J. Harrington, S.J. Collegeville, Minn.: Liturgical Press, 1991.

Johnston, Francis. *Fatima: The Great Sign*. Rockford, Ill.: Tan Books, 2010.

Jones, Stephen. "The Shroud of Turin: 3.3; The Man on the Shroud and
Jesus Were Scourged". *Shroud of Turin* (blog), July 15, 2013. http://the
Shroudofturin.blogspot.com/2013/07/the-Shroud-of-turin-33-man
-on-Shroud.html.

————. "Were Crowned with Thorns #5: Bible and the Shroud: Jesus and
the Man on the Shroud: Shroud of Turin Quotes", October 19, 2015.
https://theShroudofturin.blogspot.com/2015/10/were-crowned-with
-thorns-5-bible-and.html.

Jordan, Elaine. "The Lourdes' Miracle of Gabriel Gargam". Tradition in Action (website), April 13, 2013. http://www.traditioninaction.org /religious/h106_Lourdes.htm.

Josephus, Flavius. *Jewish Antiquities*. Translated by Louis H. Feldman. Loeb Classical Library. Cambridge, Mass.: Harvard University Press, 1965.

Jumper, Eric J., Alan D. Adler, John P. Jackson, Samuel F. Pellicori, John H. Heller, and James R. Druzik. 1984. "A Comprehensive Examination of the Various Stains and Images on the Shroud of Turin". In *Archaeological Chemistry III*. Advances in Chemistry Series, no. 205, edited by Joseph B. Lambert, pp. 447–76. Washington, D.C.: American Chemical Society, 1984.

Justin Martyr. *First Apology*. In *Ante-Nicene Fathers*. Vol. 1, *The Apostolic Fathers with Justin Martyr and Irenaeus*, edited by Philip Schaff. Grand Rapids, Mich.: Christian Classics Ethereal Library, 1985. https://archive .org/details/ante-nicene-fathers-vol-1/page/n537/mode/1up.

Keener, Craig. *Miracles: The Credibility of the New Testament Accounts*. 2 vols. Grand Rapids, Mich.: Baker Academic Publishing, 2011.

Klimiuk, Jennifer, and Kieran J. Moriarty. "The Lourdes Pilgrimage and the Impact on Pilgrim Quality of Life", *Journal of Religion and Health* 60, no. 6 (September 2021). https://www.ncbi.nlm.nih.gov/pmc/articles /PMC8428497/.

Koenig, Harold, Judith C. Hays, David B. Larson, Linda K. George, Harvey J. Cohen, Michael E. McCullough, Keith G. Meador, and Dan G. Blazer. "Does Religious Attendance Prolong Survival? A Six-Year Follow-Up Study of 3,968 Older Adults". *Journal of Gerontology, Medical Sciences* 54A, no. 7 (1999): M370.

———. "Religion, Spirituality, and Health: A Review and Update". *Advances in Mind-Body Medicine* 29, no. 3 (2015): 19–26. https://pubmed .ncbi.nlm.nih.gov/26026153/.

———. "Research on Religion, Spirituality and Mental Health: A Review". *Canadian Journal of Psychiatry* 54, no. 5 (2009): 283–91. https:// pubmed.ncbi.nlm.nih.gov/19497160/.

Koester, Helmut. "The Great Appeal: What Did Christianity Offer Its Believers That Made It Worth Social Estrangement, Hostility from Neighbors, and Possible Persecution?" Frontline, WGBH Educational Foundation, April 1998. https://www.pbs.org/wgbh/pages/frontline /shows/religion/why/appeal.html.

Köstenberger, Andreas, and Stephen Stout. "'The Disciple Jesus Loved': Witness, Author, Apostle—A Response to Richard Bauckham's Jesus and the Eyewitnesses". *Bulletin for Biblical Research* 18, no. 2 (2008): 209–31.

Lassi, Stefano, and Daniele Mugnaini. "Role of Religion and Spirituality on Mental Health and Resilience: There Is Enough Evidence". *International Journal of Emergency Mental Health and Human Resilience* 17, no. 3 (2015): 661–63. https://www.omicsonline.org/open-access/role-of-religion-and -spirituality-on-mental-health-and-resilience-there-is-enough-evidence -1522-4821-1000273.pdf.

Latourelle, Rene. *Finding Jesus through the Gospels: History and Hermeneutics.* New York: Alba House, 1979.ds

Lecky, W. E. H. *History of the Rise and Influence of the Spirit of Rationalism in Europe.* London: Longmans, Green, 1865.

Leuret, François. *Modern Miraculous Cures—A Documented Account of Miracles and Medicine in the 20th Century.* London: Peter Davies, 1957.

Lewis, C. S. *Miracles: A Preliminary Study.* New York: HarperOne, 1947.

———. *Pilgrim's Regress.* London: J. M. Dent and Sons, 1933.

Lin, Hsin-Nan. "Dealing with Depression: A Christian Perspective". *Taiwanese Journal of Psychiatry* (Taipei) 25, no. 4 (2011): 224–32. https://www.sop.org.tw/sop_journal/Upload_files/25_4/02.pdf.

Lind, Arthur C., Mark Antonacci, D. Elmore, G. Ganti, and J. Guthrie. "Production of Radiocarbon by Neutron Radiation on Linen". Proceedings of the International Workshop on the Scientific Approach to the Archeiropoietos Images, ENEA Frascati, Italy, May 4–6, 2010. http://www.acheiropoietos.info/proceedings/LindWeb.pdf.

Linoli, Edoardo. "Histological, Immunological and Biochemical Studies on the Flesh and Blood of the Eucharistic Miracle of Lanciano (8th Century)". *Quaderni Sclavo di Diagnostica Clinica e di Laboratori* 7, no. 3 (1971): 661–74.

Little, Kitty. "The Formation of the Shroud's Body Image". *British Society for the Turin Shroud Newsletter*, no. 46 (1997): 19–26. https://www.Shroud.com/bsts4607.htm.

———. "Photographic Studies of Polymeric Materials". Chapter 4 in *Photographic Techniques in Scientific Research*, vol. 3. Cambridge, Mass.: Academic Press, 1978.

Lonergan, Bernard. *Collected Works of Bernard Lonergan.* Vol. 3, *Insight: A Study of Human Understanding*, edited by Frederick E. Crowe and Robert M. Doran. Toronto: University of Toronto Press, 1992.

Lourdes Medical Bureau. *Dossier for Canonical Miracle #45: Francois Pascal.* Quoted in *The Healing Fire of Christ: Reflections on Modern Miracles—Knock, Lourdes, Fatima*, by Paul Glynn. San Francisco: Ignatius Press, 2003.

Lourdes Sanctuaire (website). "The Apparitions". Accessed April 2, 2024. https://www.lourdes-france.com/en/the-apparitions/.

———. "Miraculous Healings". Accessed February 21, 2024. https://www.lourdes-france.com/en/miraculous-healings/.

Lourenco, Inácio. "Report." Reproduced in *The True Story of Fatima*. St. Paul, Minn.: Catechetical Guild Educational Society, 1956.

Marinelli, Emanuela. "The Question of Pollen Grains on the Shroud of Turin and the Sudarium of Oviedo". Paper presented at the 1st International Congress on the Holy Shroud of Turin, Valencia, Spain, April 28–30, 2012. https://www.Shroud.com/pdfs/marinelli2veng.pdf.

Marino, Joseph G. "The Radiocarbon Dating of the Shroud of Turin in 1988: Prelude and Aftermath—an English-Language Bibliography". Academia.edu. Updated December 14, 2023. https://www.academia.edu/48831028/The_Radiocarbon_Dating_of_the_Turin_Shroud_in_1988_and_its_Aftermath_an_English_language_Bibliography.

Martindale, C.C. *The Meaning of Fatima*. New York: P.J. Kenedy & Sons, 1950.

Marto, Ti. "Testimony of Events". Reproduced in *The True Story of Fatima*, by John de Marchi. St. Paul, Minn.: Catechetical Guild Educational Society, 1956.

Mazurczak, Filip. "Ten Catholic Scientists and Inventors Everyone Should Know". *Catholic World Report*, July 21, 2022. https://www.catholicworldreport.com/2022/07/21/ten-catholic-scientists-and-inventors/.

McArthur, Harvey K. "Basic Issues: A Survey of Recent Gospel Research". In *In Search of the Historical Jesus*, edited by Harvey K. McArthur, pp. 139–44. London: Charles Scribner's Sons, 1969.

———. *In Search of the Historical Jesus*. London: Charles Scribner's Sons, 1969.

McKenzie, John L. *Dictionary of the Bible*. New York: Macmillan, 1965.

Meacham, William, James E. Alcock, Robert Bucklin, K.O.L. Burridge, John R. Cole, Richard J. Dent, and John P. Jackson. "The Authentication of the Turin Shroud: An Issue in Archaeological Epistemology". *Current Anthropology* 24, no. 3 (1983): 283–311.

Meier, John P. *A Marginal Jew: Rethinking the Historical Jesus*. Vol. 2, *Mentor, Message, and Miracles*. New York: Doubleday, 1994.

———. "The Present State of the 'Third Quest' for the Historical Jesus: Loss and Gain". *Biblica* 80 (1999): 459–87.

Moreno, Guillermo Heras, José-Delfín Villalaín Blanco, and Jorge-Manuel Rodríguez Almenar. "Comparative Study of the Sudarium of Oviedo and the Shroud of Turin". *III Congresso Internazionale di Studi sulla Sindone Turin* (1998): 1–17. Translated from the Spanish by Mark Guscin and revised by Guillermo Heras Moreno. https://www.shroud.com/heraseng.pdf.

Nagel, Thomas. *Mind and Cosmos: Why the Materialist Neo-Darwinian Conception of Nature Is Almost Certainly False*. New York: Oxford University Press, 2012.

Newman, John Henry. *An Essay in Aid of a Grammar of Assent*. Notre Dame, Ind.: University of Notre Dame Press, 1992.

————. "Proof of Theism". In *The Argument from Conscience to the Existence of God according to J. H. Newman*, edited by Adrian Boekraad and Henry Tristram, pp. 103–20. Louvain: Editions Nauwelaerts, 1961.

Noschovis, Peter P. "'Lord, I Need a Healing': The Uneasy Relationship between Faith and Medicine". *American Medical Association Journal of Ethics* 7, no. 5 (2005): 333–35.

Nuñes, Leopoldo. *Fátima—História das Aparições de Nossa Senhora do e aos Pastorinhos na Cova de Iria*. Lisbon: Tipografia Luzitania, 1927.

Oden, Thomas. *The Word of Life*. Vol. 2, *Systematic Theology*. New York: HarperOne, 1992.

Oommen, T. V. "Shroud Coins Dating by Image Extraction". Paper presented at the Shroud Science Group International Conference, "The Shroud of Turin: Perspectives on a Multifaceted Enigma", Columbus, Ohio, August 14–17, 2008. https://www.shroud.com/pdfs/ohiotvoommen.pdf.

Otto, Rudolf. *The Idea of the Holy: An Inquiry into the Non-Rational Factor in the Idea of the Divine and Its Relation to the Rational*. New York: Oxford University Press, 1958.

Our Lady of Guadalupe (website). "The Mystery in Our Lady's Eyes". Accessed March 15, 2024. http://sancta.org/eyes.html.

Our Lady of the Rosary Library (website). "Miracles of Lourdes". Accessed March 18, 2024. https://olrl.org/stories/lourdes.shtml.

Pawilkowski, Jakub, Jaroslaw Sak, Michał Wiechetek, and Marek Jarosz. "Doctors' Religiosity and Belief in Miracles". Paper presented at the 4th European Conference on Religion, Spirituality and Health, Malta, May 2014. https://www.researchgate.net/publication/311680999_Doctors'_religiosity_and_belief_in_miracles.

Peralta, Alberto. "El Códice 1548: Crítica a una Supesta Fuente Guadalupana del Siglo XVI". Proyecto Guadalupe (website), 2003. https://web.archive.org/web/20070209082837/http://www.proyectoguadalupe.com/apl_1548.html.

Petrosillo, Orazio, and Emanuela Marinelli. *The Enigma of the Shroud: A Challenge to Science*. Translated by Louis J. Scerri. Malta: Publishers Enterprise Group, 1996.

Pew Research Center. "The Global Religious Landscape", December 18, 2012. https://www.pewresearch.org/religion/2012/12/18/global-religious-landscape-exec/.

————. "Religion and Science in the United States: Scientists and Belief", November 5, 2009. https://www.pewresearch.org/religion/2022/12/21/key-findings-from-the-global-religious-futures-project/attachment/7/.

———. "Religious Composition by Country, 2010–2050", December 21, 2022. https://www.pewresearch.org/religion/interactives/religious-com position-by-country-2010-2050/.

———. "Religious Groups' Views on Evolution", February 13, 2014. https://www.pewresearch.org/religion/2009/02/04/religious-groups -views-on-evolution/.

Pilon, Roger. *Lourdes Magazine*. July 1996. Reported in *The Healing Fire of Christ: Reflections on Modern Miracles—Knock, Lourdes, Fatima*, by Paul Glynn. San Francisco: Ignatius Press, 2003.

Piotrowski, Mieczysław. "Eucharistic Miracle in Buenos Aires". *Love One Another!* 17 (2010). https://www.truechristianity.info/en/articles/article _en_0414.php.

Pius XII. Encyclical concerning Some False Opinions Threatening to Undermine the Foundations of Catholic Doctrine *Humani Generis*, August 12, 1950. https://www.vatican.va/content/pius-xii/en/encycli cals/documents/hf_p-xii_enc_12081950_humani-generis.html.

———. Encyclical on Promoting Biblical Studies *Divino Afflante Spiritu*, September 30, 1943. http://www.vatican.va/content/pius-xii/en /encyclicals/documents/hf_p-xii_enc_30091943_divino-afflante-spiritu .html.

Pohle, J. "Eucharist". In *Catholic Encyclopedia*. New York: Robert Appleton, 1909. https://www.newadvent.org/cathen/05572c.htm.

Polanyi, Michael. "Life's Irreducible Structure". *Science* 160, no. 3834 (June 21, 1968): 1308–12.

———. "Transcendence and Self-transcendence". *Soundings* 53, no. 1 (Spring 1970): 88–94.

Pontifical Academy of Sciences (website). "Nobel Laureates". Accessed March 18, 2024. https://www.pas.va/en/academicians/nobel.html.

Poole, Stafford. *The Guadalupan Controversies in Mexico*. Stanford, Calif.: Stanford University Press, 2006.

———. "History versus Juan Diego". *The Americas* 62, no. 1 (July 2005): 1–16. https://doi:10.1353/tam.2005.013.

Pylkkänen, Paavo. "Henry Stapp vs. David Bohm on Mind, Matter, and Quantum Mechanics". *Activitas Nervosa Superior* 61 (April 7, 2019): 48–50.

Real Presence Eucharistic Education and Adoration Association. *The Eucharistic Miracles of the World: Catalogue Book of the Vatican International Exhibition*. Bardstown, Ky.: Eternal Life, 2009.

Reis, Dominic. "Television Interview". Reproduced in *Meet the Witnesses*, by John Haffert. Spring Grove, Penn.: Society for DTFP, 1961.

Rigaux, Béda. "L'historicité de Jésus devant l'exégèse récente". *Revue Biblique* 68 (1958): 481–522.

Rinaudo, John-Baptiste. "Protonic Model of Image Formation on the Shroud of Turin". Paper presented at Third International Congress on the Shroud of Turin, Turin, Italy, June 5–7, 1998.

———. "A Sign for Our Time". *Shroud Sources Newsletter*, May/June 2–4, 1996, pp. 2–4.

Robinson, Kristin A., Meng-Ru Cheng, Patrick D. Hansen, and Richard Gray. "Religious and Spiritual Beliefs of Physicians". *Journal of Religion and Health* 56, no. 1 (February 2017): 205–25.

Rogers, Raymond N. "Studies on the Radiocarbon Sample from the Shroud of Turin". *Thermochimica Acta* 425, nos. 1–2 (2005): 189–94. http://www.shroud.it/ROGERS-3.PDF.

Ronenberg, Corina, Edward Alan Miller, Elizabeth Dugan, and Frank Porell. "The Protective Effects of Religiosity on Depression: A 2-Year Prospective Study". *Gerontologist* 56, no. 3 (2016): 421–31. https://academic.oup.com/gerontologist/article/56/3/421/2605601.

Sacred Congregation for the Doctrine of the Faith, *Norms Regarding the Manner of Proceeding in the Discernment of Presumed Apparitions or Revelations*. February 25, 1978. https://www.vatican.va/roman_curia/congregations/cfaith/documents/rc_con_cfaith_doc_19780225_norme-apparizioni_en.html.

Sanders, Jack T. *The New Testament Christological Hymns: Their Historical Religious Background*. Cambridge: Cambridge University Press, 1971.

Santos, Alfredo da Silva. "Interview by de Marchi". Reproduced in *The True Story of Fatima*, by John de Marchi. St. Paul, Minn.: Catechetical Guild Educational Society, 1956.

Scatizzi, Pio, S.J. *Fátima alla Luce della Fede e della Scienza*. Rome: Coletti, 1947.

———. "The Miracle of the Sun: A Critical Note". In *The True Story of Fatima*, by John de Marchi. St. Paul, Minn.: Catechetical Guild Educational Society, 1956.

Scott, Eugenie. "Antievolution and Creationism in the United States". *Annual Review of Anthropology* 26 (1997): 263–89.

Service Communication. "The Bernadette Year Has Begun". The official Lourdes website, January 17, 2009. https://www.lourdes-france.org/en/the-bernadette-year-has-begun/.

Simons, Paul. "Weather Secrets of Miracle at Fatima". *Times*, February 17, 2005.

Singh, Simon, "Even Einstein Had His Off Days". *New York Times*, January 2, 2005. www.nytimes.com/2005/01/02/opinion/02singh.html.

Smiles, Vincent. "Transcendent Mind, Emergent Universe, in the Mind of Michael Polanyi". *Open Theology* 1 (2015): 480–93. https://www.degruyter.com/document/doi/10.1515/opth-2015-0030/html?lang=en.

Smith, Jody Brant. *The Image of Guadalupe: Myth or Miracle?* Garden City, N.Y.: Doubleday, 1983.

Spitzer, Robert. *Escape from Evil's Darkness: The Light of Christ in the Church, Spiritual Conversion, and Moral Conversion.* San Francisco: Ignatius Press, 2021.

———. *Finding True Happiness: Satisfying Our Restless Hearts.* San Francisco: Ignatius Press, 2015.

———. *Five Pillars of the Spiritual Life: A Practical Guide to Prayer for Active People.* San Francisco: Ignatius Press, 2008.

———. *God So Loved the World: Clues to Our Transcendent Destiny from the Revelation of Jesus.* San Francisco: Ignatius Press, 2016.

———. *The Light Shines on in the Darkness: Transforming Suffering through Faith.* San Francisco: Ignatius Press, 2017.

———. *The Moral Wisdom of the Catholic Church.* San Francisco: Ignatius Press, 2022.

———. *New Proofs for the Existence of God: Contributions of Contemporary Physics and Philosophy.* Grand Rapids, Mich.: Eerdmans, 2010.

———. *The Sacraments, Part 1—The Sacred Eucharistic Liturgy.* Vol. 9 of the *Credible Catholic Big Book.* Garden Grove, Calif.: Magis Center, 2017. https://discover.magiscenter.com/hubfs/CC%20-%20Big%20Book%20Pdfs/BB-Vol-9-Eucharistic-Liturgy.pdf?hsLang=en.

———. *Science at the Doorstep to God: Science and Reason in Support of God, the Soul, and Life after Death.* San Francisco: Ignatius Press, 2023.

———. *The Soul's Upward Yearning.* San Francisco: Ignatius Press, 2015.

Sreenlvasan, Shoba, and Linda E. Welnberger. "Do You Believe in Miracles? Turning to Divine Intervention When Facing Serious Medical Illness". *Psychology Today,* December 15, 2017. https://www.psychologytoday.com/us/blog/emotional-nourishment/201712/do-you-believe-in-miracles#:~:text=Even%20physicians%20believe%20in%20miracles.%20A%20national%20poll,occur%20today%20%28Poll%3A%20Doctors%20Believe%20in%20Miracles%2C%202004%29.

Stoddard, John. *Rebuilding a Lost Faith: By an American Agnostic.* Tan Books and Publishers, Inc., 1990.

Swagger, Brother Leo. "Testimony". In *The Healing Fire of Christ: Reflections on Modern Miracles—Knock, Lourdes, Fatima,* by Paul Glynn. San Francisco: Ignatius Press, 2003.

Tacitus, Cornelius. *Annals.* In *The Complete Works of Tacitus.* Translated by Alfred John Church and William Jackson Brodribb. New York: Random House, 1942, edited for Perseus Digital Library. http://www.perseus.tufts.edu/hopper/text?doc=Perseus%3Atext%3A1999.02.0078%3Abook%3D15%3Achapter%3D44.

Teilhard de Chardin, Pierre. *The Divine Milieu.* New York: Harper Perennial Modern Classics, 2001.

————. *The Phenomenon of Man*. New York: Harper Perennial Modern Classics, 2008.

Thackeray, J. F. "Lepton Coin Diameters and a Circular Image on the Shroud of Turin". Shroud of Turin website, May 23, 2019. https://www.Shroud.com/pdfs/thackeray.pdf.

The Divine Mercy.org. "A Matter of Faith, a Matter of Fact", December 19, 2012. https://www.thedivinemercy.org/articles/matter-faith-matter-fact.

Topper, David. *How Einstein Created Relativity out of Physics and Astronomy*. New York: Springer, 2013.

Torroella-Bueno, Javier. 1956. "Letter Examining the Purkinje-Sanson Triple Reflection in the Guadalupe Tilma". Reported in *The Dark Virgin: The Book of Our Lady of Guadalupe—a Document Anthology*, by Donald DeMarest and Coley Taylor. Fresno, Calif.: Academy Guild Press, 1957.

Toynbee, Arnold J. *Civilization on Trial*. New York: Oxford University Press, 1948.

Udías, Augustin, and William Stauder. "The Jesuit Contribution to Seismology". *Seismological Research Letters* 67, no. 3 (May/June 1996): 10–19. https://www.seismosoc.org/inside/eastern-section/jesuit-contribution-seismology/.

United States Council of Catholic Bishops. "Catholic Education". https://www.usccb.org/offices/public-affairs/catholic-education.

van der Hoeven, Adrie. "Cold Acid Postmortem Blood Most Probably Formed Pinkish-Red Heme-Madder Lake on Madder-Dyed Shroud of Turin". *Open Journal of Applied Sciences* 5 (2015): 705–46.

Vatican Council II. Dogmatic Constitution on the Church *Lumen Gentium*, November 21, 1964. https://www.vatican.va/archive/hist_councils/ii_vatican_council/documents/vat-ii_const_19641121_lumen-gentium_en.html.

Vawter, Bruce. "The Gospel according to John". In *The Jerome Biblical Commentary*, vol. 2, edited by R. Brown, J. Fitzmyer, and R. Murphy, pp. 437–38. Englewood Cliffs, N.J.: Prentice Hall, 1968.

Wahlde, Urban C. von. "Archeology and John's Gospel". In *Jesus and Archaeology*, edited by James H. Charlesworth, pp. 523–86. Grand Rapids, Mich.: William B. Eerdmans Publishing, 2006.

Werfel, Franz. *The Song of Bernadette*. San Francisco: Ignatius Press, 2006.

Whanger, Alan D. "A Reply to Doubts concerning the Coins over the Eyes". Shroud of Turin website, August 24, 1997. https://www.Shroud.com/lombatti.htm.

Whanger, Alan D., and Mary Whanger. "Polarized Image Overlay Technique: A New Image Comparison Method and Its Applications". *Applied Optics* 24, no. 6 (March 1985): 766–72.

Wigner, Eugene. "The Unreasonable Effectiveness of Mathematics in the Natural Sciences". *Communications in Pure and Applied Mathematics* 13, no. 1 (1960): 1–14.

Wikipedia. "Herculaneum". June 10, 2013.

Williams, Jeanette. "The Amazing Science of Recent Eucharistic Miracles: A Message from Heaven?" *Ascension Press*, November 3, 2021. https://media.ascensionpress.com/2021/11/03/the-amazing-science-of-recent-eucharistic-miracles-a-message-from-heaven-%EF%BB%BF/.

Wilson, Ian. *The Shroud of Turin: The Burial Cloth of Jesus?* Rev. ed. New York: Image Books, 1979.

Wodon, Quentin. *Global Catholic Education Report 2023: Transforming Education and Making Education Transformative*. Washington, D.C.: Global Catholic Education, 2022. https://www.globalcatholiceducation.org/_files/ugd/b9597a_b54239f33dec48ddb4f2d735d10cba7c.pdf.

Woods, Thomas E. "The Church and Science". Chapter 5 in *How the Catholic Church Built Western Civilization*. Washington, D.C.: Regnery Publishing, 2005.

Wright, N. T. *Christian Origins and the Question of God*. Vol. 1, *The New Testament and the People of God*. Minneapolis: Fortress Press, 1992.

———. *Christian Origins and the Question of God*. Vol. 2, *Jesus and the Victory of God*. Minneapolis: Fortress Press, 1996.

———. *The Resurrection of the Son of God*. Minneapolis, Minn.: Fortress Press, 2003.

Wuenschel, Edward A. *Self-Portrait of Christ: The Holy Shroud of Turin*. 3rd ed. Esopus, N.Y.: Holy Shroud Guild, 1954.

ZENIT. "Science Sees What Mary Saw from Juan Diego's Tilma", 2001. https://web.archive.org/web/20100620110845/http://www.catholiceducation.org/articles/religion/re0447.htm.

Zugibe, Frederick. *The Cross and the Shroud: A Medical Examiner Investigates the Crucifixion*. Cresskill, N.J.: McDonagh, 1981.

———. *The Crucifixion of Jesus, Completely Revised and Expanded: A Forensic Inquiry*. New York: M. Evans, 2005.

———. "Testimonial Given in an Interview with Mike Willesee and Ron Tesoriero on March 15, 2005". In E-mail report. Available upon request.

SUBJECT INDEX

NAME INDEX

SCRIPTURE INDEX